T0407109

Energy Storage and Conversion Materials

This book explores the fundamental properties of a wide range of energy storage and conversion materials, covering mainstream theoretical and experimental studies and their applications in green energy. It presents a thorough investigation of diverse physical, chemical, and material properties of rechargeable batteries, supercapacitors, solar cells, and fuel cells, covering the development of theoretical simulations, machine learning, high-resolution experimental measurements, and excellent device performance.

- Covers potential energy storage (rechargeable batteries and supercapacitors) and energy conversion (solar cells and fuel cells) materials
- Develops theoretical predictions and experimental observations under a unified quasi-particle framework
- Illustrates up-to-date calculation results and experimental measurements
- Describes successful synthesis, fabrication, and measurements, as well as potential applications and near-future challenges

Promoting a deep understanding of basic science, application engineering, and commercial products, this work is appropriate for senior graduate students and researchers in materials, chemical, and energy engineering and related disciplines.

Energy Storage and Conversion Materials

Energy Storage and Conversion Materials

Properties, Methods, and Applications

Edited by

Ngoc Thanh Thuy Tran

Jeng-Shiung Jan

Wen-Dung Hsu

Ming-Fa Lin

Jow-Lay Huang

CRC Press is an imprint of the
Taylor & Francis Group, an **informa** business

First edition published 2023
by CRC Press
6000 Broken Sound Parkway NW, Suite 300, Boca Raton, FL 33487-2742

and by CRC Press
4 Park Square, Milton Park, Abingdon, Oxon, OX14 4RN

CRC Press is an imprint of Taylor & Francis Group, LLC

ISBN: 978-1-032-43421-6 (hbk)
ISBN: 978-1-032-43422-3 (pbk)
ISBN: 978-1-003-36721-5 (ebk)

DOI: 10.1201/9781003367215

Typeset in Times
by MPS Limited, Dehradun

Contents

Preface

Nowadays, green energy is the key to achieving carbon neutrality, especially with the vigorous development of electric vehicles and fifth-generation mobile communication. Safe and high-performance green energy materials will be in high demand in the future. This book presents thorough investigations of the diversified physical/chemical/material properties of rechargeable batteries, supercapacitors, solar cells, and fuel cells, covering the development of theoretical simulations, machine learning, high-resolution experimental measurements, and excellent device performances. It can significantly promote a full understanding of basic sciences, application engineering, and commercial products.

Concerning lithium-ion-based batteries, the first-principles method, molecule dynamics, and the machine learning method are utilized to fully explore the rich and unique phenomena of anodes, electrolytes, and cathodes. Specifically, the theoretical frameworks are built from them. This book provides concise physical/chemical properties, multi- and single-orbital hybridizations in chemical bonds, and spin configurations due to different atoms and orbitals. Parts of the theoretical predictions are consistent with the experimentally measured results. This is of great help in designing and screening novel green-energy materials, therefore accelerating new materials developments.

On the other hand, distinct experimental methods/techniques are developed to greatly enhance the performances of lithium-ion-based batteries, supercapacitors, solar cells, and fuel cells. In this book, you will learn about successful syntheses, fabrications, and measurements of the aforementioned energy storage and conversion materials. Moreover, the emerging open issues are currently under systematic investigation. This provides a simple, low-cost, high-efficiency, and environment-friendly approach. Detailed descriptions are provided of the potential applications, advantages, and near-future challenges. This work is appropriate for senior graduate students and scientists due to its concise pictures and content.

Acknowledgment

This work was financially supported by the Hierarchical Green-Energy Materials (Hi-GEM) Research Center, from The Featured Areas Research Center Program within the framework of the Higher Education Sprout Project by the Ministry of Education (MOE) and the Ministry of Science and Technology (MOST 111-2634-F-006-008) in Taiwan.

Editors

Ngoc Thanh Thuy Tran earned a PhD in physics in 2017 at the National Cheng Kung University (NCKU), Taiwan. Afterward, she was a postdoctoral researcher and then an assistant researcher at Hierarchical Green-Energy Materials (Hi-GEM) Research Center, NCKU. Her scientific interest is focused on the fundamental (electronic, magnetic, and thermodynamic) properties of 2D materials and rechargeable battery materials by means of first-principles calculations.

Jeng-Shiung Jan is a professor in the Department of Chemical Engineering, NCKU, Taiwan. He earned a PhD in chemical engineering in 2006 at Texas A&M University and conducted postdoctoral research at the Georgia Institute of Technology. His current research focuses on the synthesis of functional polymers and nanomaterials. He has received several awards, including the Outstanding Professor in Academic Research from LCY Education Foundation and the Lai Zaide Professor Award.

Wen-Dung Hsu is a Professor in the Department of Materials Science and Engineering, NCKU. His expertise is utilizing computational materials science methods, including first-principle calculations, molecular dynamics simulations, Monte-Carlo methods, and finite-element methods, to study materials issues. His research interests are the mechanical properties of materials from atomic to macro scale, lithium-ion batteries, solid-oxide fuel cells, ferroelectrics, solid catalyst design for biodiesel, and processing design for single-crystal growth. He earned a PhD in materials science and engineering at the University of Florida in 2007. He then served as a postdoctoral researcher in the Department of Mechanical Engineering at the University of Michigan. He joined the National Cheng Kung University in 2008.

Ming-Fa Lin is a Distinguished Professor in the Department of Physics, NCKU, Taiwan. He earned a PhD in physics in 1993 at the National Tsing-Hua University, Taiwan. His scientific interests focus on the essential properties of carbon-related materials and low-dimensional systems. He is a member of the American Physical Society, the American Chemical Society, and the Physical Society of the Republic of China (Taiwan).

Jow-Lay Huang is a Chair Professor in the Department of Materials Science and Engineering, National Cheng Kung University, Taiwan. He is the director of Hierarchical Green-Energy Materials (Hi-GEM) Research Center, NCKU, Taiwan. He earned a PhD in materials science and engineering in 1983 at the University of Utah, Salt Lake City, Utah, USA. His research interest includes the fabrication, development, and application of ceramic nanocomposites, piezo-phototronic thin films for photodetector devices, piezoelectric thin films for high-frequency devices, metal oxide/graphene, and SiCx nanocomposites as anode materials for lithium-ion batteries and 2D nanocrystal materials for the photoelectrochemical application.

Contributors

Sanjaya Brahma
Department of Materials Science and Engineering
and
Hierarchical Green-Energy Materials (Hi-GEM) Research Center
National Cheng Kung University
Tainan City, Taiwan

Chia-Chi Chang
Department of Chemical Engineering
National Cheng Kung University
Tainan City, Taiwan

Chia-Yun Chen
Department of Materials Science and Engineering
National Cheng Kung University
Tainan City, Taiwan

Shih-Hsiu Chen
Department of Materials Science and Engineering
National Cheng Kung University
Tainan City, Taiwan

Hsien-Ching Chung
RD Department
Super Double Power Technology Co. Ltd.
Changhua City, Changhua County, Taiwan

Vo Khuong Dien
Department of Physics
National Cheng Kung University
Tainan City, Taiwan

Kuan-Zong Fung
Hierarchical Green-Energy Materials (Hi-GEM) Research Center
National Cheng Kung University
Tainan City, Taiwan

Nguyen Thi Han
Department of Physics
National Cheng Kung University
Tainan City, Taiwan

Chun-Han Hsu
General Education Center
National Tainan Junior College of Nursing
Tainan City, Taiwan

Wen-Dung Hsu
Department of Materials Science and Engineering
and
Hierarchical Green-Energy Materials (Hi-GEM) Research Center
National Cheng Kung University
Tainan City, Taiwan

Yuan-Shuo Hsu
Department of Chemical Engineering
National Cheng Kung University
Tainan City, Taiwan

Chien-Ke Huang
Department of Materials Science and Engineering
and
Hierarchical Green-Energy Materials (Hi-GEM) Research Center
National Cheng Kung University
Tainan City, Taiwan

Jow-Lay Huang
Department of Materials Science and Engineering
and
Center for Micro/Nano Science and Technology
and
Hierarchical Green-Energy Materials (Hi-GEM) Research Center
National Cheng Kung University
Tainan City, Taiwan

Jeng-Shiung Jan
Department of Chemical Engineering
National Cheng Kung University
Tainan City, Taiwan

Zhe-Yun Kee
Department of Materials Science and Engineering
National Cheng Kung University
Tainan City, Taiwan

Chin-Lung Kuo
Department of Materials Science and Engineering
National Taiwan University
Taipei, Taiwan

Ting-Yuan Lee
Department of Chemical Engineering
National Cheng Kung University
Tainan City, Taiwan

Yuh-Lang Lee
Department of Chemical Engineering
and
Engineering and Hierarchical Green-Energy Materials (Hi-GEM) Research Center
National Cheng Kung University
Tainan City, Taiwan

Wei-Bang Li
Department of Physics
National Cheng Kung University
Tainan City, Taiwan

Hong-Ping Lin
Department of Chemistry
National Cheng Kung University
Tainan City, Taiwan

Ming-Fa Lin
Department of Physics
and
Hierarchical Green-Energy Material (Hi-GEM) Research Center
National Cheng Kung University
Tainan City, Taiwan

Shih-Yang Lin
Department of Physics
National Chung Cheng University
Chiayi, Taiwan

Thi Dieu Hien Nguyen
Department of Physics
National Cheng Kung University
Tainan City, Taiwan

Li-Yi Pan
Department of Materials Science and Engineering
National Taiwan University
Taipei, Taiwan

Zheng-Bang Pan
Department of Chemistry
National Cheng Kung University
Tainan City, Taiwan

Jheng-Hong Shih
Department of Physics
National Cheng Kung University
Tainan City, Taiwan

Le Vo Phuong Thuan
Department of Physics
National Cheng Kung University
Tainan City, Taiwan

Ngoc Thanh Thuy Tran
Hierarchical Green-Energy Materials (Hi-GEM) Research Center
National Cheng Kung University
Tainan City, Taiwan

Shu-Yi Tsai
Hierarchical Green-Energy Materials (Hi-GEM) Research Center
National Cheng Kung University
Tainan City, Taiwan

Nguyen Thanh Tuan
Department of Physics
National Cheng Kung University
Tainan City, Taiwan

Shanmuganathan Venkatesan
Department of Chemical Engineering
National Cheng Kung University
Tainan, Taiwan

Yu-Ming Wang
Department of Physics
National Cheng Kung University
Tainan City, Taiwan

Tsung-Yen Wu
Department of Materials Science and Engineering
National Cheng Kung University
Tainan City, Taiwan

Song-Hao Zhang
Department of Chemical Engineering
National Cheng Kung University
Tainan City, Taiwan

1 Introduction

Ngoc Thanh Thuy Tran
Hierarchical Green-Energy Materials (Hi-GEM) Research Center, National Cheng Kung University, Tainan City, Taiwan

Chin-Lung Kuo
Department of Materials Science and Engineering, National Taiwan University, Taipei, Taiwan

Ming-Fa Lin
Department of Physics and Hierarchical Green-Energy Material (Hi-GEM) Research Center, National Cheng Kung University, Tainan City, Taiwan

Wen-Dung Hsu
Department of Materials Science and Engineering and Hierarchical Green-Energy Materials (Hi-GEM) Research Center, National Cheng Kung University, Tainan City, Taiwan

Jeng-Shiung Jan
Department of Chemical Engineering, National Cheng Kung University, Tainan City, Taiwan

Hong-Ping Lin
Department of Chemistry, National Cheng Kung University, Tainan City, Taiwan

Chia-Yun Chen
Department of Materials Science and Engineering, National Cheng Kung University, Tainan City, Taiwan

Yuh-Lang Lee
Department of Chemical Engineering, and Hierarchical Green-Energy Material (Hi-GEM) Research Center, National Cheng Kung University, Tainan City, Taiwan

Shu-Yi Tsai
Hierarchical Green-Energy Materials (Hi-GEM) Research Center, National Cheng Kung University, Tainan City, Taiwan

DOI: 10.1201/9781003367215-1

Jow-Lay Huang
Department of Materials Science and Engineering, Center for Micro/Nano Science and Technology and Hierarchical Green-Energy Materials (Hi-GEM) Research Center, National Cheng Kung University, Tainan City, Taiwan

CONTENTS

1.1 INTRODUCTION

In the basic sciences and advanced engineering, it is critical to figure out how to get and use energy efficiently. To replace fossil fuels with clean, renewable energy, high-efficiency, safe, and low-cost energy conversion and storage technologies must be developed. As a result, the gap between energy supply and demand can be reduced, and energy systems can be made more reliable and efficient. The various theoretical models and experimental syntheses have been proposed to fully present the essential properties, outstanding functionalities, and commercialized products of energy storage and conversion materials. There are many rich and unique properties of these materials that can be studied by first-principles calculations, molecular dynamics, and neural networks, e.g., growth processes, optimal geometry within large unit cells, electronic energy spectra, van Hove singularities in density of states, orbital hybridizations of chemical bonds, magnetic configurations, and spectral optical absorption properties.

The organization of this book is as follows. Chapter 1 includes the general introduction, especially for the whole developments in the theoretical and experimental researches on emergent energy storage and conversion materials, namely rechargeable batteries, supercapacitors, solar cells, and fuel cells. The first-principles and molecular dynamics calculations for anodes, electrolytes, and cathodes of lithium-ion-based batteries are covered in Chapters 2–8. Chapters 2–4 focus on promising anode material, including amorphous and crystalline silicon (Chapter 2), graphite magnesium compounds (Chapter 3), and Na-intercalation compounds (Chapter 4). The high-safety solid electrolytes $LiLaTiO_4$ and Li_2S are discussed in Chapter 5 and Chapter 6, respectively. Chapter 7 and Chapter 8 investigate the high-stability $LiMnO_2$ and high-voltage $LiNiPO_4$ cathode materials. Apart from calculations, the machine learning method plays an important role in screening and accelerating novel battery materials, which is addressed in Chapter 9. On the other hand, the experimental syntheses and measurements are covered in Chapters 11–15, including Li-ion batteries (Chapter 10 and Chapter 11), supercapacitors (Chapter 12), solar cells (Chapter 13 and Chapter 14), and fuel cells (Chapter 15). Chapter 16 covers the concluding remarks of the afore-mentioned main chapters, while Chapter 17 discusses open issues and potential applications. Finally, the energy storage and conversion practice problems are provided in Chapter 18.

In Chapter 2, silicon anodes have been an important research topic due to their 10-fold theoretical capacity than the commonly used graphite anode. Silicon stores lithium by forming Li–Si alloy with a high volume expansion of 300% and pulverizes the silicon anode. Crystalline and amorphous silicon nanostructure anodes are used to solve this problem. The amorphous silicon anode, which has a two-stage lithiation behavior, is quite different from the crystalline silicon anode. Besides, it was found out that there is a critical size for cracking silicon anode. That is, silicon nanowire/nanoparticle cracks if its initial size is larger than a critical size. Amorphous silicon nanoparticle is found to have a critical size larger than 870 nm by McDowell et al., while it is only about 150 nm for crystalline silicon nanoparticles. Up until now, the atomic structure of the lithiation is still unknown, and the origin of the critical size for cracking in silicon nanostructure is still under debate. In this study, we focus on the lithiation behavior of amorphous silicon nanowires. We simulate the amorphous and crystalline silicon nanowire lithiation by our previously developed ReaxFF parameter. We reproduced the 2-stage lithiation in amorphous silicon and found out a concentration gradient in the Li–Si alloy structure. The lithiation of silicon can be modeled as a reaction-diffusion process, and the controlling factor for silicon lithiation is investigated for crystalline and amorphous silicon. Finally, the critical size issue for silicon is evaluated, and we found that it should be related to the formation of c-$Li_{15}Si_4$. The reason for the higher critical size of amorphous silicon and the origin of the critical size issue is also found out in this chapter.

The carbon atom, which belongs to group-IV element, can generate well-known organic materials with diverse physics, chemistry, optics material engineering, biomedicine, biophysics, and biochemistry. The pure carbon-based crystals cover 3D diamond, bulk AA-/AB-/ABC-stacked and turbostratic graphite systems (1–3), layered graphenes with AAA (4), ABA (5), ABC (6) and AAB stackings (7,8), the sliding/rotational bilayer graphene systems (9–11), the cylindrical and coaxial carbon nanotubes (12,13), the finite-size graphene nanoribbons with chiral or achiral edge boundaries under the planar/curved/folded/scrolled forms (14–18), and 0D carbon fullerenes/tori/onions/disks/rings/chains (19–24). Very interesting, the orbital hybridizations, without the spin configurations in most of the cases, play critical roles in determining the chemical bonding of the neighboring carbon atoms and thus dominate the high-symmetry crystal structure. The strong cooperation between the first-principles simulations (25–27) and the molecular dynamics in thoroughly understanding the gradual growth processes. The highly active environments are easily modulated by the external perturbations, being driven through the chemical, physical, engineering, biological and medical perturbations. For example, when a 3D bulk Bernal graphite (28) serves as an anode material of lithium-/sodium-/magnesium-ion-based batteries, the geometric symmetries exhibit dramatic changes. The metal-adatom intercalations are thoroughly examined to provide the n-type doping effects, such as the low-frequency plasmon modes due to free conduction electrons, and the great enhancement of the optical threshold frequency. Chapter 3 is focused on graphite magnesium intercalation compounds. The critical orbital hybridizations are thoroughly identified from the VASP calculations about the intralayer carbon-carbon and magnesium-magnesium bonds, as well as, the interlayer magnesium-carbon ones. This research strategy will be very useful in developing the enlarged quasiparticle framework, following the previously published books.

The wide use of Li-ion batteries (LIBs) in various fields, from portable products to large-scale energy storage systems, has revolutionized our daily life. The 2019 Nobel Prize in Chemistry has been awarded to John B. Goodenough, M. Stanley Whittingham, and Akira Yoshino for their contributions in developing Li-ion batteries. Although Li-ion batteries are currently on-growing research topic, lithium availability is still a problem for mass production. In contrast to lithium, sodium resources are almost unlimited on Earth, and sodium is one of the most abundant elements in the Earth's crust. Hence, Na-ion batteries as a counterpart of Li-ion batteries have the potential to serve as the next-generation batteries. In Chapter 4, a brief history and recent development of Na-ion batteries are described. The fundamental physical and electronic properties, such as geometric structures, band structure, density of states, and spatial charge distributions, of Na-intercalation compounds are discussed. The outlook of Na-ion batteries is given at the last.

The lithium- and oxygen-related materials, which belong to group I-III multi-component systems with different atoms and active orbitals in primitive unit cells, should be outstanding candidates in providing the composite quasiparticles (29), identifying their diverse behaviors and establishing a grand framework, as well as the commercial products of Li^+-based batteries. Such main-stream 3D condensed-matter systems have been successfully synthesized by various chemical techniques. In Chapter 5, to fully understand the optimal growth processes, the theoretical simulations need to be done with full cooperation between the first-principles methods (30) and molecular dynamics (31). Very interesting, both quaternary $LiLaTiO_4$ and ternary $LiTiO_2$ compounds, respectively, corresponding to electrolyte and anode (32) components of Li-ion-based battery, are chosen for a model study of composite quasiparticles (33). Such main-stream materials have covered all active orbitals: (I) [s, p_x, p_y, p_z]-I/II/III/IV/V/VI/VII/VIII [e.g., group-I alkali atoms], (II) [d_{z^2}, $d_{(x^2-y^2)}$, d_{xy}, d_{yz}, d_{xz}] - transition metal atoms [3d/4d/5d; (34)/(35)/(36)], and (III) [f_{z^3}, f_{xz^2}, f_{yz^2}, f_{xyz}, $f_{z(x^2-y^2)}$, $f_{y(3x^2-y^2)}$, $f_{x(x^2-3y^2)}$] – rare-earth metal atoms [4f/5f; (37)/(38)]. Moreover, ten/fourteen kinds of the outmost d/f orbitals are capable of creating spin-up and sin-down configurations and thus the Hubbard-like electron-electron on-site Coulomb interactions (39). The kinetic energy, exchange energy, and correlation energy (40), which are closely related to the various orbitals (41), are cooperated with the spin-dependent interactions in determining the total ground state energy. The first-principles methods, but not phenomenological models, are available in thoroughly exploring the diversified quasiparticles and delicately identifying their critical mechanisms due to orbital mixings (42) and spin magnetisms (43). The up-to-date experimental observations are only focused on chemical syntheses and X-ray patterns, while most of the examinations are absent because of intrinsic limits under the quasi-Moire superlattices and quantum confinements, how to overcome them and achieve the simultaneous progress between experimental and theoretical research will be discussed in detail.

Batteries, which store and release energy in the formation of chemical energy, have recently become a focus of research (44–46). When compared to other energy storage systems, LIBs have received a lot of attention because they process desirable features such as lightweight, long circle-life, fast charging time (47,48), and can provide a significant electronic current for electronic devices. In general, a commercial LIB is a complex combination of electrolyte (49) and negative (cathode) (50,51) and positive (anode) electrodes (52,53). Furthermore, the physical/chemical pictures in LIBs (54,55), especially at the electrode-electrolyte interface, are quite complicated and are directly related to the performance of LIBs. The theoretical framework in Chapter 6 is based on first-principles calculations (56), and delicate analyses were developed and applied for the Si-Li_2S hetero-junction, a candidate for the anode-electrolyte compound (57–59). The fundamental features, critical quasiparticle properties, and significant orbital hybridizations in various chemical bonds are thoroughly investigated. The charging and discharging of LIBs are expected to be complicated due to the variation of chemical bonds and thus orbital hybridizations. Our prediction provides meaningful information about the critical mechanisms in LIBs.

The vigorous development of energy devices promotes the use of batteries, in which Lithium batteries occupy a significant position. The main reason comes from remarkable advantages such as they can operate in a wide temperature range, low self-discharge ability, and high emitted energy. In Chapter 7, many metal oxides have been applied to produce its cathode, such as V_20_5, Li_xNiO_2, Li_xCoO_2, Li_xCrO_x, … however, compounds made from Mn atoms dominate because they are cheap and have high performance. In this chapter, to anticipate the remarkable properties of $LiMnO_2$ compounds, such as magnetic and electrical properties, we have fully calculated the characteristics based on density functional theory (DFT) using VASP software. For example, the optimized geometric structure with position-dependent chemical bonding, the atom-dominated energy spectrum at various energy ranges, the spatial charge density distribution due to different orbitals, and the atom- and orbital-projected density of states. In addition, to further understand the magnetic properties of the compound, we analyze the spin-degenerate or spin-split energy bands, the spin-decomposed van Hove singularities, and the atom- & orbital-induced magnetic moments.

Nowadays, Li-ion battery research is focusing on achieving high efficiency, energy storage systems with high capacity and high stability in electrical vehicle devices, therefore, the development of high voltage cathode materials is an important part of Li-ion battery development today. However, the current stage of olivine type high voltage cathode suffers from some issues, which is the low conductivity and ionic conductivity, this limitation results in poor cycling performance which makes them difficult to apply in the electric vehicle's system. To improve the overall performance, experimental findings observed that surface modification can improve the conductivity and ionic conductivity of the olivine type high voltage cathode, but there are few discussions about the interface properties of olivine type high voltage cathode. Therefore, in Chapter 8, first-principles calculation methods can be used in order to give a detailed vision of the surface properties of high voltage cathode, which can provide reasonable explanation for the observed result in the experiment. These can speed up the development of olivine type high voltage cathode and their applications.

Quantum mechanical simulations are highly accurate and informative in small-scale atomistic simulations; however, when it comes to studying complex electrochemical systems, they quickly reach their limit, especially in cases of structural disorder or fully amorphous phases, or when reactions occur at the electrode/electrolyte interface. Moreover, its high computational cost and time requirements make it unsuitable for high-throughput screening. As an efficient and accurate tool for atomistic simulations, machine learning (ML) has gained tremendous popularity in recent years (60–64). ML has the ability to solve complex tasks autonomously, making it a revolutionary approach to discovering material correlations, understanding materials chemistry, and accelerating materials discovery. ML has received widespread attention due to its potential to accelerate productivity in almost every field. It is worth mentioning that ML can be applied to learn the correlations between materials and predict properties and identify novel materials in the batteries area (61,65–68). In Chapter 9, we will be focusing on supervised learning as it is the main method applied in the field of LIB.

LIBs are at the forefront of interdisciplinary research due to several applications in traditional and modern electronic devices and electric/hybrid vehicles (69–73). Almost all the LIBs have graphite as the anode material but the low capacity (372 mAh g−1) (72) is a major concern for applications in high-capacity devices. Metal oxides have higher theoretical capacity and are considered as an alternative to graphite (74). Out of a large variety of metal oxides, SnO_2 is considered a key material due to its high capacity (782 mAhg−1), low cost and non-toxicity. However, low electric conductivity and large capacity fading during charge/discharge cycles are serious issues to overcome. In combination with carbon-based materials, such as graphene, it delivers better electrochemical properties. In Chapter 10, we demonstrate the room temperature chemical reduction procedure for the growth of SnOx quantum dots over RGO sheets and the composite has delivered very good capacity after 100 charge-discharge cycles. Furthermore, doing with Mo has led to enhancing the capacity, cyclic stability, and rate capability of the as-prepared composite.

The demand for secondary batteries with superior capacity, larger energy density, and better safety has sharply risen due to the recent development of electric/hybrid electric vehicles, energy storage devices, portable devices, and household appliances. Among the recently developed secondary batteries, LIBs have received the most attention. The overall performance and safety of LIBs are crucially determined by the electrolytes, carrying Li ions between the electrodes to complete a circuit of charges. The introduction of poly(ionic liquid) (PILs) can render the as-prepared polymer electrolyte (PEs) displaying the intrinsic properties of ILs including negligible flammability, extremely low vapor pressure, and broad electrochemical window. In Chapter 11, the thiol-ene chemistry was chosen to prepare PEs based on imidazolium-based IL and flexible PEG via in-situ solvent-less photopolymerization by depositing precursors directly onto the lithium anode. The combination of thiol-ene click chemistry and in-situ solvent-less polymerization strategy could facilitate the formation of a seamless contact at the PE/electrode interface.

In Chapter 12, we successfully synthesized multiporous carbons through a physical blending method by using water chestnut shell biochar (WCSB) as a carbon source and nano-$CaCO_3$ nanoparticles and K_2CO_3 as activating agents. The multiporous carbons with different specific surface areas and pore sizes could be obtained by modulating the weight ratios of the nano-$CaCO_3$ and K_2CO_3 activating agents to the WCSB. In addition, we also used wasted eggshells as both $CaCO_3$-activating agents and nitrogen dopants to synthesize nitrogen-doped multiporous carbons. Because of high specific surface areas, the assembled supercapacitor with the multiporous carbon has a high-power density and capacitance and a retention rate of nearly 97% after 10,000 cycles. When assembled into a capacitive deionization device, the multiporous carbon electrodes have a large electrosorption capacity for the removal of NaCl salt from aqueous solution.

Confronting the decreasing natural reserved fossil fuels and severe global climate change, renewable energies have been developed for decades before it becomes too late. Solar power holds the potential to offer an unlimited energy supply, and low-dimensional materials seem to stand for the promising opportunity to further improve the existing commercialized silicon-based solar cells. In Chapter 13, recent research of promising methods toward more effective and reliable fabrication processes for low dimensional nanomaterials, and the development of nanostructured high-performance solar cells are comprehensively discussed.

Dye-sensitized solar cells (DSSCs) are highly efficient in producing the small energy needed for electronic applications such as wearable devices, wireless sensors, portable electronics, and the internet of things (IoT) devices. Chapter 14 reviews the electrode and electrolyte materials developed for indoor light DSSCs applications. The introduction section describes the overall development of photovoltaics and their related performance under one sun and room light conditions. Following this section, the importance of the optimization of DSSCs components and their various effects on the performance of the room light DSSCs are briefly described. Thereafter, the effect of titanium dioxide (TiO_2) compact layer, TiO_2 particle size, TiO_2 layer thicknesses, TiO_2 layer structure, and various dyes on the performance of room light DSSCs are elaborately described in the next section. In section 4, the importance of the iodide, cobalt, and copper redox system-based liquid, polymer gel, and printable electrolytes on the efficiency and stability of the laboratory and sub-module DSSCs under ambient lighting are reviewed. Finally, the performance of the counter electrodes fabricated using platinum and poly (3, 4-ethylenedioxythiophene) catalyst layers in the room light DSSCs are explained.

Fuel cells are power supply devices that have advantages such as high efficiency, low noise, and pollutant output. Within several types of fuel cells, the solid oxide fuel cell has shown its excellent performances such as high conversion efficiency and reduced overall cost drastically due to being free of noble metal catalysts. SOFCs can work in both fuel cell mode and electrolysis mode. Excess energy can be stored in chemicals (eg. H2, CO, CH4), and when needed, can be converted to electricity again. The anode materials should have high electrical conductivity, high catalytic efficiency, and a stable reducing environment due to the fuel gas directly contacting the anode materials. The electrolyte is a dense ceramic which is chemically stable in both reducing and oxidizing environments. And to avoid a short circuit across the cell, it must be a sufficiently high ionic and an electronic insulator. For the cathode material, it should have high oxygen reduction reaction (ORR) site. The improvement in the cathode kinetics can directly increase the operating efficiency of SOFCs. For future environmentally friendly energy resources, fuel cells are one of the indispensable empowered technologies for next-generation hydrogen energy production as demonstrated in Chapter 15.

1.2 METHODS

1.2.1 Simulations

1.2.1.1 First-Principle Calculations

Besides theoretical and experimental studies, computer modeling and simulation play an indispensable role in fundamental scientific research. In any system, the presence of the complicated many-body effects is attributed to the electron-electron Coulomb interactions and the electron-ion crystal potential. Apparently, it is rather difficult to accurately deal with the many-particle Schrodinger equation. The evaluation difficulties will be greatly enhanced when the unusual geometric structures and the strong chemical adsorptions need to be taken into consideration. Some approximation methods have been developed to obtain a reliable electronic structure. Till now, the first-principles calculations become an efficient numerical method for studying the essential properties. In this book, the essential properties of emerging materials are investigated by first-principle calculations under the Vienna Ab initio Simulation Package (VASP) codes. In detail, VASP finds the approximate solution within the DFT by solving the Kohn-Sham equation within the Hartree–Fock approximation. Obviously, compared to the phenomenological methods (e. g. tight-binding model), such numerical calculation is very effective in finding the optimal structure as well as the complicated orbital hybridizations in various chemical bonds in many body-particles systems. Thus, many essential properties could be fully explored such as the magnetic quantization phenomena, the quantum spin Hall effect, Coulomb excitation, and magneto-optical properties.

1.2.1.2 Molecular Dynamics Calculations

Molecular dynamics (MD) simulates the motions, interactions, and dynamics at the atomic level of a complex system by choosing a force field that describes all the interatomic interactions and integrating Newtonian equations to determine the position and velocity of atoms over time. A force field describes the interaction energy between atoms/ions in a simulation and is parameterized based on data from experiments and DFT. Currently, simulations using MD can be performed on systems with about 10 million atoms for 1 ns. In this book, the MD calculations are done by means of Large-scale Atomic/Molecular Massively Parallel Simulator (LAMMPS), a widely used MD package.

1.2.2 Machine Learning

The concept of "machine learning" first came up in 1952; however, the lack of a database significantly impeded its development in the material science field. Recently, a large amount of data including material structures and their corresponding properties has been accumulated, not only from experimental measurements but also from first-principles calculations. Therefore, the power of machine learning has gained a lot of attention from researchers. A large number of studies are dedicated to implementing machine learning to learn the relationship between material structures between their properties from existing data in recent years. Gradually, machine learning becomes a powerful method for investigating materials at a large scale in the initial stage of material design without doing time-consumption calculations and costly experiments.

1.2.3 Experimental Measurements

1.2.3.1 X-ray Diffraction (XRD) Spectroscopy

XRD is one of the very efficient methods in measuring the various crystal structures, initiated by Bragg and Laue et al. The high-resolution measurements, which mainly arise from the elastic scatterings between charge densities and electromagnetic waves, are available for sufficiently large samples with observable responses. Without energy transfers, the prominent peak structures could be detected only under the Laue conditions, in which the charge of wave vectors is identical to the reciprocal lattice vectors. The number, intensity and form of them are very sensitive to the incident angles and wavelengths. The XRD diffraction patterns should be attributed to the chemical bonds/the orbital hybridizations in a primitive unit cell. In general, the initial and final lattice symmetries could be identified from the up-to-date XRD measurements, such as the highly non-uniform environments due to various chemical bonds and observable bond-length fluctuations.

1.2.3.2 Scanning Tunneling Microscopy (STM) and Tunneling Electron Microscopy (TEM)

Both STM and TEM are very powerful in detecting nano-scaled geometries with/without local defects. Their high-precision measurements directly provide an active environment of chemical bonds. As for STM, a gate voltage is applied between the nanoproble and material surface, so that a very weak quantum tunneling is greatly enhanced under the increment of voltage. The different probe heights, which correspond to identical currents, are detected at various positions. In the case of TEM, it is probed by an electron beam with suitable kinetic energy, which is available in clarifying the side-view structures of the few-layer/very thin materials. Only the elastic scattering events, being distinguished from the inelastic Coulomb excitation spectra, belong to the diffraction patterns. Apparently, STM and TEM can present the diversified crystal symmetries of low-dimensional materials due to the physical and chemical modifications. Their combinations are able to present the full top and side views of various geometric structures.

1.2.3.3 Scanning Tunneling Spectroscopy (STS) and Angle-Resolved Photoemission Spectroscopy (ARPES)

How to examine and verify electronic properties is one of the mainstream research. STS and ARPES are two kinds of experimental tools that are frequently used to detect basic quasiparticle properties. However, they have advantages and disadvantages. For STS, its high-precision measurements are done on the whole surface through the position-dependent tunneling quantum currents at the same probe height. By the delicate analyses on the differential conductance, the strong dependence on gate voltage is assumed to be roughly proportional to the density of states. That is to say, the prominent structures in the measured results mainly arise from the van Hove singularities. Only the band-edge state energies of valence and conduction energy spectra could be tested in the STS experiments, while the information about the wave-vector dependences cannot be clarified through them. On the other hand, ARPES can directly examine the wave-vector-dependent energy spectra and lifetimes of the occupied quasiparticle states, but not for the unoccupied ones. Part of the information about the latter could be supported by the STS measurements.

REFERENCES

1. Chiu C-W, Shyu F-L, Lin M-F, Gumbs G, Roslyak O. Anisotropy of π-plasmon dispersion relation of AA-stacked graphite. *Journal of the Physical Society of Japan*. 2012;81(10):104703.
2. Ho JH, Lu CL, Hwang CC, Chang CP, Lin MF. Coulomb excitations in AA- and AB-stacked bilayer graphites. *Physical Review B*. 2006;74(8).
3. Lu CL, Chang CP, Huang YC, Ho JH, Hwang CC, Lin MF. Electronic properties of AA-and ABC-stacked few-layer graphites. *Journal of the Physical Society of Japan*. 2007;76(2):024701.
4. Chiu C-W, Chen S-C, Huang Y-C, Shyu F-L, Lin M-F. Critical optical properties of AA-stacked multilayer graphenes. *Applied Physics Letters*. 2013;103(4):041907.
5. Lin C-Y, Ho C-H, Wu J-Y, Lin M-F. Unusual electronic excitations in ABA trilayer graphene. *Scientific Reports*. 2020;10(1):11106.
6. Lin Y-P, Lin C-Y, Ho Y-H, Do T-N, Lin M-F. Magneto-optical properties of ABC-stacked trilayer graphene. *Physical Chemistry Chemical Physics*. 2015;17(24): 15921–15927.
7. Do T-N, Lin C-Y, Lin Y-P, Shih P-H, Lin M-F. Configuration-enriched magneto-electronic spectra of AAB-stacked trilayer graphene. *Carbon*. 2015;94:619–632.

8. Lin C-Y, Huang B-L, Ho C-H, Gumbs G, Lin M-F. Geometry-diversified Coulomb excitations in trilayer AAB stacking graphene. *Physical Review B*. 2018;98(19): 195442.
9. Dodaro JF, Kivelson SA, Schattner Y, Sun X-Q, Wang C. Phases of a phenomenological model of twisted bilayer graphene. *Physical Review B*. 2018;98(7):075154.
10. Stauber T, Kohler H. Quasi-flat plasmonic bands in twisted bilayer graphene. *Nano Letters*. 2016;16(11):6844–6849.
11. Huang Y-K, Chen S-C, Ho Y-H, Lin C-Y, Lin M-F. Feature-rich magnetic quantization in sliding bilayer graphenes. *Scientific Reports*. 2014;4(1):1–10.
12. Martel R, Schmidt T, Shea H, Hertel T, Avouris P. Single-and multi-wall carbon nanotube field-effect transistors. *Applied Physics Letters*. 1998;73(17):2447–2449.
13. Iijima S. Helical microtubules of graphitic carbon. *Nature*. 1991;354(6348):56–58.
14. Savoskin MV, Mochalin VN, Yaroshenko AP, Lazareva NI, Konstantinova TE, Barsukov IV, et al. Carbon nanoscrolls produced from acceptor-type graphite intercalation compounds. *Carbon*. 2007;45(14):2797–2800.
15. Zhang J, Xiao J, Meng X, Monroe C, Huang Y, Zuo J-M. Free folding of suspended graphene sheets by random mechanical stimulation. *Physical Review Letters*. 2010; 104(16):166805.
16. Terrones M, Botello-Méndez AR, Campos-Delgado J, López-Urías F, Vega-Cantú YI, Rodríguez-Macías FJ, et al. Graphene and graphite nanoribbons: Morphology, properties, synthesis, defects and applications. *Nano Today*. 2010;5(4):351–372.
17. Kumar P, Panchakarla L, Rao C. Laser-induced unzipping of carbon nanotubes to yield graphene nanoribbons. *Nanoscale*. 2011;3(5):2127–2129.
18. Kosynkin DV, Higginbotham AL, Sinitskii A, Lomeda JR, Dimiev A, Price BK, et al. Longitudinal unzipping of carbon nanotubes to form graphene nanoribbons. *Nature*. 2009;458(7240):872–876.
19. Yannouleas C, Bogachek EN, Landman U. Collective excitations of multishell carbon microstructures: Multishell fullerenes and coaxial nanotubes. *Physical Review B*. 1996;53(15):10225.
20. Lu X, Chen Z. Curved Pi-conjugation, aromaticity, and the related chemistry of small fullerenes (<C60) and single-walled carbon nanotubes. *Chemical Reviews*. 2005; 105(10):3643–3696.
21. Golden MS, Knupfer M, Fink J, Armbruster J, Cummins T, Romberg H, et al. The electronic structure of fullerenes and fullerene compounds from high-energy spectroscopy. *Journal of Physics: Condensed Matter*. 1995;7(43):8219.
22. Meunier V, Lambin P, Lucas AA. Atomic and electronic structures of large and small carbon tori. *Physical Review B*. 1998;57(23):14886–14890.
23. Shyu FL, Tsai CC, Lin MF, Hwang CC. Electronic properties of carbon tori in external fields. *Journal of the Physical Society of Japan*. 2006;75(10):104710.
24. Shyu F-L, Tsai C-C, Lee C, Lin M-F. Magnetoelectronic properties of chiral carbon nanotubes and tori. *Journal of Physics: Condensed Matter*. 2006;18(35):8313.
25. Hafner J. Ab-initio simulations of materials using VASP: Density-functional theory and beyond. *Journal of Computational Chemistry*. 2008;29(13):2044–2078.
26. Hafner J. Materials simulations using VASP—a quantum perspective to materials science. *Computer Physics Communications*. 2007;177(1-2):6–13.
27. Hafner J, Kresse G. The vienna ab-initio simulation program VASP: An efficient and versatile tool for studying the structural, dynamic, and electronic properties of materials. *Properties of Complex Inorganic Solids*: Springer. 1997. p. 69–82.
28. Ho Y-H, Chiu Y-H, Su W-P, Lin M-F. Magneto-absorption spectra of Bernal graphite. *Applied Physics Letters*. 2011;99(1):011914.
29. Jeon GS, Jain JK. Nature of quasiparticle excitations in the fractional quantum Hall effect. *Physical Review B*. 2003;68(16):165346.

30. Behler J. First principles neural network potentials for reactive simulations of large molecular and condensed systems. *Angewandte Chemie International Edition*. 2017;56(42):12828–12840.
31. Rapaport DC, Rapaport DCR. *The art of molecular dynamics simulation*. Cambridge University Press; 2004.
32. Durmus YE, Zhang H, Baakes F, Desmaizieres G, Hayun H, Yang L, et al. Side by side battery technologies with lithium-ion based batteries. *Advanced Energy Materials*. 2020;10(24):2000089.
33. Mazets I, Kurizki G, Katz N, Davidson N. Optically induced polarons in Bose-Einstein condensates: monitoring composite quasiparticle decay. *Physical Review Letters*. 2005;94(19):190403.
34. Kaloni TP. Tuning the structural, electronic, and magnetic properties of germanene by the adsorption of 3d transition metal atoms. *The Journal of Physical Chemistry C*. 2014;118(43):25200–25208.
35. Cox A, Louderback J, Apsel S, Bloomfield L. Magnetism in 4d-transition metal clusters. *Physical Review B*. 1994;49(17):12295.
36. Beljakov I, Meded V, Symalla F, Fink K, Shallcross S, Ruben M, et al. Spin-crossover and massive anisotropy switching of 5d transition metal atoms on graphene nanoflakes. *Nano Letters*. 2014;14(6):3364–3368.
37. Gerken F. Calculated photoemission spectra of the 4f states in the rare-earth metals. *Journal of Physics F: Metal Physics*. 1983;13(3):703.
38. Mayer MG. Rare-earth and transuranic elements. *Physical Review*. 1941;60(3):184.
39. Watson MD, Backes S, Haghighirad AA, Hoesch M, Kim TK, Coldea AI, et al. Formation of Hubbard-like bands as a fingerprint of strong electron-electron interactions in FeSe. *Physical Review B*. 2017;95(8):081106.
40. Perdew JP, Constantin LA. Laplacian-level density functionals for the kinetic energy density and exchange-correlation energy. *Physical Review B*. 2007;75(15):155109.
41. Edmiston C, Ruedenberg K. Localized atomic and molecular orbitals. *Reviews of Modern Physics*. 1963;35(3):457.
42. Nguyen TDH, Tran NTT, Lin M-F. Open Issues and Potential Applications. *Lithium-Ion Batteries and Solar Cells*: CRC Press; 2021. p. 261–277.
43. Tsymbal EY, Zutic I. *Spin transport and magnetism*. CRC Press; 2012.
44. Liu C, Neale ZG, Cao G. Understanding electrochemical potentials of cathode materials in rechargeable batteries. *Materials Today*. 2016;19(2):109–123.
45. Schlögl R. The role of chemistry in the energy challenge. *ChemSusChem: Chemistry & Sustainability Energy & Materials*. 2010;3(2):209–222.
46. Zu C-X, Li H. Thermodynamic analysis on energy densities of batteries. *Energy & Environmental Science*. 2011;4(8):2614–2624.
47. Li F, Qu Y, Zhao M. Germanium sulfide nanosheet: A universal anode material for alkali metal ion batteries. *Journal of Materials Chemistry A*. 2016;4(22):8905–8912.
48. Zhao G. *Reuse and recycling of lithium-ion power batteries*. John Wiley & Sons; 2017.
49. Zhang SS. A review on electrolyte additives for lithium-ion batteries. *Journal of Power Sources*. 2006;162(2):1379–1394.
50. Aravindan V, Lee YS, Madhavi S. Best practices for mitigating irreversible capacity loss of negative electrodes in Li-ion batteries. *Advanced Energy Materials*. 2017;7(17):1602607.
51. Zhao X, Wang J, Yu R, Wang D. Construction of multishelled binary metal oxides via coabsorption of positive and negative ions as a superior cathode for sodium-ion batteries. *Journal of the American Chemical Society*. 2018;140(49):17114–17119.

52. Dugas R, Zhang B, Rozier P, Tarascon J-M. Optimization of Na-ion battery systems based on polyanionic or layered positive electrodes and carbon anodes. *Journal of The Electrochemical Society*. 2016;163(6):A867.
53. Liu D-H, Lü H-Y, Wu X-L, Wang J, Yan X, Zhang J-P, et al. A new strategy for developing superior electrode materials for advanced batteries: Using a positive cycling trend to compensate the negative one to achieve ultralong cycling stability. *Nanoscale Horizons*. 2016;1(6):496–501.
54. Lin M-F, Hsu W-D, Huang J-L. *Lithium-ion Batteries and Solar Cells: Physical, Chemical, and Materials Properties*: CRC Press; 2021.
55. Han NT, Dien VK, Lin M-F. Excitonic Effects in the Optical Spectra of Lithium metasilicate (Li2SiO3). arXiv preprint arXiv:201011621. 2020.
56. Scheidemantel T, Ambrosch-Draxl C, Thonhauser T, Badding J, Sofo JO. Transport coefficients from first-principles calculations. *Physical Review B*. 2003;68(12): 125210.
57. Winter M, Appel WK, Evers B, Hodal T, Möller K-C, Schneider I, et al. Studies on the anode/electrolyte interface in lithium ion batteries. *Electroactive Materials*: Springer; 2001. pp. 53–66.
58. Liu B, Du M, Chen B, Zhong Y, Zhou J, Ye F, et al. A simple strategy that may effectively tackle the anode-electrolyte interface issues in solid-state lithium metal batteries. *Chemical Engineering Journal*. 2022;427:131001.
59. Shen Z, Zhang W, Zhu G, Huang Y, Feng Q, Lu Y. Design principles of the anode–electrolyte interface for all solid-state lithium metal batteries. *Small Methods*. 2020;4(1):1900592.
60. Shimizu R, Kobayashi S, Watanabe Y, Ando Y, Hitosugi T. Autonomous materials synthesis by machine learning and robotics. *APL Materials*. 2020;8(11):111110.
61. Tong Q, Gao P, Liu H, Xie Y, Lv J, Wang Y, et al. Combining machine learning potential and structure prediction for accelerated materials design and discovery. *The Journal of Physical Chemistry Letters*. 2020;11(20):8710–8720.
62. Cai J, Chu X, Xu K, Li H, Wei J. Machine learning-driven new material discovery. *Nanoscale Advances*. 2020;2(8):3115–3130.
63. Schmidt J, Marques MR, Botti S, Marques MA. Recent advances and applications of machine learning in solid-state materials science. *npj Computational Materials*. 2019;5(1):1–36.
64. Raccuglia P, Elbert KC, Adler PD, Falk C, Wenny MB, Mollo A, et al. Machine-learning-assisted materials discovery using failed experiments. *Nature*. 2016;533(7601): 73–76.
65. Deringer VL. Modelling and understanding battery materials with machine-learning-driven atomistic simulations. *Journal of Physics: Energy*. 2020;2(4):041003.
66. Shandiz MA, Gauvin R. Application of machine learning methods for the prediction of crystal system of cathode materials in lithium-ion batteries. *Computational Materials Science*. 2016;117:270–278.
67. Barrett DH, Haruna A. Artificial intelligence and machine learning for targeted energy storage solutions. *Current Opinion in Electrochemistry*. 2020;21:160–166.
68. Houchins G, Viswanathan V. An accurate machine-learning calculator for optimization of Li-ion battery cathodes. *The Journal of Chemical Physics*. 2020;153(5): 054124.
69. Deng D. Li-ion batteries: basics, progress, and challenges. *Energy Science & Engineering*. 2015;3(5):385–418.
70. Nitta N, Wu F, Lee JT, Yushin G. Li-ion battery materials: Present and future. *Materials Today*. 2015;18(5):252–264.
71. Wu S, Xu R, Lu M, Ge R, Iocozzia J, Han C, et al. Graphene-containing nanomaterials for lithium-ion batteries. *Advanced Energy Materials*. 2015;5(21):1500400.

72. Goriparti S, Miele E, De Angelis F, Di Fabrizio E, Zaccaria RP, Capiglia C. Review on recent progress of nanostructured anode materials for Li-ion batteries. *Journal of Power Sources*. 2014;257:421–443.
73. Etacheri V, Marom R, Elazari R, Salitra G, Aurbach D. Challenges in the development of advanced Li-ion batteries: A review. *Energy & Environmental Science*. 2011;4(9):3243–3262.
74. Zhao Y, Li X, Yan B, Xiong D, Li D, Lawes S, et al. Recent developments and understanding of novel mixed transition-metal oxides as anodes in lithium ion batteries. *Advanced Energy Materials*. 2016;6(8):1502175.

2 Molecular Dynamics Simulation of Amorphous Silicon Anode in Li-Ion Batteries

Li-Yi Pan and Chin-Lung Kuo
Department of Materials Science and Engineering,
National Taiwan University, Taipei, Taiwan

CONTENTS

2.1 INTRODUCTION

Silicon anodes have attracted more interest nowadays due to their about 10-fold specific capacity than the commonly used graphite anode. Silicon anode stores lithium by forming lithium-silicon alloys, called the lithiation of the silicon anode. The lithiation process starts from Si to amorphous lithium silicon alloy (a-Li_xSi) and then crystalline $Li_{15}Si_4$ (c-$Li_{15}Si_4$). The formation of this phase resulted in the 3,579 mAh/g specific capacity for the silicon anode. Nonetheless, silicon anode's high lithium capacity suffers from a high volume expansion of 300% which causes its pulverization. The nanostructured silicon is used to solve the problem.

Experimentally, both crystalline and amorphous silicon (denoted as c-Si and a-Si) have been utilized as anode material. However, their lithiation behavior is quite different. c-Si undergoes a simple one-stage lithiation, forming a crystalline silicon core and a lithium-silicon alloy with a clear phase boundary between the two phases. This behavior is called 2-phase lithiation. In contrast, amorphous silicon undergoes a two-stage lithiation [1]. The lithiation starts by

DOI: 10.1201/9781003367215-2

forming a 2-phase lithiation when the silicon core is fully consumed. After that, it undergoes a 1-phase lithiation, with only the lithium-silicon alloy phase lithiated to its final stage.

In addition to the difference in 2-stage and 1-stage lithiation between *c*-Si and *a*-Si, it is found that crystalline silicon nanowire (*c*-SiNW) and nanoparticle (*c*-SiNP) undergoes anisotropic lithiation. The anisotropic lithiation is a critical problem for the crystalline silicon nanowire/nanoparticle that causes the stress concentration on the concave region of Li–Si alloy. However, amorphous silicon undergoes isotropic lithiation due to its lack of order. There is also a "critical size" found for silicon nanoparticles or nanowires that they crack if their initial diameter is larger than this size. *c*-SiNW is reported to have a critical size of about 300 nm [2], while *c*-SiNP has about 150 nm [3]. In contrast, amorphous silicon nanoparticle [4] is reported to have a critical size larger than 870 nm, which is much greater than the *c*-Si. It is suggested that the volume expansion is lower in the Li–Si alloy generated by amorphous silicon in McDowell *et al.*'s discussion.

In this study, we focused on the lithiation behavior of amorphous silicon. We used the ReaxFF that we developed previously to simulate the lithiation of amorphous silicon and compare it with crystalline silicon. We tried to simulate the 2-stage lithiation of amorphous silicon and find out its mechanism. The general lithiation mechanism of crystalline/amorphous silicon and its controlling factors are revealed. Finally, we tried to find out the reason for the critical size issues for crystalline/amorphous silicon.

2.2 COMPUTATIONAL DETAILS

In this study, we used molecular dynamics to simulate the lithiation behavior of amorphous/crystalline silicon nanowires. The Reactive Force Field (ReaxFF) proposed by van Duin *et al.* [5] is used for describing the Li–Si interaction, while the ReaxFF Li–Si parameter was developed by ourselves in a previous study. The package LAMMPS [6] with the ReaxFF implementation with OpenMP by Aktulga *et al.* [7 and 8] is used. We constructed amorphous silicon and Si[111] nanowire with a diameter of 8 nm for the silicon structure. Li surrounds silicon structures with a ratio of roughly Li:Si = 9:1. The initial structure is presented in Figure 2.1. In order to reduce the high silicon diffusion rate, we used a regional temperature control scheme. The inner core region, which is mainly composed of amorphous silicon core and some lithium for the alloying reaction, is set as 1,100 K. The outer Li–Si alloy and pure lithium are set at 700 K to prevent its diffusion. A buffer layer between the two regions with a length of 10 Å is set as NVE. This regional temperature control scheme will be denoted as NVT 1,100 K/NVE/NVT 700 K. The cell volume is kept fixed during the simulation process.

Two kinds of amorphous silicon are constructed in this study. The first one (highly defected amorphous silicon) is equilibrated at 1,200 K for 100 ps and then relaxed. The second one (weakly defective) used the relaxed structure from the first

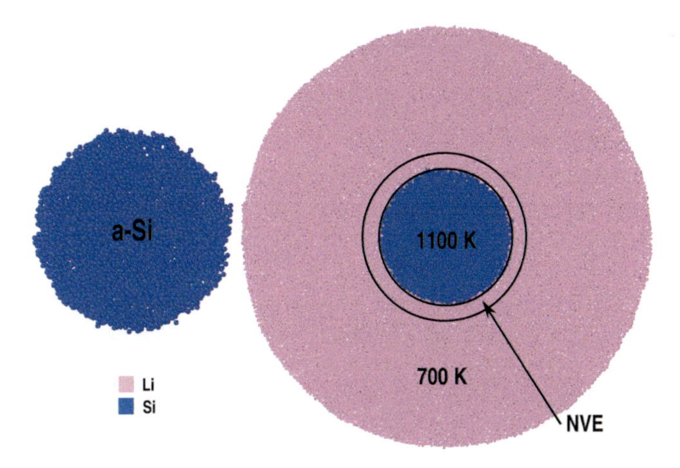

FIGURE 2.1 The initial structure of the amorphous silicon nanowire used in this study.

one and was further equilibrated at 2,500 K 500 ps. It is later quenched to 300 K with a quenching rate of 10^{13}K/s. The lithiated structure of the amorphous silicon for highly/weakly defected amorphous silicon is compared to find out the effect of the defect concentration of amorphous silicon on the Li–Si alloy concentration.

To clarify the lithiation mechanism difference between *c*-Si and *a*-Si, we performed slab analysis in Si(112) and *a*-Si slabs with a similar temperature control scheme as NVT 1,100 K/NVE/NVT 700 K. These slabs are also used for later analysis at NVT 300 K and 500 K for comparing the relative barrier between lithium insertion and Si–Si bond-breaking reaction. The controlling factor is analyzed by the concentration gradient, which is done by dividing the number of lithium by the total number of atoms within a 2 Å shell. To calculate the Si–Si bond breaking rate and the Si diffusion rate, we calculate the Si diffusivity and the corresponding Li diffusivity at 700 K (our temperature for Li–Si alloy) with different compositions as *a*-$Li_{0.25}Si$, *a*-$Li_{0.5}Si$, *a*-LiSi, *a*-Li_2Si, *a*-Li_3Si, *a*-Li_4Si, *a*-Li_5Si by sampling the means square displacement (MSD) for Si and Li for 500 ps in NVT (These structures are pre-equilibrated in NPT for 500 ps).

In order to find out the strength of the *a*-Li_xSi alloy and *c*-$Li_{15}Si_4$, we performed tensile tests for *a*-LiSi, *a*-$Li_{2.5}Si$, *a*-$Li_{3.75}Si$, and *c*-$Li_{15}Si_4$. The structure is constructed by randomly distributing the corresponding Li, Si atoms and equilibrating it at z-directional NPT 1,200 K for 500 ps. It is later equilibrated at NVT 1,200 K for 500 ps, quenched to 300 K with a quench rate of 10^{13}K/s, equilibrated at 300 K by z-NPT for 200 ps and NVT for another 300 ps. It is later quenched in NVT to 1 K for 50 ps and equilibrated at 1 K for 10 ps. The x, y vacuum is added for simulating the nanowire structure and equilibrated for 300 K for 200 ps. For the tensile test, the engineering strain rate is set as 10^8 K/s, and the z-directional stress is extracted and scaled back by the actual area of the Li–Si alloy nanowire used in the tensile test. The toughness is calculated by integrating the area below the engineering stress-strain curve.

2.3 RESULTS AND DISCUSSIONS

2.3.1 Microscopic Evolution of Amorphous Silicon Lithiation

The lithiation snapshot of the amorphous silicon nanowire is shown in Figure 2.2. Its lithiation follows a two-stage lithiation with an amorphous silicon core to Li–Si alloy with lower lithium concentration until 250 ps. The amorphous silicon core is fully consumed after this point, but lithium continues to flow into the Li–Si alloy, which increases its lithium content. This is the second-stage lithiation with only one phase as Li–Si alloy without the silicon core. This result is compared with the Si[111] lithiation case as in Figure 2.3 with only two-phase lithiation. The lithiation rate of amorphous silicon is much higher than its crystalline silicon counterpart, causing its lower Li concentration of 0.77 (about a-$Li_{2.5}Si$) than crystalline 0.85 (about a-$Li_{3.75}Si$) when the silicon core is fully consumed. The lithiation mechanism of amorphous silicon is compared with the crystalline silicon by Si(112) slab as in Figure 2.4. Si(112) shows a ledge peeling-off mechanism as observed in the experiment, while a-Si shows a diffusion-like silicon lithiation due to the lack of structural order and periodicity.

The effect of the defect concentration in the amorphous silicon is presented by the Li–Si alloy concentration in weakly/highly defected amorphous silicon, as in Figure 2.5. The highly defected amorphous silicon shows a much higher lithiation rate (silicon flux from the core) and lower lithium concentration. This indicates that the higher defect concentration will increase the lithiation rate and the relative number of silicon in Li–Si alloy.

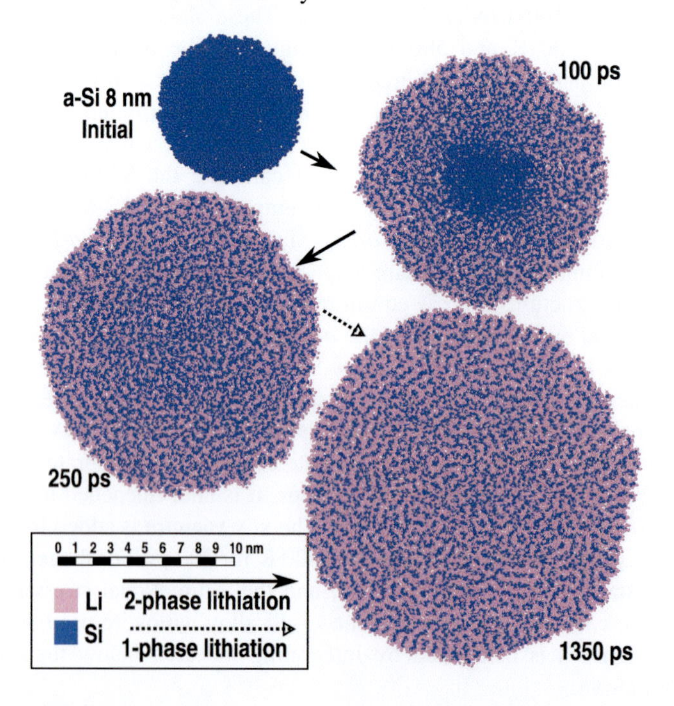

FIGURE 2.2 The lithiation snapshot of amorphous silicon nanowire with diameter 8 nm.

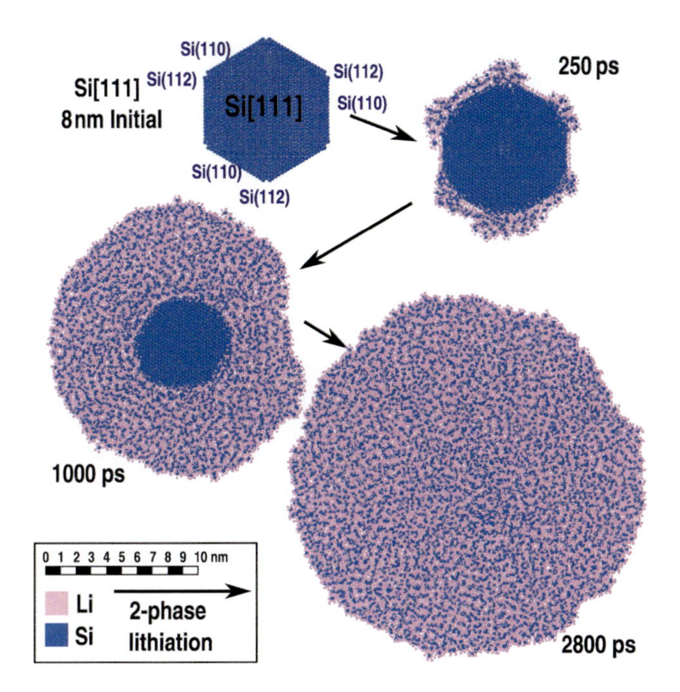

FIGURE 2.3 The lithiation snapshot of Si[111]nanowire with diameter 8 nm.

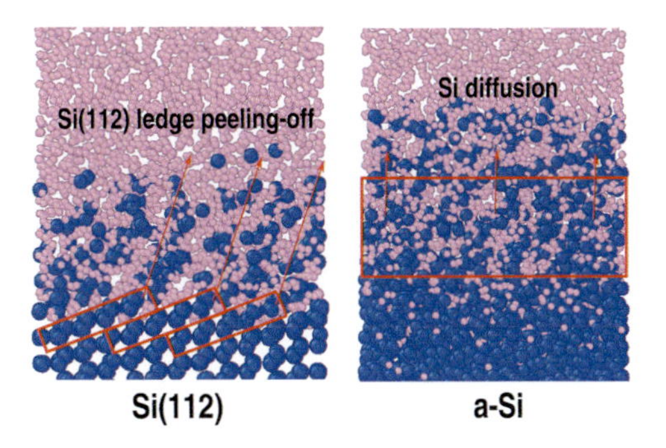

FIGURE 2.4 Comparison of the lithiation behavior for *a*-Si with Si(112) slab.

2.3.2 The Kinetic Process of Silicon Lithiation from a Macroscopic Point of View

To study the macroscopic lithiation process, we need to model the whole lithiation process in regular kinetics terms. The general silicon lithiation process can be viewed as a reaction-diffusion process, as in Figure 2.6. The Si–Si bond-breaking reaction (denoted as "reaction" in our later analysis), and the silicon diffusion in Li–Si alloy (denoted as "diffusion") are the two major processes in the whole lithiation. Adding the lithium insertion in the reaction front (denoted as "insertion"), it becomes an insertion-reaction-diffusion system.

We calculated the concentration gradient for crystalline/amorphous silicon nanowires, while the crystalline silicon nanowire is the Si[111] nanowire, which undergoes isotropic lithiation. The phase region and its corresponding concentration profile are shown in Figure 2.7. The lithiation of a crystalline silicon nanowire, as in Figure 2.8

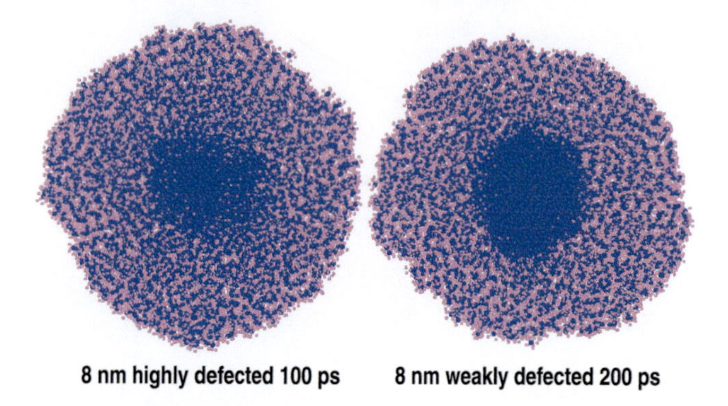

FIGURE 2.5 The lithiation snapshot of highly defected and weakly defected amorphous silicon. The lithium concentration is higher in the weakly defected case.

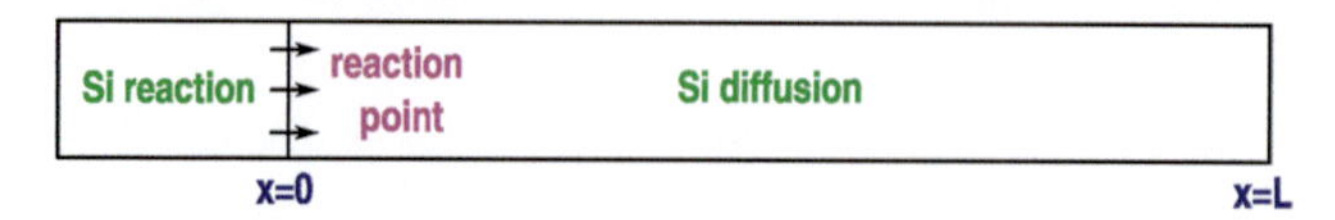

FIGURE 2.6 The reaction-diffusion system in crystalline and amorphous silicon.

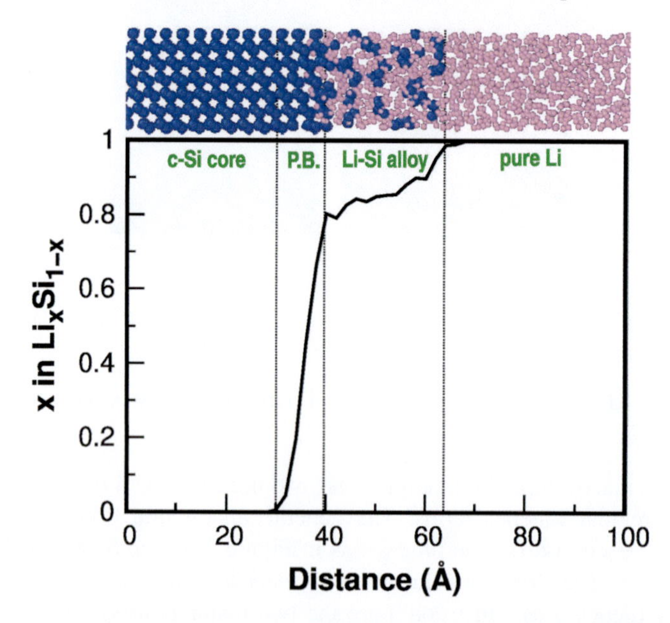

FIGURE 2.7 Illustration of concentration profile of Si[111] 8 nm silicon nanowire at 500 ps with its corresponding phase region. P.B. denotes the phase boundary between Si and Li-Si alloy.

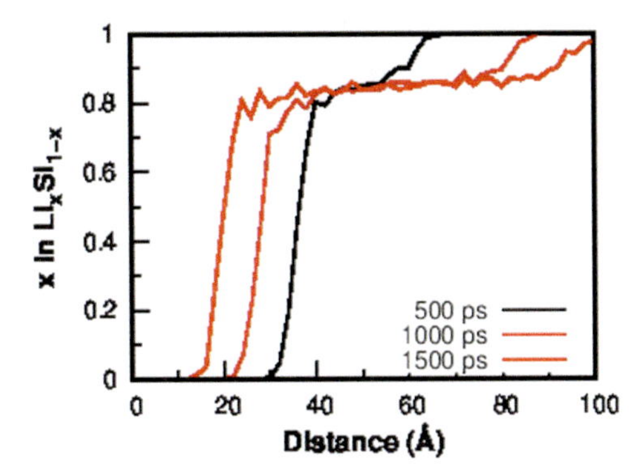

FIGURE 2.8 The concentration profile of Si[111] 8 nm silicon nanowire.

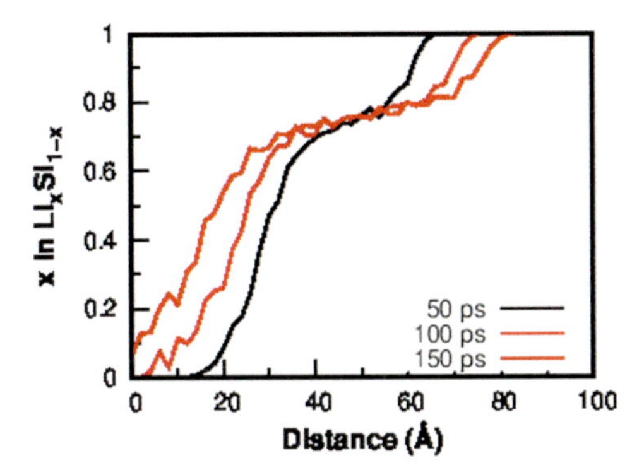

FIGURE 2.9 Concentration of amorphous silicon nanowire 8 nm.

shows a sharp concentration gradient in the phase boundary. In addition, the concentration gradient is very low in the Li–Si alloy layer. In contrast, amorphous silicon, as in Figure 2.9 shows a much higher concentration gradient in Li–Si alloy layer. This result confirms that the 2-phase lithiation in amorphous silicon is not a homogeneous process, which has been suggested by a previous study [4]. Besides, the silicon diffusion rate in amorphous silicon is much slower than that in crystalline silicon. This result also suggests that crystalline silicon lithiation is a reaction-controlled process, while amorphous silicon lithiation is also reaction-controlled with a slower diffusion process.

In the following discussion, we tried to determine the controlling reaction (the bottleneck reaction) in crystalline/amorphous silicon lithiation. We argue that the Si-Si bond-breaking reaction is the slowest among the insertion-reaction-diffusion system. First, we argue that the Si–Si bond-breaking reaction is slower than the lithium insertion process. Figure 2.10 and 2.11 shows the lithiation reaction at different temperature for *a*-Si and Si(112) slab, respectively. Lithium insertion occurs at a much lower temperature as 300 K for both *a*-Si and Si(112) slab, while Si-Si bond breaking is activated at 500 K only for amorphous silicon and not for Si (112). This result shows that the Si-Si bond-breaking reaction is much harder

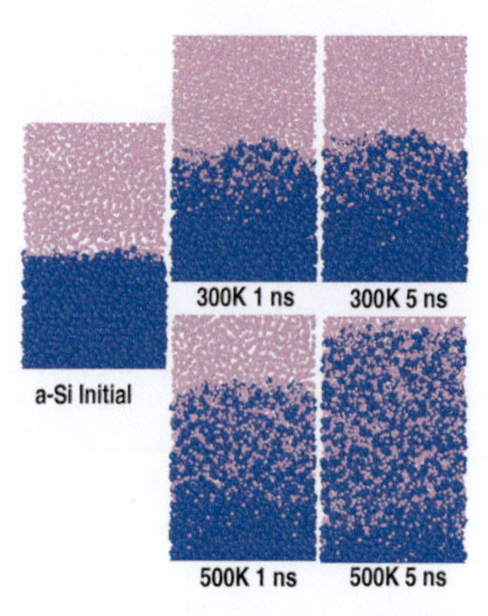

FIGURE 2.10 Amorphous silicon slab lithiation at 300 K and 500 K.

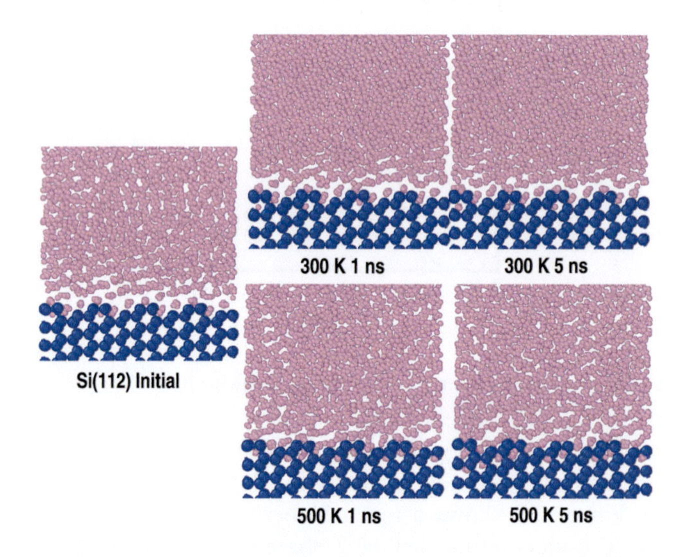

FIGURE 2.11 Si(112) slab lithiation at 300 K and 500 K.

and slower than the lithium insertion rate. In addition, the Si-Si bond breaking is much easier in amorphous than in crystalline silicon.

Second, we argue that Si-Si bond-breaking reaction is slower than the silicon diffusion rate in Li–Si alloy. This can be inferred from the silicon diffusivity result in Figure 2.12. We will first discuss the case for amorphous silicon lithiation. Its Si–Si bond-breaking reaction rate near the boundary should be directly related to Si diffusivity, due to the diffusion-like nature of the amorphous silicon lithiation. In addition, since our lithiation is a reaction-diffusion system, the Li concentration in the Li–Si alloy layer will always be higher than at the phase boundary. From Figure 2.12, the silicon diffusion becomes faster in the higher lithium-concentrated Li–Si alloy. This means that the silicon diffusion rate in Li–Si alloy will always be higher than the reaction rate at the boundary and proves our statement for amorphous

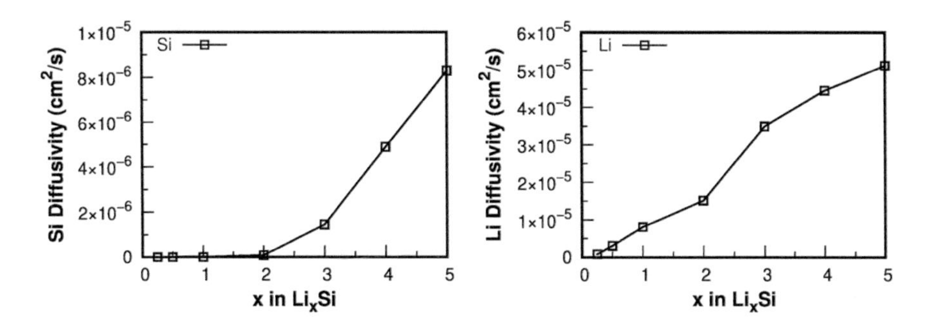

FIGURE 2.12 The diffusivity of Si and Li at 700 K in different Li/Si concentrations.

silicon. That is, the Si-Si bond-breaking reaction rate is slower than the Si diffusion rate in Li–Si alloy for amorphous silicon. Nonetheless, the concentration of Li–Si alloy will not differ much in the phase boundary and the Li–Si alloy layer. Therefore, the Si diffusion rate will only be slightly faster than the reaction rate in amorphous silicon.

The same argument can be applied to crystalline silicon. Continuing from our result that amorphous silicon has a slower reaction rate than diffusion rate. Crystalline silicon is found to have a much slower Si–Si bond-breaking reaction rate than amorphous silicon. Besides, the lithium concentration in Li–Si alloy is much higher for crystalline silicon than for amorphous silicon, which implies a higher diffusion rate in Li–Si alloy than amorphous silicon. This means that the Si diffusion rate is much higher than the Si–Si bond-breaking reaction rate in crystalline silicon, and our statement for crystalline silicon is also proved.

Following these discussions, we found that the Si–Si bond-breaking reaction is the slowest and is the controlling factor for crystalline/amorphous silicon lithiation. The much higher diffusion rate for crystalline silicon than the Si–Si bond breaking rate makes it a reaction-controlled process. The close diffusivity for amorphous silicon than the Si–Si bond-breaking rate makes it a **mixed-controlled process with a more reaction-controlled feature**.

Here, we shall shortly summarize the reason for the two-stage lithiation of amorphous silicon. The faster Si-Si bond-breaking rate of amorphous silicon makes Li–Si alloy with lower lithium concentration, slowing down the silicon diffusivity and further impeding the silicon diffusion in Li–Si alloy as in Figure 2.12. This makes a different Li–Si alloy concentration when crystalline/amorphous silicon core is fully consumed. The lower lithium concentrated Li–Si alloy for amorphous silicon will still accommodate more lithium after the silicon core is fully consumed, which causes the second-stage lithiation in amorphous silicon.

2.3.3 Origin of the Critical Size in Crystalline/Amorphous Silicon

Hoop stress on Li–Si alloy layer is a major reason for the cracking of silicon nanowires [9], while stress concentration makes the status worse. Many theoretical analyses state that the hoop stress does not increase during the 2-phase lithiation process [9]. It does not even increase with the initial size of the silicon nanowire/nanoparticle by the same argument. Therefore, it is crucial to find out the real reason for cracking the silicon anode. In this part, we tried to find out the reason for the higher critical size for amorphous silicon (nanoparticle larger than 870 nm by McDowell et al. [4]) than crystalline silicon (nanoparticle as 150 nm by Liu et al. [3]).

Experimentally, *c*-$Li_{15}Si_4$ appears when *a*-Li_xSi reaches x = 15/4 [10]. This process occurs very fast without long-range diffusion [11]. Furthermore, the formation of *c*-$Li_{15}Si_4$ is detrimental to the performance of silicon anode [12]. Combining these studies, we argue that the formation of *c*-$Li_{15}Si_4$ is the main reason for the critical size in silicon nanowire/nanoparticle. First, we claim that the formation of *c*-$Li_{15}Si_4$ hampers the mechanical properties, which is proven by tensile tests on various *a*-Li_xSi and *c*-$Li_{15}Si_4$ as in Figure 2.13. The toughness is calculated by integrating the area below the stress-strain curve in Table 2.1. Although *c*-$Li_{15}Si_4$ shows higher ultimate tensile strength, its toughness is much lower than its amorphous counterpart, even at a similar composition as *a*-$Li_{3.75}Si$. This might result from the ordered crystal structure and the ionic character of *c*-$Li_{15}Si_4$. This means that *c*-$Li_{15}Si_4$ is much more brittle when compared with *a*-Li_xSi when $x < 3.75$ and contributes to the cracking of the silicon anode.

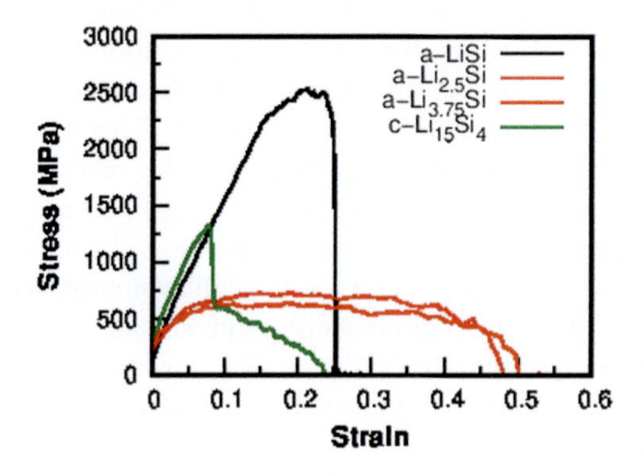

FIGURE 2.13 The tensile test of *a*-LiSi, *a*-$Li_{2.5}Si$, *a*-$Li_{3.75}Si$, and *c*-$Li_{15}Si_4$.

TABLE 2.1
The Calculated Toughness of Some Amorphous and Crystalline Li–Si alloy by Integrating the Area below the Engineering Stress/Strain Curve

Sample	Toughness (MPa)
a-LiSi	426
a-$Li_{2.5}Si$	289
a-$Li_{3.75}Si$	263
c-$Li_{15}Si_4$	133

Since the brittle c-$Li_{15}Si_4$ formation cracks the silicon anode, the reason why c-$Li_{15}Si_4$ forms become a critical issue in crystalline/amorphous silicon. The reason will be clear from our previous discussion of the insertion-reaction-diffusion nature of silicon lithiation as in Figure 2.6. Amorphous silicon, which has a higher silicon lithiation rate and lower lithium concentration in a-Li_xSi, tends to have a longer time to reach the concentration of $x = 15/4$ for c-$Li_{15}Si_4$ than crystalline silicon. This leads to a higher critical size for amorphous silicon than its crystalline silicon counterpart. In addition, a longer silicon diffusion path for Li–Si alloy exists in a larger silicon nanowire. This makes the formation of c-$Li_{15}Si_4$ easier on the outer surface of Li–Si alloy, where higher lithium flux and lower silicon flux than the inner region of Li-Si alloy. Therefore, the size-dependent cracking of the silicon anode is related to the length of diffusion path, which causes c-$Li_{15}Si_4$ formation. As a result, if one wants to increase the critical size of silicon nanostructures, the silicon bond-breaking rate should be increased. Therefore, it will be beneficial to increase the defect concentration of amorphous silicon or decrease the Si-Si bond strength for crystalline silicon to improve the performance of the silicon anode.

2.4 CONCLUSIONS

In this study, we studied amorphous silicon lithiation by our previously developed ReaxFF model. We reproduced the two-stage lithiation of amorphous silicon. The lithiation of amorphous silicon shows a more diffusion-like behavior, different from the ledge mechanism observed in crystalline silicon. The general lithiation behavior of silicon is revealed, while crystalline silicon lithiation is reaction-controlled, and amorphous silicon lithiation is mixed control with a more reaction-controlled feature. The two-stage lithiation of amorphous silicon is caused by the higher Si-Si bond-breaking reaction rate. Besides, the high Si content in Li–Si alloy further impeded silicon diffusion and caused an unsaturated Li–Si alloy. Therefore, the Li–Si alloy further lithiates after the silicon cores are fully consumed, causing the second-stage lithiation. Finally, the cracking and critical size issues for amorphous and crystalline silicon are related to the formation of c-$Li_{15}Si_4$. Amorphous silicon, which has a smaller x for a-Li_xSi, is less prone to form the c-$Li_{15}Si_4$, and has a larger critical size than crystalline silicon. In addition, the larger nanoparticle/nanowire tends to have a longer silicon diffusion path and a smaller silicon flux at outer surface of Li–Si alloy. This makes c-$Li_{15}Si_4$ formation easier and crack the Li-Si alloy layer. These phenomena makes the Li/Si concentration in Li-Si alloy closer to c-$Li_{15}Si_4$ and contribute to the critical size phenomenon for silicon nanowire/nanoparticle.

REFERENCES

1. J. W. Wang, Y. He, F. Fan, X. H. Liu and S. Xia *et al.* 2013 Two-phase electrochemical lithiation in amorphous silicon. *Nano Letters*, **13** (2), 709–715. doi:10.1021/nl304379k.
2. I. Ryu, J. W. Choi, Y. Cui and W. D. Nix. 2011 Size-dependent fracture of Si nanowire battery anodes. *Journal of the Mechanics and Physics of Solids* **59**(9), 1717–1730. doi:10.1016/j.jmps.2011.06.003.

3. X. H. Liu and J. Y. Huang. 2011 In situ TEM electrochemistry of anode materials in lithium ion batteries. *Energy & Environmental Science*, **4** (10), 3844. doi:10.1039/c1ee01918j.
4. M. T. McDowell, S. W. Lee, J. T. Harris, B. A. Korgel and C. Wang *et al.* 2013 In situ TEM of two-phase lithiation of amorphous silicon nanospheres. *Nano Letters*, **13** (2), 758–764. doi:10.1021/nl3044508.
5. A. C. T. van Duin, S. Dasgupta, F. Lorant and W. A. Goddard. 2001 ReaxFF: a reactive force field for hydrocarbons. *Journal of Physical Chemistry A*, **105** (41), 9396–9409. doi:10.1021/jp004368u.
6. S. Plimpton. 1995 Fast parallel algorithms for short-range molecular dynamics. *Journal of Computational Physics*, **117** (1), 1–19. doi:10.1006/jcph.1995.1039.
7. H. M. Aktulga, J. C. Fogarty, S. A. Pandit and A. Y. Grama. 2012 Parallel reactive molecular dynamics: Numerical methods and algorithmic techniques. *Parallel Computing*, **38**, 245–259.
8. H. M. Aktulga, C. Knight, P. Coffman, K. A. OHearn and T. R. Shan *et al.* 2019 Optimizing the performance of reactive molecular dynamics simulations for multi-core architectures. *International Journal of High Performance Computing Applications*, **33** (2), 304–321.
9. Z. Jia and T. Li. 2016 Intrinsic stress mitigation via elastic softening during two-step electrochemical lithiation of amorphous silicon. *Journal of the Mechanics and Physics of Solids*, **91**, 278–290. doi: 10.1016/j.jmps.2016.03.014.
10. M. N. Obrovac and L. Christensen. 2004 Structural changes in silicon anodes during lithium insertion/extraction. *Electrochemical and Solid-State Letters*, **7** (5), A93. doi:10.1149/1.1652421.
11. M. Gu, Z. Wang, J. G. Connell, D. E. Perea and L. J. Lauhon *et al.* 2013 Electronic origin for the phase transition from amorphous LixSi to crystalline $Li_{15}Si_4$. *ACS Nano*, **7** (7), 6303–6309. doi:10.1021/nn402349j.
12. D. Iaboni and M. Obrovac. 2015 $Li_{15}Si_4$ formation in silicon thin film negative electrode. *Journal of The Electrochemical Society*, **163**, 2. doi:10.1149/2.0551602jes.

3 Rich Intercalations in Graphite Magnesium Compounds

Yu-Ming Wang, Jheng-Hong Shih, Wei-Bang Li, and Thi Dieu Hien Nguyen
Department of Physics, National Cheng Kung University, Tainan City, Taiwan

Ming-Fa Lin
Department of Physics and Hierarchical Green-Energy Material (Hi-GEM) Research Center, National Cheng Kung University, Tainan City, Taiwan

CONTENTS

PREFACE

Carbon atoms, which belong to group-IV elements, have four active orbitals of [2s, 2pz, 2py] in creating the various crystal phases. Furthermore, the up-to-date successful syntheses clearly show the existence of organic and non-organic materials. For example, the former include 3D diamond [1,2], bulk simple hexagonal/Bernal/rhombohedral/turbostratic graphite systems [3–7], the 2D layered graphenes with AAA [8], ABA [9], ABC [10] and AAB stackings [11], the sliding/rotational bilayer graphene systems [12,13], the 1D cylindrical and coaxial carbon nanotubes [14], the finite-size graphene nanoribbons with chiral/achiral open edges under the planar/curved/folded/scrolled structures [15–19], and 0D carbon fullerenes/onions/disks/tori/rings/chains [20–25]. The diverse chemical bondings in C-C bonds, sp^3, sp^2, and sp [26,27], are

DOI: 10.1201/9781003367215-3

thoroughly identified to be responsible for the low-energy quasiparticle properties. This clearly illustrates represents the successful cooperation between the theoretical [28,29] and experimental science researches [30–32]. The pure orbital hybridizations will become richer and/or complicated by the chemical modifications [33–35], the physical perturbations [36–39], and material engineering, e.g., donor/acceptor-type graphite intercalation compounds [40,41], magnetic quantization phenomena of layered bulk graphite systems [42], and graphite-carbon-nanotubes composites/carbon nanotube bundles [43]. Such critical modulations can greatly generate graphene-related materials in developing the basic [31,32,44] and applied sciences [33].

The bulk graphite systems are layered semi-metals [7,29,45,46], being totally different from the insulating diamond [47]. This indicates the significant differences between sp^2 and sp^3 chemical bondings. The former has a 2D honeycomb structure with the strongest mechanical property among any condensed-matter systems [48], in which the interlayer interactions are delicately identified to be van der Waals of $2p_z$-$2p_z$ orbital mixings [49,50]. The stacking configurations can present the well-characterized AA [8], AB [8], and ABC ones [51,52]. Such graphitic spacing is responsible for the unusual quasiparticle phenomena and the high applications, e.g., the semi-metallic behaviors sensitive to stacking configurations [53], as well as, the high performance of anode material about the ion-based batteries [54]. The first/third one, with the highest/lowest symmetry, possesses the largest free carrier densities for the band-overlap-induced valence holes and conduction electrons [55]. However, the Bernal graphite frequently appears in the natural statuses [56,57], e.g., more than 95% and below 5%, respectively, corresponding to AB and ABC stacking configurations according to the up-to-date experimental examinations [58]. Very interesting, each planar carbon honeycomb provides a super-active chemical/physical/material environment under strong anisotropy. The unusual quasiparticle behaviors are further by the strong modifications, e.g., the great enhancement of superconductivity temperature in the presence of donor- or acceptor-type dopings [59]. Obviously, the chemical bonds and their orbital hybridizations will dramatically change after the adatom intercalations How to obtain the critical orbital hybridization. In addition to the intralayer and interlayer carbon-carbon, there exist the carbon-intercalant and intercalant-intercalant ones. The rich chemical bondings will be achieved by thoroughly exploring the crystal symmetries and electronic properties under the first-principles simulations. How to verify the theoretical predictions from the high-precision measurements needs to provide full discussions.

As to the phenomenological modes, many previous studies have successfully illustrated the diverse quantum quasiparticles in 3D layered graphite systems [details in books, [42,45,60]]. A planar honeycomb crystal, with the same heights of A and B sublattices, possesses the perpendicular π- and σ-electronic bondings [$2p_z$ and ($2s$, $2p_x$, $2p_y$)]. The latter present at the deeper energies, so their contributions to the low-energy quasiparticle phenomena are negligible. The carbon-$2p_z$ orbital hybridizations are sufficient in observing almost all of composite quasiparticles. The up-to-date theoretical and experimental researches clearly show that the single-particle Hamiltonians are very successful in fully understanding the stacking-enriched phenomena [11,45,61]. For example, the AA- [8], AB- [8] and ABC-stacked [51] bulk graphites present the different interlayer $2p_z$ orbital hybridizations

[49,50] and thus the diversified band structures/optical absorption spectra/plasmon modes [62–67]. The electronic/optical/Coulomb-excitation properties are calculated by the tight-binding model/the dynamic Kubo formula/the random phase approximations. As to graphite interaction compounds, the superlattice model is roughly utilized to investigate the optical plasmon modes due to the intercalation-induced free carriers [68], as well as, the quasiparticle energy spectra and lifetimes [69–71]. This qualitative model is worthy of further modifications. The ignored mechanisms need to be recovered by the necessary addition of the interlayer carbon intercalant interactions and intercalant-intercalant hybridizations. Certain intrinsic interactions will be thoroughly discussed in the modified Hamiltonian [29,72].

The first-principles methods are very suitable for studying the rich intercalation phenomena in layered graphitic systems [73–75], The composite quasiparticle will be greatly diversified by the guest-atom configurations and their weak, but significant interlayer interactions. The critical mechanisms, being accompanied by magnesium guest adatoms, will be thoroughly examined from all consistent quantities [76]. The VASP simulations will be done through the various Mg concentrations with the neighboring layers of graphene and intercalant [Figures 3.1(a)–3.1(f)]. The calculation results cover the various crystal symmetries with the top and side views, the featured band structures in terms of carbon and magnesium dominances [76], the spatial charge density distributions in the presence of intralayer and interlayer orbital hybridizations

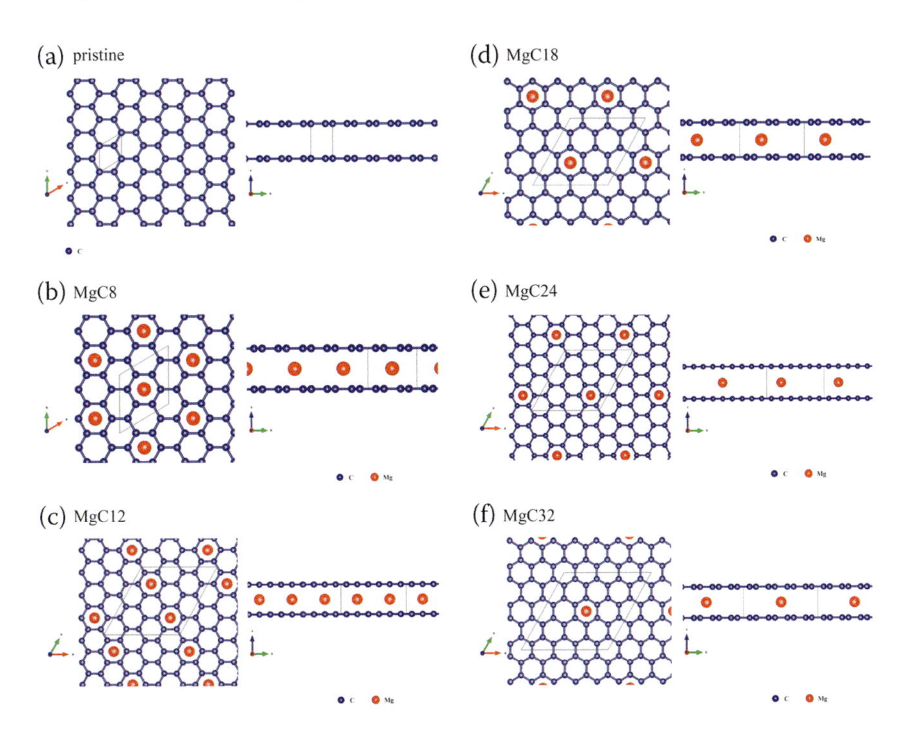

FIGURE 3.1 The optimal crystal structures of graphite magnesium-adatom compounds through the top and side views under the various intercalation concentrations: (a) a pristine system, (b) MgC8, (c) MgC12, (d) MgC18, (e) MgC24 and (f) MgC32.

[49,50], and the unusual van Hove singularities from the atom- and orbital-projected contributions [77]. The dramatic transformation of chemical bondings should be responsible for the semimetal-metal transitions and thus play critical roles in the low-energy physics behaviors, e.g., the plasmon modes due to free conduction electrons [donor-type dopings; [59]], but independent of weak band overlaps [78].

How to build the phenomenological models with the important host-guest interlayer hybridizations is expected to be the next studying focus. This is very useful in developing the quasiparticle framework, as clearly illustrated in the previously published books. Moreover, the important differences among the lithium-, sodium-, magnesium-, iron- and aluminum-ion-based batteries are discussed in detail, such as a lot of merits and drawbacks.

3.1 LAYERED GRAPHITE GROUP-II-RELATED COMPOUNDS

The unusual features of nitrogen substitutions in silicene are investigated in detail by using the VASP code [79,80]. The exchange and correlation energies due to many-particle electron–electron interactions are calculated with the use of the Perdew–Burke– Ernzerhof (PBE) functional under the generalized gradient approximation [81]. Furthermore, the electron–ion interactions can be characterized by the projector-augmented wave (PAW) pseudopotentials [82]. A space of 15 A width is inserted between the silicene planes to avoid their interaction. In general, a planewave basis set with a maximum kinetic energy cutoff of 500 eV is chosen to expand the wave function. The first Brillouin zone is sampled by $5 \times 5 \times 5$ and $40 \times 40 \times 9$ k-point meshes within the Monkhorst–Pack scheme for further calculations of the electronic and magnetic properties. The energy convergence is set to be 10^{-8} eV between two consecutive steps and the maximum Hellmann–Feynman force acting on each atom is less than 0.01 eV Å-1 during the ionic relaxations.

The 3D bulk graphite systems have the most stable honeycomb structures, but very weak van der Waals interlayer interactions [49,50], The graphitic spacings are very useful in the basic [83] and applied sciences [84]. They are available in greatly diversifying the various crystal phases by the adatom intercalations, e.g., the donor-type intercalations of group-I and group-II metal adatoms, as well as, the acceptor-type ones about $AlCl_4/Al_2Cl_7$ [85] and $FeCl_3$ [86]. Furthermore, graphite in the former [29] and the latter [87], respectively, serves as the commercial cathode and anode of ion-based batteries. In general, the intercalant layers, which appear in the planar forms, play the critical roles in the operation stability of charging and discharging processes through the stationary ion transports. Their concentrations and distributions will determine the diversified crystal symmetries under the intralayer/interlayer interactions of honeycomb and intercalant lattices [88,89] and the host-quest interlayer hybridizations [90]. The very active environments can provide rich chemical bondings in observing and identifying the composite quasiparticles and thus developing an enlarged framework [the details in the previously published books].

The neutral magnesium atoms [Mg], but not ionic configurations [Mg^{2+}], are chosen as quasiparticles into graphite during the charging and discharging processes, as shown by the alkali atoms [e.g., Li]. When Mg^{2+} ions are initiated from the cathode material, the negative charges of electrons circulate an external lead line and quickly accumulate in the graphitic anode. All Fermions are very light, display the super-screening ability and thus enclose each ionic Mg^{2+}. Such guest intercalants should be close to the neutral status, since the extra chemical bonings between the guest and host atoms can greatly strengthen the geometric stability. On the other side, Mg^{2+} ions hardly create significant couplings with the neighboring graphitic sheets; that is, the n-type doping effects almost disappear under this chemical case. Such configurations are predicted to survive in electrolytes [91] and separators, in which they experience/generate the macroscopic Coulomb fields during the stationary ion flows. Very interesting, a lot of intermediate configurations come to exist in battery operations [92]. The initial or final stable crystal structures are frequently illustrated in the VASP simulations. How to combine the first-principles method and the chemical reactions can present the time-dependent diverse crystal phases [93].

Each carbon honeycomb structure, which presents a very active environment of chemistry, physics, and material engineering, has the perpendicular π-electronic bondings of the half occupied. The pristine AA- [8,62], AB- [8,65], and ABC-stacked [51,94] bulk graphites clearly show the different interlayer C-$2p_z$ orbital hybridizations [the distinct van der Waal interactions]. Their significant difference could be distinguished from the well-known position-dependent interaction formula. The Bernal graphite, with an optimal interlayer distance of 3.35 Å [the top view and the side view in Figure 3.1(a)], shows the lowest ground state, being consistent with the experimental observations [56,57]. The layered graphene structures are very suitable for intercalations/de-intercalations, substitutions [33,95,96], and adsorptions [few layers; [97,98]]. For example, graphite magnesium compounds, as clearly illustrated in Figures 3.1(b)-3.1(f), display the various crystal structures through the distinct Mg-adatom concentrations. The intercalant Mg-layer could survive with the neighboring graphitic sheet so that the total intercalation density of guest atoms is highest during the charging/discharging processes of the stationary ion transport [99]. By the delicate VASP simulations, the host-intercalant distances/Mb-C bond lengths [Figure 3.1 and Table 3.1] are, respectively, 3.48/1.46, 3.26/1.41, 3.32/1.46, 3.33/1.41 for MgC8, MgC12, MGC18, MgC24m, and MgC32. Apparently, the crystal

TABLE 3.1
The Geometric Parameters about (a) Pristine Systems, (b) MgC8, (c) MgC12, (d) MgC18, (e) MgC24, and (f) MgC32

	C-C Bond(Å)	Interlayer Distance(Å)	Distance between Mg and Mg(Å)
(a)pristine	1.41	3.24	–
(b)MgC8	1.46	3.48	4.98
(c)MgC12	1.41	3.26	4.27, 7.39, 8.54
(d)MgC18	1.46	3.32	7.4
(e)MgC24	1.41	3.33	8.54
(f)MgC32	1.42	3.41	9.87

symmetries are mainly determined by the intralayer host and quest-atom orbital hybridizations, as well as, their interlayer interactions. In addition, the chemical substitution of Mg-enriched few-layer graphene systems will be totally different from graphite magnesium compounds, mainly owing to more complicated chemical bondings in the latter [100].

3.2 FEATURED HOLE AND ELECTRON STATES WITH HOST AND GUEST ATOM DOMINANCES

By the delicate VASP simulations and analyses, the critical Mechanisms, the intralayer and interlayer orbital hybridizations due to the host and guest atoms, are responsible for a lot of significant features in quasiparticle energy spectra and wave functions. Apparently, the quasiparticle behaviors are greatly diversified by the stacking configurations with the distinct interaction interactions between neighboring layers [49,50,101,102], the zone folding effects associated with the diluter distributions along the different/same directions, and the semimetal-metal transitions arising from electron affinities. The theoretical predictions on occupied valence hole and conduction electron energy spectra could be verified from the high-precision measurements of angle-resolved photoemission spectroscopy [ARPES; [103–107]], The main reason is that the concise band structures exhibit the well-defined low quasiparticle energy spectra near the K and H valleys. This is capable of ignoring the intrinsic limit of the non-conserved momentum transfer under a sample surface.

The diverse quasiparticle energy spectra and wave functions are clearly revealed in graphite-related systems. For exemplum, graphite magnesium compounds [Figures 3.2(b)–3.2(k)], pristine bulk graphite [Figure 3.2(a)], and monolayer graphene [108], respectively, belong to metallic, semi-metallic, and zero-gap semiconducting systems. The third 2D material has only one hexagonal crystal structure so that three nearest-neighboring carbon atoms can create the linear and anisotropic Dirac cone from the K/K′ valleys [details [109]]. Furthermore, the linear valence and conduction subband just intersect at the Fermi level, leading to a zero density of states there [the absence of free carriers and their 2D plasmons at very low temperatures [110]]. The interlayer van der Waals further creates the weak, but important overlap near the stable valleys [49], in which free valence holes and conduction electrons appear simultaneously with the same low-density cases [Figure 3.2(a)]. Both free carriers make the same contributions to the low-energy physical phenomena, such as the temperature- and polarization-dependent plasmon modes of about 100 meV in Bernal graphite. Very interesting, the donor-/acceptor-type intercalants can create the large electron/hoe transfer; that is, the interlayer guest-host interactions induce the blue/red shift of the Fermi level relative to the Dirac-cone-like valleys [discussed later in the density of states in Figures 3.4(a)–3.4(p)]. The strong intercalations effects are verified to exhibit the prominent 3D plasmon modes from the high-resolution measurements of electron energy loss spectra [EELS; [111]], as well as, the high threshold absorption frequency by measuring the reflectance spectra [112]. The observable modulations of the Fermi level provide an outstanding research strategy for the further development

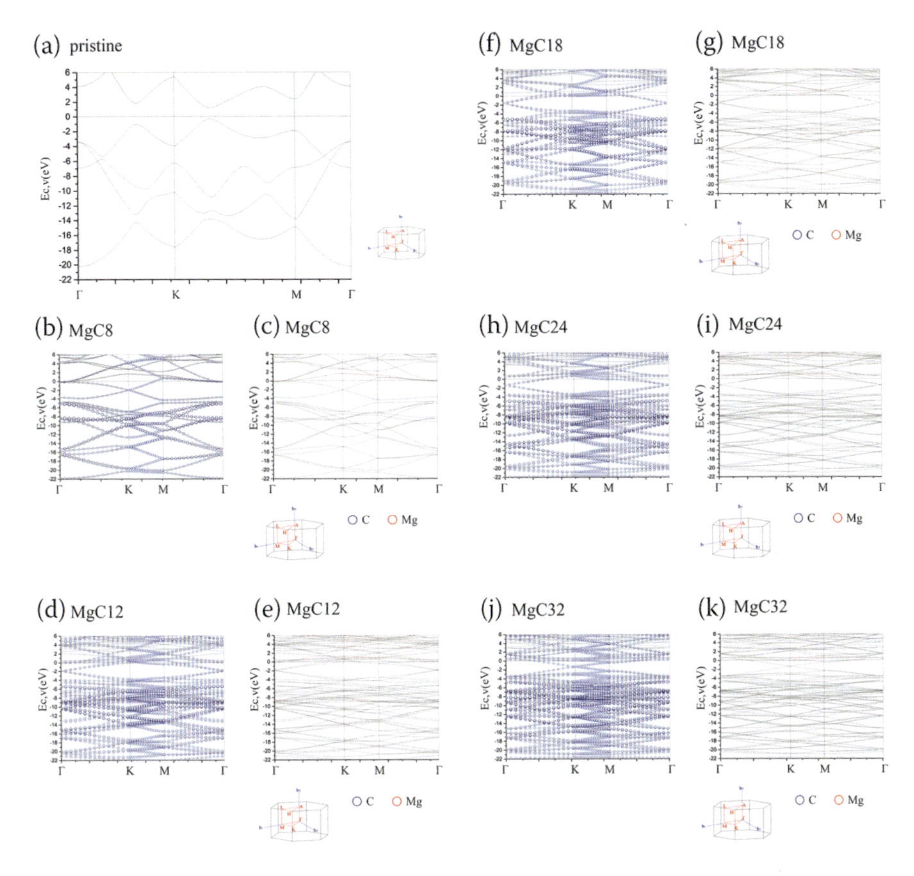

FIGURE 3.2 The featured band structures along the high-symmetry points for (a) a pristine system, (b)/(c) MgC8, (d)/(e) MgC12, (f)/(g) MgC18, (h)/(i) MgC24 and (j)/(k) MgC32, in which the C and Mg atom dominances by the red and blue balls, respectively. Also shown is that of the first Brillouin zone.

of an enlarged quasiparticle framework [113]. For example, the time-dependent intercalations/de-intercalations in graphite anode, which occur during the charging/discharging processes of magnesium- [114], sodium- [115], or lithium-ion-based batteries [116], can create the continuous variations of intermediate crystal structures. This will lead to a lot of Moire superlattices [117], being worthy of systematic investigations. 3D band structures, which correspond to pristine [Figure 3.2(a)], MgC8 [Figure 3.2(b)(c)], MgC12 [Figure 3.2(d)(e)], MgC18 [Figure 3.2(f)(g)], (e) MgC24 [Figure 3.2(h)(i)], and MgC32 [Figure 3.2(j)(k)], are clearly shown along the high-symmetry points of the first Brillouin zones. The planar hexagonal symmetry of non-buckled carbon-honeycomb crystal [118], the optimal stacking configuration [119], the zone folding associated with the intercalant distributions and concentrations [120], and the host-host, host-intercalant, and intercalant-intercalant orbital hybridizations are responsible the diversified quasiparticle behaviors. The dramatic changes among the different graphite-related layered systems cover a great enhancement about the asymmetric hole and electron spectra about the Fermi level after the Mg-adatom interactions [the significant C- and Mg-layer interactions; [99]], E_F far away the K or Γ valleys [its obvious blue shift due to charge transfer from the guest atoms; the higher electron affinity of host atoms, the semimetal-metal transitions; [121]], the observable variations about subband number, energy dispersions, the initial energies and stable valleys [49], the well-characterized π- and σ-electronic states [the non-deformed crystals; [122]], and the distinct carbon and magnesium dominances at the specific energy ranges. Specifically, the drastic changes in the low-lying energy subbands mainly arise from the intercalation effects, being sensitive to experimental syntheses [123]. Such electronic states will be very useful in establishing the reliable tight-binding model/ the generalized tight-bonding model and thus promote the simultaneous progress of the VASP simulations [124], phenomenological models [125], and experimental observations [57]. This is under the current investigations, as clearly illustrated in the previously published books.

3.3 UNUSUAL INTRALAYER AD INTERLAYER CHARGE DENSITY DISTRIBUTIONS

The graphite-related materials [3], which posse the planar honeycomb carbon lattices [118], exhibit concise covalent bondings due to the active orbital of host and guest atoms. Such orbital hybridizations are thoroughly examined from the spatial charge densities before and after the formation of the modified crystal structures, For example, a 2D monolayer graphene [108], a Bernal graphite [top- and side-view in Figures 3.3(a), respectively] and the well-defined graphite magnesium compounds [Figures 3.3(b), 3.3(l) under the different concentrations]. Both π- and σ-electronic states, respectively, arise from the linear superposition of $2p_z$ and [2s, $2p_x$, $2p_y$], in which they are responsible for the lower and deeper-energy quasiparticle phenomena [above and below the Fermi level about 3 eV and beyond it; [42]]. Apparently, the latter generates the most stable systems and the best mechanic properties. The featured covalent bonding, which is associated with the top-view charge distribution, clearly reveal the highest charge density between three pair of neighboring carbon atoms. In general, the planar σ bondings, being perpendicular to the π bondings, almost keep unchanged after the participation of interlayer van der Waals interactions or host-guest ones [35]. The π-electronic bonding, which comes to exist in each planar honeycomb crystal, clearly displays the wave-like distribution at the outer regions [the green/yellow ones in Figure 3.3(b)] about the [x, z]- and [y, z]- plane side views. Obviously, there exist the observable charge density variations, such as the great enhancement of anisotropic environments and effective distribution widths, being very sensitive to the intercalant concentrations and diistribut5io configurations [Figures 3.3(b)(c)/3.3(d)(e)(f)/3.3(g)(h)/3.3(i)(j)/3.3(k)(l)]. The interlayer orbital hybridizations are deduced to mainly arise from C-$2p_z$ and Mg-[3s, $3p_z$]. It should be noticed that this concise picture is fully assisted by the merged can Hove singularities [details in Figures 3.4(a)-3.4(p)]. Of course, the [3s, 3pz]-[3s, $3p_z$] hybridizations play critical roles in generating the intercalant layers. Successful works have been done for

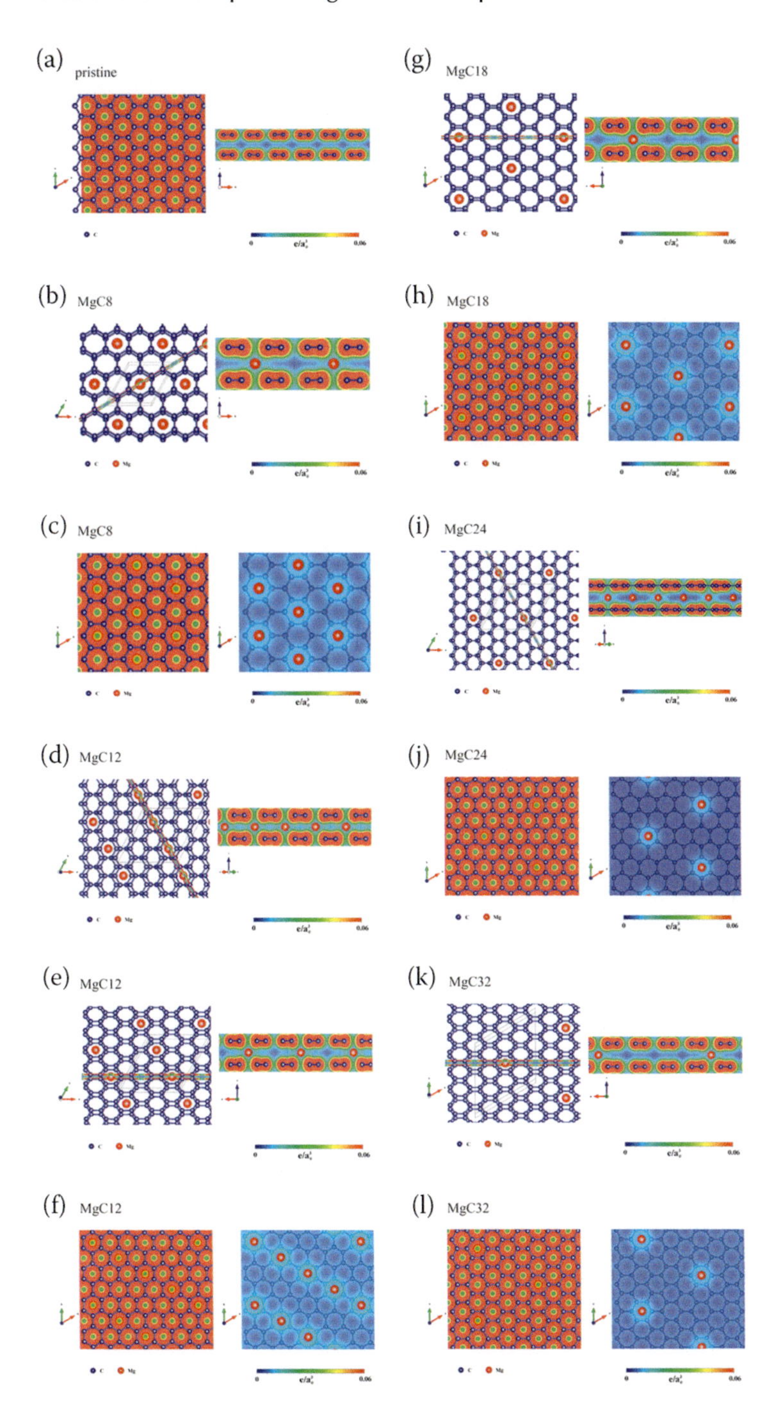

FIGURE 3.3 The spatial charge density distributions about graphite magnesium-adatom compounds projected along the top and side views under the various intercalation concentrations: (a) a pristine system, (b) (c) MgC8, (d)(e)(f) MgC12, (g)(h) MgC18, (i)(j) MgC24 and (k)(l) MgC32.

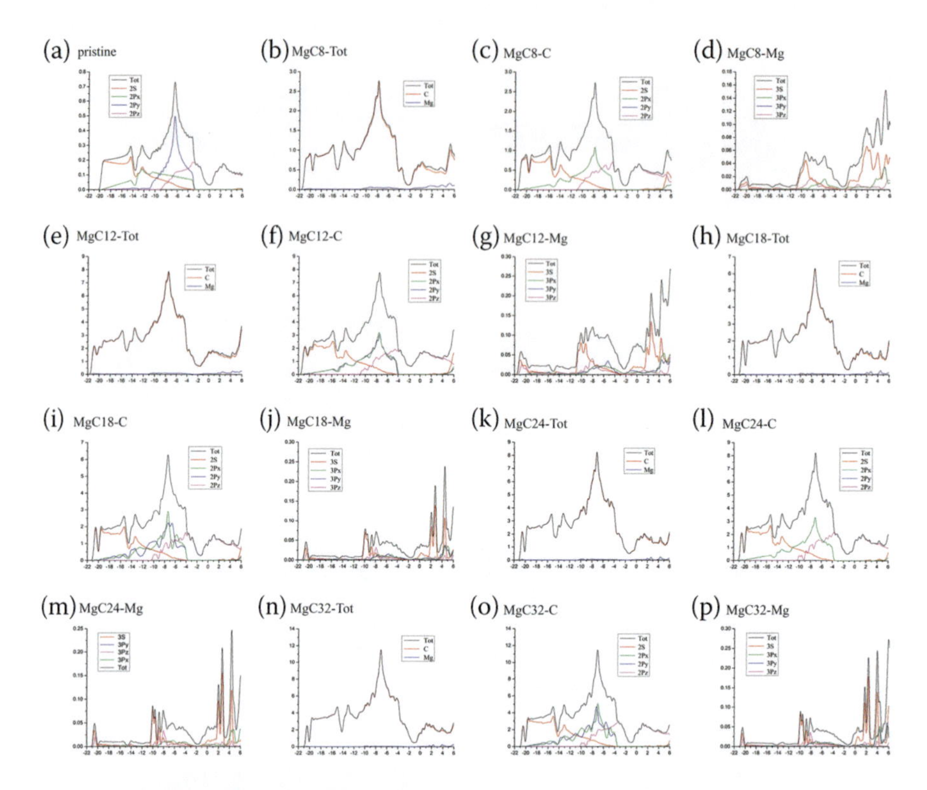

FIGURE 3.4 The atom- and orbital-decomposed density for (a) a pristine system, (b)/(c)/(d) MgC8, (e)/(f)/(g) MgC12, (h)/(i)/(j) MgC18, (k)/(l)/(m) MgC24 and (n)/(o)/(p) MgC32.

graphite alkali compounds, and similar studies could be generalized to more complicated atoms and molecules in the presence of strong host-guest interactions. In addition, while the ionic Mg^{+2} is intercalated into graphite, its closed-shell configurations are expected to effectively terminate the chemical bonds. The absence of active graphite-Mg^{+2} interactions will lead to simple quasiparticles in a 2D monolayer graphene [108]. That is to say, a 3D semimetal [28] is dramatically changed into a 2D zero-gap semiconductor [61]. Compared with intercalations/de-intercalations, the chemical adsorptions and substitutions in graphene-related materials are capable of creating richer and more complicated orbital hybridizations, e.g., the hydrogenations/oxidizations/halogenations of few-layer graphene systems [126–128], as well as, 2D binary SiC/GeC/SnC/PbC compounds. Graphane and graphone, which, respectively, correspond to the full double- and single-side hydrogen adsorptions on a monolayer graphene, clearly illustrate an observable deformation about the honeycomb lattice. This is attributed to the significant sp^3 bondings of 1s-[2s, $2p_x$, $2p_y$, $2p_z$] in H-C bonds. The very strong orbital mixings even lead to the dramatic transition from a zero-gap semiconductor into a wide-gap insulator [E_g higher than 4 eV; [61]]. Also, all active orbitals of carbon atoms take part in the significant hybridizations with those of oxygen [129] and halogen adatoms [130], as done in the previously published books. Very interesting, a planar

SiC can show the perpendicular π and σ bondings, in which both of them belong to the impure orbital bondings of $2p_z$-$3p_z$ and [2s, $2p_x$, $2p_y$]-[3s, $3p_x$, $3p_y$]. The unusual orbital hybridizations, which account for the diverse composite quasiparticles [131,132], frequently come to exist through chemical modifications. Furthermore, the geometric symmetries play critical roles in greatly diversifying quasiparticle behaviors, e.g., the stacking configurations and layer numbers of layered graphene systems [133,134], the cylindrical, chiral, and coaxial structures of carbon nanotubes [29], and the finite-width quantum confinement and the chiral/achiral open edges in graphene nanoribbons [16]. The strong relationships among the geometric, electronic [135], optical [33], and Coulomb-excitations properties have been successfully established in the theoretical works [40].

3.4 ATOM- AND ORBITAL-DECOMPOSED VAN HOVE SINGULARITIES

The density of states can well characterize the energy-dependent energy spectra, being useful in understanding the atom- and orbital-decomposed contributions [63]. Its definition is the integration of the inverse of group velocity on the constant-energy closed surfaces/loops/dots for 2D/1D/0D condensed-matter systems [136–138]. Of course, the quasi-0D dispersionless states exhibit delta-function-like behaviors, e.g., the discrete states in quantum dots [139] and the quantized Landau levels [12]. The zero or irregular group velocities, which mainly come from the critical points in the energy-wave-vector space, are able to generate the well-known van Hove singularities. For example, the extreme points of 3D, 2D, and 1D parabolic dispersions are able to, respectively, exhibit diverse energy dependences in the square root [140], plateau [141], and inverse form of square root [142]. Very interestingly, the 2D systems clearly present the shoulders [143], the logarithmic ally divergent peaks [144], the inverse of square-root divergences [145], the delta-function-like prominent peaks [146], and the V-shape valleys [147], respectively, corresponding to the extreme points [minima or maxima [the saddle points, [148]], the constant-energy loops [149], the partially flat subbands [150], and the Dirac-cone band structures [151]. The 2D, 1D, and 0D density of states, being closely related to the unique geometries, could be examined simultaneously by the direct combinations of high-resolution STS and STM [152] measurements, e.g., the geometry-determined band properties of few-layer graphene systems with the distinct layer numbers and stacking configurations [semimetals due to band overlaps; [78]], the radius- and chirality-dependent metals, narrow- and middle-gap semiconductors for the single-walled carbon nanotubes in the presence of periodical boundary conditions and curvature effects, and the narrow- and middle-gap graphene nanoribbons under the quantum confinement and open edges along the transverse and longitudinal directions, respectively [16]. The simultaneous progress between the theoretical predictions and the STS measurements is urgently requested under the current investigations.

The low-energy van Hove singularities, which cross the Fermi level, are very useful in characterizing the band properties of a zero-/narrow-/middle-/or wide-gap wide semiconductor, a semimetal [153] and a metal [154]. For example, a 2D monolayer graphene clearly shows a V-shape density of states with a zero value at E_F because of only two intersecting Dirac points there [155]. The π bondings further display the logarithmic peak at ~−2. 5 eV and generate its bandwidth of more than 7 eV without any anti-crossings with σ bondings. That is, both kinds of chemical bondings do not have any relations and thus cannot show the merged van Hove singularities. The van Der Waals interactions, the interlayer caron-$2p_z$ orbital hybridizations, have strong effects on E_F-dependent free carriers due to valence and conduction subband overlaps. The semi-metallic behaviors are well characterized by the temperature-dependent valence holes and conduction electrons with the same low densities, e.g., their small densities sensitive to AA [61], AB [13] and ABC stackings [12]. After further intercalations. the donor-type graphite intercalation compounds clearly display the blue shift about the Fermi level relative to the left-hand dip structure, the great enhancements of density of states at E_F under the various Mg-atom intercalation cases [Figures 3.4(b)-3.4(p)]. These intercalation-induced free conduction electrons can further generate the composite quasiparticles, e.g., the ~−1eV optical plasmon modes in electron energy loss functions [156] and the drastic plasmon edges of optical reflectance spectra [157]. The well-defined and merged van Hove singularities are available to fully understand the concise pictures in all active chemical bonds. According to the delicate VASP calculations and analyses, the atom- and orbital-decomposed van Hove singularities have successfully the most important information in the C-C, C-Mg, and Mg-Mg bonds: (I) with the perpendicular $2p_z$-$2p_z$ and [2s, $2p_x$. $2p_y$]-[2s. $2p_x$. $2p_y$] [the intralayer atomic interactions in each honeycomb; Refs], (II) $2p_z$-[3s, $3p_z$] [the host-intercalant orbital hybridizations; Refs], and (III) and 3s-3s and [2s. $2p_x$. $2p_y$]-[2s. $2p_x$. $2p_y$] [the atomic interactions in the intercalant layer]. It should be noticed that only the σ bondings on carbon-honeycomb crystals are independent of intercalation effects. The concise orbital hybridizations should play critical roles in establishing the tight-binding model/the generalized tight-binding model [72,158], never appearing in the previously published works. The close partnerships among the first-principles methods [53], phenomenological models [28], and experimental examinations [159] are able to clearly reveal the diverse composite quasiparticles. Apparently, rich research strategies can be freely developed and presented through a lot of science channels.

3.5 LITHIUM-, LITHIUM-SULFUR-, SODIUM-, MAGNESIUM-, ALUMINUM-, AND IRON-RELATED BATTERIES

Rechargeable batteries have been developed for many significant applications, e.g., mobile devices, vehicles, and other electronic equipment. The demand of the industrial market provides the chances and challenges for scientists to find diverse methods to enhance the efficiency of these secondary batteries. Certain important types of batteries have been built for the past decades such as lithium-ion-, lithium-sulfur-, sodium-, magnesium-, aluminum- and iron- batteries [Figures 3.5(a–f)]. Principally, a typical cell consists of three components, the anode, the cathode, and the electrolyte. For example, lithium-ion batteries can be formed by graphite (anode), $LiFePO_4$, and liquid or solid electrolyte parts as shown in Figure 3.5(a). Lithium-ion batteries (LIBs) possess several rich features, such as an average

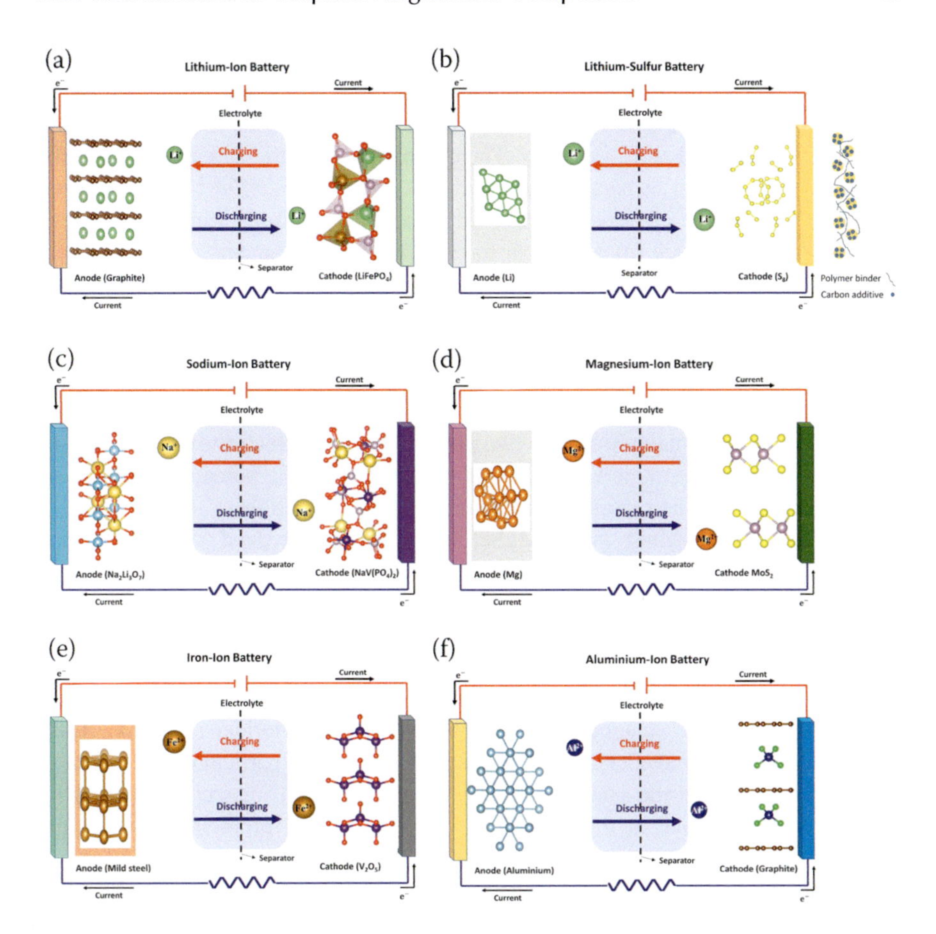

FIGURE 3.5 The (a) lithium-, (b) lithium-sulfur-, (c) sodium-, (d) magnesium-, (e) iron-, and (f) aluminum-ion-related batteries.

operating voltage of 3.7 V [160], high theoretical energy densities, low self-discharge index, and a wide range of temperature use. Many compounds can be served as anode, cathode, and electrolyte. The diverse combinations of them will play a critical role in the current experimental and theoretical research. On the other hand, lithium-ion batteries also have several disadvantages, e.g., volume expansions, the safety problems related to the flammable liquid materials for an electrolyte. Consequently, some kinds of batteries are taken into consideration for the next generation of secondary batteries such as lithium-sulfur, sodium, magnesium, or iron-ion batteries.

Lithium-Sulfur batteries (LISs) also contribute to the secondary batteries, which are shown certain merits such as the theoretical specific capacity of about 1,675 m Ah g-1, acceptable cost, and environmental friendliness. The charging and discharging process will determine the essential mechanism of LIBs (detailed in Figure 3.5(b)). As the same principle with lithium-ion batteries (LIBs), Lithium-ion will transport inside the battery cell. The different materials, which are used for creating electrodes and electrolytes of lithium-sulfur batteries, govern some specific changes in the charging and discharging mechanism. For example, sulfur will serve as the cathode side, and it will diminish by combination with lithium ions from the anode by the discharging process. The compatibility of the core materials in the anode, cathode, binder, separator, and electrolyte of lithium-sulfur batteries plays a vital role to improve electrochemical performance. Therefore, lithium-sulfur batteries also require much improvement to meet industrial demands.

As some potential candidates of rechargeable generations, sodium-ion and magnesium-ion batteries [Figures 3.5(c-d)] have been produced with alternative technologies [161,162]. The remarkable solid-state electrolyte development is one of the important methods, but it is difficult to arrange all available components. Magnesium batteries possess some specific merits, for example, abundance, and safety. Additionally, magnesium has two positive charges which can produce higher energy stored with the same volume as lithium batteries. The potential use of magnesium batteries can be reported with diverse cathode materials (V_2O_5 [163], TiS_2, ZnS_2, CO_2O_3, WSe_2, MnO_2 [164], $MgFeSiO_4$ [165], graphite, and so on [166]), and anode compounds (Mg metal [164], Bi [164], Sn [164,167], Mg_2Sn [168,169]). Since the first $Mg(BPh_2Bu_2)_2$ electrolyte compound is used in magnesium battery, however, it showed some limitations, particularly in electrochemical oxidation. Currently, some other kinds of electrolytes materials are studied such as gel polymers containing $Mg(AlCl_2EtBu)_2$ in tetraglyme/PVDF [164], Bis(trimethylsilyl)amine) [170], carborane salts, magnesium chloride, aluminum chloride, etc [164]. Each kind of material will have its advantages and disadvantages, particularly magnesium metal tends to create the con conducting layer which can interrupt the charging process. Therefore, one of the approaches for solving this problem is to mix the Mg metal anode with sulfur or carbon cathode. Our current research mainly discusses the theoretical analysis of the Mg-intercalation compound with a rich and unique hybridization theoretical framework.

Additionally, iron-ion batteries are successfully synthesized by an Indian group in 2019 [163], which is characterized to reach a specific capacity of 207 mA h g-1 within a specific current of 30 mA g-1. In this work, the intercalation and deintercalation process of iron ions into the electrode materials through the electrolyte is achieved based on ex situ measurement. Remarkably, the anode and cathode materials are composed of mild steel and V_2O_5, respectively. The significant cycling stabilities can be maintained at 54.5% (after 50 cycles) and high coulombic efficiency (approximately 70 %) shows that iron-ion cells might have the potential development for the post-batteries. The current endeavor is enhancing the electrochemical performance by understanding the mechanism of the charging and discharging process, also the ion transport properties inside the electrolyte and batteries. Solving this problem might be promising and fascinating research of theories and experiments.

3.6 SUMMARIES

The VASP simulations clearly illustrate develop the composite quasiparticles in graphite and graphite intercalation compounds [65,112], being very useful in a series of published books [30,33,165]. By the delicate calculations and analyses, the critical mechanisms are deduced to be the hexagonal symmetry of a planar honeycomb lattice, the rich stacking configurations [AA/AB/ABC/turbostratic ones; [39,49,53,61,166,167]], the zone folding of the host and quest lattices, the dynamic intercalations/de-intercalations [20,65–67,112], and the semimetal-metal transitions due to n-/p-type doping cases [the high density of conduction electrons/valence holes; donor-/acceptor graphite intercalation compounds; [28,29]]. Apparently, the concise pictures, which mainly arise from the orbital mixings in the intralayer and interlayer chemical bonds, are accurately achieved from the detailed examinations of the spatial charge density distributions [details in Fig. W3] and the merged structures of the atom-/orbital-decomposed van Hove singularities [69,168]. The simulation development is also successful for the large ionic/molecular intercalants, e.g., graphite $AlCl_4$ and Al_2Cl_7 compounds [77], as well as, graphite $FeCl_3$ ones [78]. The low-lying valence and conduction energy subbands, crossing the Fermi level, are initiated from the very stable K/H and G/A valleys. Whether they will be well-characterized by the suitable Hamiltonians with the significant hopping integrals and on-site Coulomb potential energies can greatly strengthen the prominent relations among the first-principles methods [116,169,170], the phenomenological models [3,64,117,152, 171–173], and experimental observations, Their merits are combined together, but drawbacks need to be solved urgently under the current investigations. As to 2D few-layer graphene systems, the chemical adsorptions and substitutions dramatically change are expected to dramatically change the crystal phases, orbital hybridizations, and composite quasiparticle behaviors. These will be fully investigated and finished in the near-future science researches.

Graphite magnesium compounds, which belong to 3D donor-type metals, present diverse quasiparticle phenomena. For example, there exist certain important differences among group-II [Be/Mg/Ca/Sr/Ba] and group-I Li/Na/K/Rb/Cs] by modulating the distinct guest atoms [174], the intercalant distributions and concentrations, as well as, their normal/ irregular stacking orderings [175]. The delicate calculations are consistent with one another through the obvious effects due to the well-behaved carbon- and magnesium layer, the zone folding [176,177], and the interlayer charge transfers [178]. The 3D layered structures clearly indicate the highly anisotropic and active environments of chemistry, physics, and material engineering. The various crystals can be generated by different methods, being very helpful in establishing an enlarged quasiparticle framework. The featured quasiparticle states are clearly shown in 3D band structures along the high-symmetry points of the first Brillouin zone, e.g., the great enhancement of asymmetric hole and electron spectra about the Fermi level after the Mg-adatom interactions [the significant C- and Mg-layer interactions; [179]], E_F far away the K or G valleys [the charge transfer from the guest atoms; the higher electron affinity of host atoms; Refs], the observable variations of subband number, energy dispersions, their initial energies and stable valleys, and π- and σ- electronic states. Obviously, each carbon honeycomb lattice displays the perpendicular π and σ bodings, in which the former and the latter have very high and middle carrier densities. It keeps a planar structure after the prominent intercalations, clearly illustrating the almost unchanged σ bondings and the π-induced interlayer interactions. The side-view charge densities of host and guest layers present important variations. The above-mentioned features are further reflected in the atom and orbital-decomposed density of states, such as the blue shift of the Fermi level under the n-type dopings, and the merged van Hove singularities due to the orbital hybridizations of active orbitals.

REFERENCES

1. Wallace, P.R., The band theory of graphite. *Physical Review*, 1947. **71**(9): p. 622–634.
2. Li, G., A. Luican, and E.Y. Andrei, Scanning tunneling spectroscopy of graphene on graphite. *Physical Review Letters*, 2009. **102**(17).
3. Reich, S., et al., Tight-binding description of graphene. *Physical Review B*, 2002. **66**(3).
4. Geim, A.K. and K.S. Novoselov, The rise of graphene. *Nature Materials*, 2007. **6**(3): p. 183–191.
5. Novoselov, K.S., Electric field effect in atomically thin carbon films. *Science*, 2004. **306**(5696): p. 666–669.
6. Kuila, T., et al., Chemical functionalization of graphene and its applications. *Progress in Materials Science*, 2012. **57**(7): p. 1061–1105.
7. Gorjizadeh, N. and Y. Kawazoe, Chemical functionalization of graphene nanoribbons. *Journal of Nanomaterials*, 2010. **2010**: p. 513501.
8. Bragg, W.H. and W.L. Bragg, The structure of the diamond. *Proceedings of the Royal Society of London. Series A, Containing Papers of a Mathematical and Physical Character*, 1913. **89**(610): p. 277–291.
9. Sinnott, S.B., Chemical functionalization of carbon nanotubes. *Journal of Nanoscience and Nanotechnology*, 2002. **2**(2): p. 113–123.
10. Wu, J.-Y., et al., The effect of perpendicular electric field on temperature-induced plasmon excitations for intrinsic silicene. *RSC Advances*, 2015. **5**(64): p. 51912–51918.
11. Slonczewski, J.C. and P.R. Weiss, Band structure of graphite. *Physical Review*, 1958. **109**(2): p. 272–279.
12. Yuan, S., R. Roldán, and M.I. Katsnelson, Landau level spectrum of ABA-and ABC-stacked trilayer graphene. *Physical Review B*, 2011. **84**(12): p. 125455.
13. Do, T.-N., et al., Configuration-enriched magneto-electronic spectra of AAB-stacked trilayer graphene. *Carbon*, 2015. **94**: p. 619–632.
14. Gopinadhan, K., et al., Giant magnetoresistance in single-layer graphene flakes with a gate-voltage-tunable weak antilocalization. *Physical Review B*, 2013. **88**(19): p. 195429.
15. Dresselhaus, G., M.S. Dresselhaus, and R. Saito, Physical properties of carbon nanotubes. 1998: World scientific.
16. Kumar, P., L. Panchakarla, and C. Rao, Laser-induced unzipping of carbon nanotubes to yield graphene nanoribbons. *Nanoscale*, 2011. **3**(5): p. 2127–2129.
17. Zhang, J., et al., Free folding of suspended graphene sheets by random mechanical stimulation. *Physical Review Letters*, 2010. **104**(16): p. 166805.
18. Terrones, M., et al., Graphene and graphite nanoribbons: Morphology, properties, synthesis, defects and applications. *Nano Today*, 2010. **5**(4): p. 351–372.
19. Kosynkin, D.V., et al., Longitudinal unzipping of carbon nanotubes to form graphene nanoribbons. *Nature*, 2009. **458**(7240): p. 872–876.
20. Savoskin, M.V., et al., Carbon nanoscrolls produced from acceptor-type graphite intercalation compounds. *Carbon*, 2007. **45**(14): p. 2797–2800.

21. Lu, X. and Z. Chen, Curved Pi-conjugation, aromaticity, and the related chemistry of small fullerenes (<C60) and single-walled carbon nanotubes. *Chemical Reviews*, 2005. **105**(10): p. 3643–3696.
22. Yannouleas, C., E.N. Bogachek, and U. Landman, Collective excitations of multishell carbon microstructures: Multishell fullerenes and coaxial nanotubes. *Physical Review B*, 1996. **53**(15): p. 10225.
23. Golden, M.S., et al., The electronic structure of fullerenes and fullerene compounds from high-energy spectroscopy. *Journal of Physics: Condensed Matter*, 1995. **7**(43): p. 8219.
24. Shyu, F.-L., et al., Magnetoelectronic properties of chiral carbon nanotubes and tori. *Journal of Physics: Condensed Matter*, 2006. **18**(35): p. 8313.
25. Shyu, F.L., et al., Electronic properties of carbon tori in external fields. *Journal of the Physical Society of Japan*, 2006. **75**(10): p. 104710.
26. Wong, J.-H., B.-R. Wu, and M.-F. Lin, Strain effect on the electronic properties of single layer and bilayer graphene. *The Journal of Physical Chemistry C*, 2012. **116**(14): p. 8271–8277.
27. Tsai, S.-J., et al., Gate-voltage-dependent Landau levels in AA-stacked bilayer graphene. *Chemical Physics Letters*, 2012. **550**: p. 104–110.
28. Guerard, D., et al., Electronic structure of donor-type graphite intercalation compounds. *Il Nuovo Cimento B (1971-1996)*, 1977. **38**(2): p. 410–417.
29. Sorokina, N., et al., Acceptor-type graphite intercalation compounds and new carbon materials based on them. *Russian Chemical Bulletin*, 2005. **54**(8): p. 1749–1767.
30. Lin, C.-Y., et al., Diverse quantization phenomena in layered materials. 2019: CRC Press.
31. Tan, P.H., et al., Photoluminescence spectroscopy of carbon nanotube bundles: Evidence for exciton energy transfer. *Physical Review Letters*, 2007. **99**(13).
32. Gusynin, V.P. and S.G. Sharapov, Unconventional Integer Quantum Hall Effect in Graphene. *Physical Review Letters*, 2005. **95**(14).
33. Lin, C.-Y., et al., Electronic and optical properties of graphite-related systems. 2017: CRC Press.
34. Wong, J.-H., B.-R. Wu, and M.-F. Lin, Electronic properties of rhombohedral graphite. *Computer Physics Communications*, 2011. **182**(1): p. 77–80.
35. Saslow, W., T. Bergstresser, and M.L. Cohen, Band structure and optical properties of diamond. *Physical Review Letters*, 1966. **16**(9): p. 354.
36. Akinwande, D., et al., A review on mechanics and mechanical properties of 2D materials—Graphene and beyond. *Extreme Mechanics Letters*, 2017. **13**: p. 42–77.
37. Charlier, J.-C., X. Gonze, and J.-P. Michenaud, Graphite interplanar bonding: Electronic delocalization and van der Waals interaction. *EPL (Europhysics Letters)*, 1994. **28**(6): p. 403.
38. Holstein, B.R., The van der Waals interaction. *American Journal of Physics*, 2001. **69**(4): p. 441–449.
39. Lu, C.L., et al., Electronic properties of AA-and ABC-stacked few-layer graphites. *Journal of the Physical Society of Japan*, 2007. **76**(2): p. 024701.
40. Zhou, X., et al., Thermoelectrics of two-dimensional conjugated benzodithiophene-based polymers: Density-of-states enhancement and semi-metallic behavior. *Journal of Materials Chemistry A*, 2019. **7**(17): p. 10422–10430.
41. Zhou, X., et al., Strategies towards low-cost dual-ion batteries with high performance. *Angewandte Chemie International Edition*, 2020. **59**(10): p. 3802–3832.
42. Tang, K., et al., Electric-field-induced energy gap in few-layer graphene. *The Journal of Physical Chemistry C*, 2011. **115**(19): p. 9458–9464.

43. Bernal, J.D., The structure of graphite. *Proceedings of the Royal Society of London. Series A, Containing Papers of a Mathematical and Physical Character*, 1924. **106**(740): p. 749–773.
44. Lipson, H.S. and A. Stokes, The structure of graphite. *Proceedings of the Royal Society of London. Series A. Mathematical and Physical Sciences*, 1942. **181**(984): p. 101–105.
45. Lui, C.H., et al., Imaging stacking order in few-layer graphene. *Nano Letters*, 2011. **11**(1): p. 164–169.
46. Poverenov, E., et al., Unusual doping of donor–acceptor-type conjugated polymers using lewis acids. *Journal of the American Chemical Society*, 2014. **136**(13): p. 5138–5149.
47. Lin, C.-Y., et al., Coulomb excitations and decays in graphene-related systems. 2019: CRC Press.
48. Lin, C.-Y., et al., Geometry-diversified Coulomb excitations in trilayer AAB stacking graphene. *Physical Review B*, 2018. **98**(19): p. 195442.
49. Chiu, C.-W., et al., Anisotropy of π-plasmon dispersion relation of AA-stacked graphite. *Journal of the Physical Society of Japan*, 2012. **81**(10): p. 104703.
50. Mak, K.F., J. Shan, and T.F. Heinz, Electronic structure of few-layer graphene: Experimental demonstration of strong dependence on stacking sequence. *Physical Review Letters*, 2010. **104**(17).
51. Ho, Y.-H., et al., Magneto-absorption spectra of Bernal graphite. *Applied Physics Letters*, 2011. **99**(1): p. 011914.
52. Lu, C.-L., et al., Absorption spectra of trilayer rhombohedral graphite. *Applied Physics Letters*, 2006. **89**(22): p. 221910.
53. Chang, C.-P., et al., Magnetoelectronic properties of the AB-stacked graphite. *Carbon*, 2005. **43**(7): p. 1424–1431.
54. Lin, Y.-P., et al., Magneto-optical properties of ABC-stacked trilayer graphene. *Physical Chemistry Chemical Physics*, 2015. **17**(24): p. 15921–15927.
55. Do, T.-N., et al., Rich magneto-absorption spectra of AAB-stacked trilayer graphene. *Physical Chemistry Chemical Physics*, 2016. **18**(26): p. 17597–17605.
56. Alvarez, M.M., et al., Optical absorption spectra of nanocrystal gold molecules. *The Journal of Physical Chemistry B*, 1997. **101**(19): p. 3706–3712.
57. Yacobi, B., F. Boswell, and J. Corbett, Intercalation-induced shift of the absorption edge in ZrS2 and HfS2. *Journal of Physics C: Solid State Physics*, 1979. **12**(11): p. 2189.
58. Giuliani, G.F. and J.J. Quinn, Lifetime of a quasiparticle in a two-dimensional electron gas. *Physical Review B*, 1982. **26**(8): p. 4421.
59. Bostwick, A., et al., Quasiparticle dynamics in graphene. *Nature physics*, 2007. **3**(1): p. 36–40.
60. Bena, C. and S.A. Kivelson, Quasiparticle scattering and local density of states in graphite. *Physical Review B*, 2005. **72**(12): p. 125432.
61. Ho, J.H., et al., Coulomb excitations in AA- and AB-stacked bilayer graphites. *Physical Review B*, 2006. **74**(8): p. 085406.
62. Ferro, S., Synthesis of diamond. *Journal of Materials Chemistry*, 2002. **12**(10): p. 2843–2855.
63. Kelly, B.T., Physics of graphite. 1981.
64. Goringe, C.M., D.R. Bowler, and E. Hernández, Tight-binding modelling of materials. *Reports on Progress in Physics*, 1997. **60**(12): p. 1447–1512.
65. Li, W.-B., et al., Essential electronic properties on stage-1 Li/Li+-graphite-intercalation compounds for different concentrations. arXiv preprint arXiv:2006. 12055, 2020.

66. Dresselhaus, M. and G. Dresselhaus, Intercalation compounds of graphite. *Advances in Physics*, 1981. **30**(2): p. 139–326.
67. Lin, M., C. Huang, and D. Chuu, Plasmons in graphite and stage-1 graphite intercalation compounds. *Physical Review B*, 1997. **55**(20): p. 13961.
68. Huang, C.-L. and E. Kuo, Mechanism of hypokalemia in magnesium deficiency. *Journal of the American Society of Nephrology*, 2007. **18**(10): p. 2649–2652.
69. Li, G., et al., Observation of Van Hove singularities in twisted graphene layers. *Nature Physics*, 2010. **6**(2): p. 109–113.
70. Andersen, D.R. and H. Raza, Plasmon dispersion in semimetallic armchair graphene nanoribbons. *Physical Review B*, 2012. **85**(7): p. 075425.
71. Kresse, G. and J. Furthmüller, Efficient iterative schemes for ab initio total-energy calculations using a plane-wave basis set. *Physical Review B*, 1996. **54**(16): p. 11169.
72. Kresse, G. and D. Joubert, From ultrasoft pseudopotentials to the projector augmented-wave method. *Physical Review B*, 1999. **59**(3): p. 1758.
73. Perdew, J.P., K. Burke, and M. Ernzerhof, Generalized gradient approximation made simple. *Physical Review Letters*, 1996. **77**(18): p. 3865.
74. Blöchl, P.E., Projector augmented-wave method. *Physical Review B*, 1994. **50**(24): p. 17953.
75. Steward, E., B. Cook, and E. Kellett, Dependence on temperature of the interlayer spacing in carbons of different graphitic perfection. *Nature*, 1960. **187**(4742): p. 1015–1016.
76. Zou, G., et al., Controllable interlayer spacing of sulfur-doped graphitic carbon nanosheets for fast sodium-ion batteries. *Small*, 2017. **13**(31): p. 1700762.
77. Rolland, P. and G. Mamantov, Electrochemical reduction of Al2Cl7– ions in chloroaluminate melts. *Journal of the Electrochemical Society*, 1976. **123**(9): p. 1299.
78. Niemi, V., et al., Polymerization of 3-alkylthiophenes with FeCl3. *Polymer*, 1992. **33**(7): p. 1559–1562.
79. Taft, E. and H. Philipp, Optical properties of graphite. *Physical Review*, 1965. **138**(1A): p. A197.
80. Becerril, D., G. Pirruccio, and C. Noguez, Optical band engineering via vertical stacking of honeycomb plasmonic lattices. *Physical Review B*, 2021. **103**(19): p. 195412.
81. Chen, C., et al., Probing interlayer interaction via chiral phonons in layered honeycomb materials. *Physical Review B*, 2021. **103**(3): p. 035405.
82. Chen, Y. and S.Y. Quek, Tunable bright interlayer excitons in few-layer black phosphorus based van der Waals heterostructures. *2D Materials*, 2018. **5**(4): p. 045031.
83. Wolynes, P.G., Dynamics of electrolyte solutions. *Annual Review of Physical Chemistry*, 1980. **31**(1): p. 345–376.
84. Jossen, A., et al., Reliable battery operation—a challenge for the battery management system. *Journal of Power Sources*, 1999. **84**(2): p. 283–286.
85. Belcher, A.M., et al., Control of crystal phase switching and orientation by soluble mollusc-shell proteins. *Nature*, 1996. **381**(6577): p. 56–58.
86. Chiu, C.-W., et al., Excitation spectra of ABC-stacked graphene superlattice. *Applied Physics Letters*, 2011. **98**(26): p. 261920.
87. Lv, R., et al., Nitrogen-doped graphene: Beyond single substitution and enhanced molecular sensing. *Scientific Reports*, 2012. **2**(1).
88. Panchakarla, L.S., et al., Synthesis, structure, and properties of boron- and nitrogen-doped graphene. *Advanced Materials*, 2009: p. NA-NA.

89. Huang, H.-C., et al., Configuration- and concentration-dependent electronic properties of hydrogenated graphene. *Carbon*, 2016. **103**: p. 84–93.
90. Jin, K.-H., S.-M. Choi, and S.-H. Jhi, Crossover in the adsorption properties of alkali metals on graphene. *Physical Review B*, 2010. **82**(3): p. 033414.
91. Du, Z., D.L. Wood III, and I. Belharouak, Enabling fast charging of high energy density Li-ion cells with high lithium ion transport electrolytes. *Electrochemistry Communications*, 2019. **103**: p. 109–113.
92. Nilsson, A., L.G. Pettersson, and J. Norskov, Chemical bonding at surfaces and interfaces. 2011: Elsevier.
93. Ryu, Y.K., R. Frisenda, and A. Castellanos-Gomez, Superlattices based on van der Waals 2D materials. *Chemical Communications*, 2019. **55**(77): p. 11498–11510.
94. Berland, K., et al., van der Waals forces in density functional theory: A review of the vdW-DF method. *Reports on Progress in Physics*, 2015. **78**(6): p. 066501.
95. Korenman, V. and R. Prange, Local-band-theory analysis of spin-polarized, angle-resolved photoemission spectroscopy. *Physical Review Letters*, 1984. **53**(2): p. 186.
96. Damascelli, A., Probing the electronic structure of complex systems by ARPES. *Physica Scripta*, 2004. **T109**: p. 61.
97. Ohta, T., et al., Interlayer interaction and electronic screening in multilayer graphene investigated with angle-resolved photoemission spectroscopy. *Physical Review Letters*, 2007. **98**(20): p. 206802.
98. Cattelan, M. and N. Fox, A perspective on the application of spatially resolved ARPES for 2D materials. *Nanomaterials*, 2018. **8**(5): p. 284.
99. Sobota, J.A., Y. He, and Z.-X. Shen, Angle-resolved photoemission studies of quantum materials. *Reviews of Modern Physics*, 2021. **93**(2).
100. Lee, C., et al., Measurement of the elastic properties and intrinsic strength of monolayer graphene. *Science*, 2008. **321**(5887): p. 385–388.
101. Park, J., et al., Anisotropic Dirac fermions in a Bi square net of SrMnBi 2. *Physical Review Letters*, 2011. **107**(12): p. 126402.
102. Lau, A., D. Levine, and P. Pincus, Novel electrostatic attraction from plasmon fluctuations. *Physical Review Letters*, 2000. **84**(18): p. 4116.
103. Verbeeck, J. and S. Van Aert, Model based quantification of EELS spectra. *Ultramicroscopy*, 2004. **101**(2-4): p. 207–224.
104. Vrhel, M.J., R. Gershon, and L.S. Iwan, Measurement and analysis of object reflectance spectra. *Color Research & Application*, 1994. **19**(1): p. 4–9.
105. Tselyaev, V., Quasiparticle time blocking approximation within the framework of generalized Green function formalism. *Physical Review C*, 2007. **75**(2): p. 024306.
106. Huie, M.M., et al., Cathode materials for magnesium and magnesium-ion based batteries. *Coordination Chemistry Reviews*, 2015. **287**: p. 15–27.
107. Jiang, C., et al., A multi-ion strategy towards rechargeable sodium-ion full batteries with high working voltage and rate capability. *Angewandte Chemie*, 2018. **130**(50): p. 16608–16612.
108. Durmus, Y.E., et al., Side by side battery technologies with lithium-ion based batteries. *Advanced Energy Materials*, 2020. **10**(24): p. 2000089.
109. Moon, P. and M. Koshino, Electronic properties of graphene/hexagonal-boron-nitride moiré superlattice. *Physical Review B*, 2014. **90**(15): p. 155406.
110. Hong, Y.J. and T. Fukui, Controlled van der Waals heteroepitaxy of InAs nanowires on carbon honeycomb lattices. *ACS Nano*, 2011. **5**(9): p. 7576–7584.
111. Caillet, J. and P. Claverie, Differences of nucleotide stacking patterns in a crystal and in binary complexes—the case of adenine. *Biopolymers: Original Research on Biomolecules*, 1974. **13**(3): p. 601–614.

112. Li, W.-B., et al., Essential geometric and electronic properties in stage-n graphite alkali-metal-intercalation compounds. *RSC Advances*, 2020. **10**(40): p. 23573–23581.
113. Liu, G., et al., Strain-induced semimetal-metal transition in silicene. *EPL (Europhysics Letters)*, 2012. **99**(1): p. 17010.
114. Abdelaziz, K.B., et al., A broad omnidirectional reflection band obtained from deformed Fibonacci quasi-periodic one dimensional photonic crystals. *Journal of Optics A: Pure and Applied Optics*, 2005. **7**(10): p. 544.
115. Moissette, A., et al., Sulfate graphite intercalation compounds: New electrochemical data and spontaneous intercalation. *Carbon*, 1995. **33**(2): p. 123–128.
116. Hafner, J., Ab-initio simulations of materials using VASP: Density-functional theory and beyond. *Journal of computational chemistry*, 2008. **29**(13): p. 2044–2078.
117. Dodaro, J.F., et al., Phases of a phenomenological model of twisted bilayer graphene. *Physical Review B*, 2018. **98**(7): p. 075154.
118. Peng, Q., et al., A theoretical analysis of the effect of the hydrogenation of graphene to graphane on its mechanical properties. *Physical Chemistry Chemical Physics*, 2013. **15**(6): p. 2003–2011.
119. Sun, L., Structure and synthesis of graphene oxide. *Chinese Journal of Chemical Engineering*, 2019. **27**(10): p. 2251–2260.
120. Karlicky, F., et al., Halogenated graphenes: rapidly growing family of graphene derivatives. *ACS Nano*, 2013. **7**(8): p. 6434–6464.
121. Khazraie, A., et al., Oxygen holes and hybridization in the bismuthates. *Physical Review B*, 2018. **97**(7): p. 075103.
122. Scheiner, S., Halogen bonds formed between substituted imidazoliums and n bases of varying n-hybridization. *Molecules*, 2017. **22**(10): p. 1634.
123. Sen, S., K.S. Gupta, and J. Coey, Mesoscopic structure formation in condensed matter due to vacuum fluctuations. *Physical Review B*, 2015. **92**(15): p. 155115.
124. Meunier, V., P. Lambin, and A.A. Lucas, Atomic and electronic structures of large and small carbon tori. *Physical Review B*, 1998. **57**(23): p. 14886–14890.
125. Lin, C.-Y., et al., Unusual electronic excitations in ABA trilayer graphene. *Scientific Reports*, 2020. **10**(1): p. 1–9.
126. Cazalilla, M., et al., One dimensional bosons: From condensed matter systems to ultracold gases. *Reviews of Modern Physics*, 2011. **83**(4): p. 1405.
127. Slusher, R. and C. Weisbuch, Optical microcavities in condensed matter systems. *Solid State Communications*, 1994. **92**(1–2): p. 149–158.
128. Bimberg, D., M. Grundmann, and N.N. Ledentsov, Quantum dot heterostructures. 1999: John Wiley & Sons.
129. Bonča, J. and T. Pruschke, Van Hove singularities in the paramagnetic phase of the Hubbard model: DMFT study. *Physical Review B*, 2009. **80**(24): p. 245112.
130. Wu, S., et al., Chern insulators, van Hove singularities and topological flat bands in magic-angle twisted bilayer graphene. *Nature materials*, 2021. **20**(4): p. 488–494.
131. Alabugin, I.V., S. Bresch, and G. dos Passos Gomes, Orbital hybridization: A key electronic factor in control of structure and reactivity. *Journal of Physical Organic Chemistry*, 2015. **28**(2): p. 147–162.
132. Popov, I.V., et al., Relative stability of diamond and graphite as seen through bonds and hybridizations. *Physical Chemistry Chemical Physics*, 2019. **21**(21): p. 10961–10969.
133. Stauber, T. and H. Kohler, Quasi-flat plasmonic bands in twisted bilayer graphene. *Nano Letters*, 2016. **16**(11): p. 6844–6849.
134. Huang, Y.-K., et al., Feature-rich magnetic quantization in sliding bilayer graphenes. *Scientific Reports*, 2014. **4**(1): p. 1–10.
135. Shyu, F.L. and M.F. Lin, Low-frequency π-electronic excitations of simple hexagonal graphite. *Journal of the Physical Society of Japan*, 2001. **70**(3): p. 897–901.

136. Hodges, C., Van Hove singularities and continued fraction coefficients. *Journal de Physique Lettres*, 1977. **38**(9): p. 187–189.
137. Lawrence, R.L., A.M. Ellingson, and P.M. Ludewig, Validation of single-plane fluoroscopy and 2D/3D shape-matching for quantifying shoulder complex kinematics. *Medical Engineering & Physics*, 2018. **52**: p. 69–75.
138. Lau, H.W. and P. Grassberger, Information theoretic aspects of the two-dimensional Ising model. *Physical Review E*, 2013. **87**(2): p. 022128.
139. Bonča, J., Kondo effect in the presence of Rashba spin-orbit interaction. *Physical Review B*, 2011. **84**(19): p. 193411.
140. Zhu, B., X. Chen, and X. Cui, Exciton binding energy of monolayer WS2. *Scientific Reports*, 2015. **5**(1): p. 1–5.
141. Rycerz, A., J. Tworzydło, and C. Beenakker, Valley filter and valley valve in graphene. *Nature Physics*, 2007. **3**(3): p. 172–175.
142. Chu, H. and Y.-C. Chang, Saddle-point excitons in solids and superlattices. *Physical Review B*, 1987. **36**(5): p. 2946.
143. Owerre, S., Dirac magnon nodal loops in quasi-2D quantum magnets. *Scientific Reports*, 2017. **7**(1): p. 1–9.
144. Aoki, H., Theoretical possibilities for flat band superconductivity. *Journal of Superconductivity and Novel Magnetism*, 2020. **33**(8): p. 2341–2346.
145. Dolui, K. and S.Y. Quek, Quantum-confinement and structural anisotropy result in electrically-tunable dirac cone in few-layer black phosphorous. *Scientific Reports*, 2015. **5**(1): p. 1–12.
146. Hansma, P.K. and J. Tersoff, Scanning tunneling microscopy. *Journal of Applied Physics*, 1987. **61**(2): p. R1–R24.
147. Young, S.M., et al., Dirac semimetal in three dimensions. *Physical review letters*, 2012. **108**(14): p. 140405.
148. Mead, C. and W. Spitzer, Fermi level position at metal-semiconductor interfaces. *Physical Review*, 1964. **134**(3A): p. A713.
149. Deshpande, A., et al., Spatially resolved spectroscopy of monolayer graphene on SiO 2. *Physical Review B*, 2009. **79**(20): p. 205411.
150. De Abajo, F.G., Optical excitations in electron microscopy. *Reviews of Modern Physics*, 2010. **82**(1): p. 209.
151. Kurokawa, Y. and H.T. Miyazaki, Metal-insulator-metal plasmon nanocavities: Analysis of optical properties. *Physical Review B*, 2007. **75**(3): p. 035411.
152. Hancock, Y., et al., Generalized tight-binding transport model for graphene nanoribbon-based systems. *Physical Review B*, 2010. **81**(24): p. 245402.
153. Meskine, H., et al., Examination of the concept of degree of rate control by first-principles kinetic Monte Carlo simulations. *Surface Science*, 2009. **603**(10-12): p. 1724–1730.
154. Burheim, O.S., Engineering energy storage. 2017: Academic press.
155. Moury, R., A. Gigante, and H. Hagemann, An alternative approach to the synthesis of NaB3H8 and Na2B12H12 for solid electrolyte applications. *International journal of hydrogen energy*, 2017. **42**(35): p. 22417–22421.
156. Roedern, E., et al., Magnesium ethylenediamine borohydride as solid-state electrolyte for magnesium batteries. *Scientific Reports*, 2017. **7**(1): p. 1–6.
157. Nóvák, P., V. Shklover, and R. Nesper, Magnesium insertion in vanadium oxides: A structural study. *Zeitschrift für Physikalische Chemie*, 1994. **185**(1): p. 51–68.
158. Mohtadi, R. and F. Mizuno, Magnesium batteries: Current state of the art, issues and future perspectives. *Beilstein Journal of Nanotechnology*, 2014. **5**(1): p. 1291–1311.
159. Orikasa, Y., et al., High energy density rechargeable magnesium battery using earth-abundant and non-toxic elements. *Scientific Reports*, 2014. **4**(1): p. 1–6.

160. Bella, F., et al., An overview on anodes for magnesium batteries: Challenges towards a promising storage solution for renewables. *Nanomaterials*, 2021. **11**(3): p. 810.
161. Nguyen, D.T., et al., Magnesium storage performance and surface film formation behavior of tin anode material. *ChemElectroChem*, 2016. **3**(11): p. 1813–1819.
162. Nguyen, D.-T. and S.-W. Song, Magnesium stannide as a high-capacity anode for magnesium-ion batteries. *Journal of Power Sources*, 2017. **368**: p. 11–17.
163. Singh, N., et al., A high energy-density tin anode for rechargeable magnesium-ion batteries. *Chemical Communications*, 2013. **49**(2): p. 149–151.
164. Zhao-Karger, Z., et al., A new class of non-corrosive, highly efficient electrolytes for rechargeable magnesium batteries. *Journal of Materials Chemistry A*, 2017. **5**(22): p. 10815–10820.
165. Tran, N.T.T., et al., Geometric and electronic properties of graphene-related systems: Chemical bonding schemes. 2017: CRC Press.
166. Ho, C.-H., C.-P. Chang, and M.-F. Lin, Evolution and dimensional crossover from the bulk subbands in ABC-stacked graphene to a three-dimensional Dirac cone structure in rhombohedral graphite. *Physical Review B*, 2016. **93**(7): p. 075437.
167. Koshino, M., Interlayer screening effect in graphene multilayers with A B A and A B C stacking. *Physical Review B*, 2010. **81**(12): p. 125304.
168. Van Hove, L., The occurrence of singularities in the elastic frequency distribution of a crystal. *Physical Review*, 1953. **89**(6): p. 1189–1193.
169. Hafner, J., Materials simulations using VASP—a quantum perspective to materials science. *Computer Physics Communications*, 2007. **177**(1-2): p. 6–13.
170. Hafner, J. and G. Kresse, The vienna ab-initio simulation program VASP: An efficient and versatile tool for studying the structural, dynamic, and electronic properties of materials, *in* Properties of Complex Inorganic Solids. 1997: Springer. p. 69–82.
171. Kliros, G.S., A phenomenological model for the quantum capacitance of monolayer and bilayer graphene devices. arXiv preprint arXiv:1105.5827, 2011.
172. Slater, J.C. and G.F. Koster, Simplified LCAO method for the periodic potential problem. *Physical Review*, 1954. **94**(6): p. 1498–1524.
173. Grüneis, A., et al., Tight-binding description of the quasiparticle dispersion of graphite and few-layer graphene. *Physical Review B*, 2008. **78**(20), 205425.
174. Daley, A.J., Quantum computing and quantum simulation with group-II atoms. *Quantum Information Processing*, 2011. **10**(6): p. 865–884.
175. Yuan, L., et al., Photocarrier generation from interlayer charge-transfer transitions in WS2-graphene heterostructures. *Science Advances*, 2018. **4**(2): p. e1700324.
176. Boykin, T.B. and G. Klimeck, Practical application of zone-folding concepts in tight-binding calculations. *Physical Review B*, 2005. **71**(11): p. 115215.
177. Sato, K., et al., Zone folding effect in RamanG-band intensity of twisted bilayer graphene. *Physical Review B*, 2012. **86**(12): p. 125414.
178. Khusayfan, N.M. and H.K. Khanfar, Impact of Mg layer thickness on the performance of the Mg/Bi2O3 plasmonic interfaces. *Thin Solid Films*, 2018. **651**: p. 71–76.
179. Ban, X., et al., Enhanced electron affinity and exciton confinement in exciplex-type host: Power efficient solution-processed blue phosphorescent OLEDs with low turn-on voltage. *ACS Applied Materials & Interfaces*, 2016. **8**(3): p. 2010–2016.

4 Na-Intercalation Compounds and Na-Ion Batteries

Hsien-Ching Chung
RD Department, Super Double Power Technology Co. Ltd., Changhua City, Changhua County, Taiwan

Wei-Bang Li and Yu-Ming Wang
Department of Physics, National Cheng Kung University, Tainan City, Taiwan

Ming-Fa Lin
Department of Physics and Hierarchical Green-Energy Material (Hi-GEM) Research Center, National Cheng Kung University, Tainan City, Taiwan

CONTENTS

In this chapter, a brief history and recent developments in Na-ion batteries are described. The fundamental physical and electronic properties, such as geometric structure, band structure, density of states, and spatial charge distribution, of Na-intercalation compounds are discussed. The outlook of Na-ion batteries is given at the last.

DOI: 10.1201/9781003367215-4

4.1 INTRODUCTION

The wide use of lithium-ion (Li-ion) batteries in various fields, from 3C products to MWh-class grid-scale energy storage systems (ESSs), has revolutionized our daily life [1,2]. The 2019 Nobel Prize in Chemistry has been awarded to John B. Goodenough, M. Stanley Whittingham, and Akira Yoshino for their contributions to developing Li-ion batteries. The story can go back to the 1970s, Whittingham demonstrated the prototype framework of Li-ion batteries [3,4]. However, the commercialization is not smooth. The dendrite of lithium metal leads to short-circuiting and thermal runaway. After a lot of effort, stable full-cell Li-ion batteries are made. In 1985 at Asahi Kasei Corporation, Akira Yoshino [5,6] assembled a full rechargeable battery including the petroleum coke anode with Goodenough's $LiCoO_2$ cathode [7–13] based on Whittingham's framework [3,4], preventing the risk of dendrite-formation-induced thermal runaway and ensuring the stability for the commercial market. This battery was later commercialized by Sony in 1991 with a gravimetric energy capacity of 80 Wh/kg and volumetric energy capacity of 200 Wh/L [14].

Currently, Li-ion batteries have exhibited many branches based on their active materials, performing various applications [15]. They are commonly named by the cathode (positive) materials, such as lithium cobalt oxide ($LiCoO_2$ or LCO) [9,14,16], lithium manganese oxide ($LiMnO_2$ or LMO) [17–21], lithium iron phosphate ($LiFePO_4$ or LFP) [22–25], lithium nickel cobalt manganese oxide ($Li(Ni_xCo_yMn_z)O_2$ or NCM) [26–28], and lithium nickel cobalt aluminum oxide ($LiNi_{0.8}Co_{0.15}Al_{0.05}O_2$ or NCA) [29,30]. Some are named by their anode (negative) materials, such as lithium titanium oxide ($Li_4Ti_5O_{12}$ or LTO) [31–46]. LCO and LMO-based batteries with medium energy capacities and high nominal voltages (3.6~3.7 Volt) are usually used in portable devices, such as smartphones and laptops. NCM and NCA-based batteries with high energy capacities and high nominal voltages (3.6~3.7 Volt) are usually used in power applications, such as electric vehicles. LFP-based batteries with low energy capacities, medium nominal voltages (3.2 Volt), and high safety features are usually used in large-scale power and stationary application, such as electric buses and grid-scale energy storage systems. LTO-based batteries with low energy capacities, low nominal voltages (2.4 Volt), and high safety features are usually used in power applications, such as electric vehicles and uninterruptible power supply (UPS) [1,2].

In 2019, the battery market is USD 25 billion and showing a growing tendency. By 2030, the battery market will be worth USD 116 billion annually according to BloomNEF's forecasts [47], and this doesn't include investment in the supply chain. A decade ago, the Li-ion batteries were a pricey proposition. In 2010, the Li-ion battery packs cost 1,183/kWh. Nine years later, the price had decreased nearly tenfold to USD 156/kWh in 2019, falling 87% in real terms, driving a fast-expanding market for electric vehicles (EVs). Manufacturers are closing in on a point where EVs will approach cost parity with their fossil fuel-powered cousins at around USD 100/kWh. That price is widely seen as a sweet point in the sector, where consumers will no longer regard EVs as pricey options. Cost reductions in 2019 are attributed to increasing order size, growth in EV sales, and the continued penetration of high-energy-density cathodes. The advance in pack designs and falling manufacturing costs will further drive prices down. BloombergNEF forecast that Li-ion battery costs will fall under 100 USD/kWh in 2024 and hit around 60 USD/kWh by 2030 [47].

Although Li-ion batteries currently give an appropriate solution to overcome the challenges in realizing sustainable energy development, reflecting on the growing EV market and emerging ESS market, the lithium reserves must be taken into consideration [48]. According to the U.S. Geological Survey (USGS), lithium has historically been acquired from either continental brines or hard-rock minerals. Chile has been a leading producer of lithium carbonate (Li_2CO_3) for a long time, with production from two Salar de Atacama (Atacama Salt Flat) brine operations next to the Andes Mountains in South America [49]. Although lithium markets vary by location, global end-use markets are estimated as follows: batteries are the largest (74%), ceramics and glass (14%) and the rest are lubricating greases (3%), continuous casting mold flux powders (2%), polymer production (2%), air treatment (1%), and other uses (4%). Lithium consumption for batteries has largely grown in recent years owing to rechargeable Li-ion batteries used extensively in the raising market for EVs, portable electronic devices, and grid-scale storage applications [50]. Excluding U.S. production, worldwide lithium production in 2021 increased by 21% to approximately 100,000 tons from 82,500 tons in 2020 reflecting strong demand from the Li-ion battery market and increased prices of lithium. Global consumption of lithium in 2021 was estimated to be 93,000 tons, a 33% increase from 70,000 tons in 2020. Lithium resources are unevenly distributed, the mine production capacities of the first two countries are Australia (55,000 tons) and Chile (26,000 tons) [50]. In contrast to lithium, sodium resources are almost unlimited on Earth, and sodium is one of the most abundant elements in the Earth's crust. Sodium resources can be easily found in the ocean. Sodium can easily be obtained by evaporation of seawater (11,000 mg/L in seawater) where the lithium content in seawater is much lower than that of sodium (0.18 mg/L) [51]. Additionally, sodium is the second-lightest and -smallest alkali metal below lithium in the periodic table. Based on material abundance and standard electrode potential, rechargeable sodium-ion batteries (or Na-ion batteries) are the potential alternative to Li-ion batteries [52–54].

Na-ion batteries are operable at ambient temperature, and metallic sodium is not used as the anode (negative) electrode, which is different from other commercialized high-temperature sodium-based technology, such as sodium-sulfur batteries (Na/S batteries) [55] and $Na/NiCl_2$ [56] batteries. These batteries apply alumina-based solid (ceramic) electrolyte under high operation temperatures (250~350 °C) for maintaining the electrodes in the liquid state to ensure good contact with the solid electrolyte. Because molten sodium and sulfur are adopted as active materials at such high temperatures, safety issues remain a critical problem for consumer appliances. On the contrary, Na-ion batteries composed of sodium insertion materials with polar aprotic solvent as an electrolyte are free from metallic sodium and safety issues at high temperatures. The structures, components, systems, and charge storage mechanisms of Na-ion batteries are essentially analogous to those of Li-ion batteries [54]. Na-ion batteries are composed of two sodium insertion materials, cathode (positive) and anode (negative) electrodes, which are electronically separated by electrolyte as a pure ionic conductor. The battery performance relies on selected battery components, and many different Na-ion batteries for different purposes can be fabricated.

The abundance of materials is a straightforward reason for considering sodium ions as the charge carriers in secondary rechargeable batteries. There is an obvious disadvantage when compared between Li and Na metal electrodes, i.e., the theoretical gravimetric capacities of Na (1,166 mAh/g) are much less than that of Li (3,861 mAh/g), also the volumetric capacities of Na (1,131 mAh/g) are much less than that of Li (2,062 mAh/g) [57]. It seems that if treated Na^+/Na as an electrochemical equivalent of Li^+/Li, Na is more than three times heavier than Li. However, when compared to the layered cobalt oxides of Na and Li (i.e., $NaCoO_2$ and $LiCoO_2$), the difference between their theoretical capacity becomes smaller. The theoretical capacity is 235 and 274 mAh/g respectively for $NaCoO_2$ and $LiCoO_2$, as one-electron redox of the cobalt ion is assumed to happen (Co^{3+}/Co^{4+} redox). In this case, the capacity is decreased by an acceptable 14%. Another sacrifice can be found in the difference in working voltage range. The working voltage range is 2~3.5 V [58] for $NaCoO_2$, which is much lower than the 3~4.2 V for $LiCoO_2$ [9]. If the final goal is to establish the energy storage technology based on sodium ions instead of sodium metal, the sacrifice in energy density can be potentially eliminated. Hence, Na-ion batteries are expected to be the alternative battery system for Li-ion batteries.

Early-stage studies of Li^+ and Na^+ ions as charge carriers for electrochemical energy storage at ambient temperature began in the 1970s. In 1980, lithium cobalt oxide ($LiCoO_2$), which is a lithium-containing layered structure, demonstrated the electrode performance for high-energy positive electrode materials in Li-ion batteries [9]. On the other side, sodium cobalt oxide (Na_xCoO_2) as sodium-containing layered oxides was also reported [58]. The early history of sodium intercalation materials was reviewed and published in 1982 [59,60]. However, in the past three decades, important research efforts have been conducted only for Li-ion batteries, and studies on the counterpart sodium intercalation materials for energy storage once almost disappeared.

Two reasons stand for the disappear studies. The first reason is the available energy density. Li-ion batteries were believed to possess higher energy densities than their Na-ion counterpart. Although both systems have similar crystal lattice construction by sheets of edge-sharing CoO_6 octahedra, the voltage of $NaCoO_2$ at the end of discharge is lower than that of $LiCoO_2$. When both systems are charged to >100 mAh/g, the voltage difference is dropped to approximately 0.4 V [9,58], similar to the difference in the standard electrochemical potential of Li (3.040 V) [61–63] and Na (2.71 V). The voltage difference becomes more critical as Na is a major component in the structure. Consequently, the available energy density of the Na system is much lower than that of the Li system. Especially, when the same chemistry is used, such as redox species and host crystal structures. The comparison between $NaCoO_2$ and $LiCoO_2$ is the typical case.

The second reason was the lack of appropriate anode (negative) electrodes for a long period of time. Since the 1980s, carbonaceous materials have been found to be potential candidates for Li intercalation hosts. Carbonaceous materials are currently widely adopted as anode (negative) electrode materials for Li-ion batteries, such as carbon fiber [64], pyrolytic carbon [65], and graphite [66,67]. The research interest in Li-ion batteries is further accelerated by the application of graphite, resulting in high capacity (theoretically 372 mAh/g). Unfortunately, graphite is not suitable for an intercalation host of sodium ions, e.g., Ge and Fouletier (35 mAh/g) [68], Doeff et al. (93 mAh/g) [69,70], Thomas et al. (55 mAh/g) [71]. Before the 1990s, several studies were available as potential anode (negative) electrode materials for Na-ion batteries, e.g., Na-Pb alloy [72,73] and disordered carbon [70]. However, it was obvious that the energy density of Na-ion batteries with these anode materials was inferior to that of Li-ion batteries with graphite anodes.

As the first turning point in the Na-ion battery research field, in 2000, Stevens and Dahn reported a high capacity of 300 mAh/g for Na-ion batteries with hard carbon [74], close to that for Li-ion batteries. Hard carbon is now extensively studied as a potential candidate for anode (negative) electrode material for Na-ion batteries. As the second important finding, in 2006, Okada et al. reported that $NaFeO_2$ is electrochemically active in Na-ion batteries based on the Fe^{3+}/Fe^{4+} redox couple [75]. The capacity of $NaFeO_2$ is 80 mAh/g, and the 3.3 V flat discharge profile is similar to that of isostructural $LiCoO_2$ in Li-ion batteries. The Fe^{3+}/Fe^{4+} redox is unique chemistry for the Na-ion batteries and never been demonstrated as active for $LiFeO_2$ in the Li-ion batteries (since layered rocksalt $LiFeO_2$ is metastable in Li-ion batteries).

4.2 RECENT DEVELOPMENT

The Na-ion batteries with organic electrolytes are state-of-the-art technology. The development of renewable and environment-friendly organic-electrolyte-based Na-ion batteries has attracted researchers' attention in the past decade [51,76]. In organic-electrolyte-based Na-ion batteries, the electrolytes are synthesized by dissolving one of the ionic sodium salts (e.g., $NaClO_4$, $NaPF_6$, and NaTFSI) in non-aqueous solvents (e.g., EC, PC, DMC, DEC, DME), like in Li-ion batteries [77]. The ionic conductivity of an electrolyte synthesized with 1 M $NaClO_4$ in EC:DME (50:50 wt%) approaches 12.5 ms/cm (almost the same value as 1 M $LiPF_6$ in EC:DMC (50:50 wt%)) [77].

The energy and power density of the Na-ion batteries is directly related to the cathode materials. Analogous to Li-ion batteries, sodium cathodes can be divided into three major groups: layered, olivine (NASICON), and spinel. Currently, layered structure oxides [78–82] and NASICON structure phosphates have demonstrated potential performance [83]. Recently, Prussian blue analogs (PBAs) ($Na_2M[Fe(CN)_6]$, where M = Co, Mn, Ni, Cu, Fe, etc.) have been extensively studied based on their large vacancies in lattice space structure, providing many sites and transport channels for reversible Na-ion deintercalation [84,85].

Layered oxides are possible candidates as cathode materials for Na-ion batteries, possessing a common formula Na_xXO_2, where X is one or several transition metals (Mn, Fe, Co, Ni, Ti, V, Cr, etc). 2D layered sodium transition metal oxides are extensively studied based on their electrochemical features. The 2D layered sodium transition metal oxides are classified as an O3 type and P2 type structure, where the O and P indicate the location of Na ion in the crystal structure (i.e., P-prismatic site and O-octahedral site), and the numbers indicate the transition metal layer in the repeating unit cell in the structure [86]. O3- and P2-types are the common structural polymorphs of layered transition metal oxides.

In the P2-type materials, manganese-based ($Na_{2/3}MnO_2$) cathode has attracted much attention based on the low cost of manganese, and it gives high discharge capacity (>150 mAh/g) [87]. However, manganese leads to structural distortions as Mn^{3+} is dominant in the structure. This phenomenon can be attributed to Jahn-Teller distortion, causing elongation or compression in the z-axis. The anisotropic changes in the lattice parameters during charge and discharge result in fast capacity fading [81]. One of the possible routes to overcome the problem is to dilute the Mn^{3+} concentration in the crystal structure and decrease the anisotropic change by substituting it with other metal cations [88–91].

Clément et al. [92] studied the effect of Mg doping on the P2-type $Na_{2/3}MnO_2$ compound by varying the amount of dopant, i.e., $Na_{2/3}Mn_{1-y}Mg_yO_2$ (y = 0, 0.05, and 0.1). They demonstrated that Mg substitution results in smoother electrochemistry, with fewer distinct electrochemical processes, improving rate performance, and better capacity retention. Mg doping reduces the number of Mn^{3+} Jahn-Teller centers and delays the high voltage phase transition occurring in P2-$Na_{2/3}MnO_2$. The 5% Mg-doped phase exhibited the highest capacity and rate performance. Kang et al. [93] demonstrated copper-substituted P2-type $Na_{0.67}Cu_x Mn_{1-x}O_2$ (x = 0, 0.14, 0.25, and 0.33) compound to enhance the rate performance of P2-type $Na_{0.7}MnO_2$. The materials show excellent stability, retaining more than 70% of the initial capacity after 500 cycles at 1,000 mA/g.

In NASICON type of cathodes, $Na_3V_2(PO_4)_3$ and its derivatives demonstrated great electrochemical performance [90]. Although the capacity is less than that of the layered oxides, the operating potential makes it more promising. However, the cost and toxicity of the vanadium restrict its next step and widespread to the market. Many research works [85,88,94] have reported systematic studies of the electrochemical performance of various cathode materials. Considerable comparison can be obtained between the cycling profiles for these materials, realizing the cathode behavior.

4.3 FUNDAMENTAL PHYSICAL AND ELECTRONIC PROPERTIES OF NA-INTERCALATION COMPOUNDS

The fundamental physical and electronic properties, such as geometric structure, band structure, density of states, and spatial charge distribution, of Na-intercalation compounds (NaC_8, NaC_{18}, NaC_{24}, and NaC_{32}) are discussed.

4.3.1 Geometric Structure

Generally, it is well-known that graphite intercalation compounds exhibit planar carbon-honeycomb symmetries and well-characterized intercalant distributions, especially for metal atoms or even for large molecule ones [95]. The graphitic layers are attracted by the weak but significant van der Waals interactions, leading to easy modulations by the adatom [96] or large-molecule intercalations/de-intercalations [97]. This clearly indicates strong σ bondings of graphitic layers, as well as the non-covalent interlayer and intra-intercalant interactions. From the delicate analyses of VASP simulations, Na-intercalations could be classified into stable and quasi-stable configurations according to their ground state energies, E_{gs}.

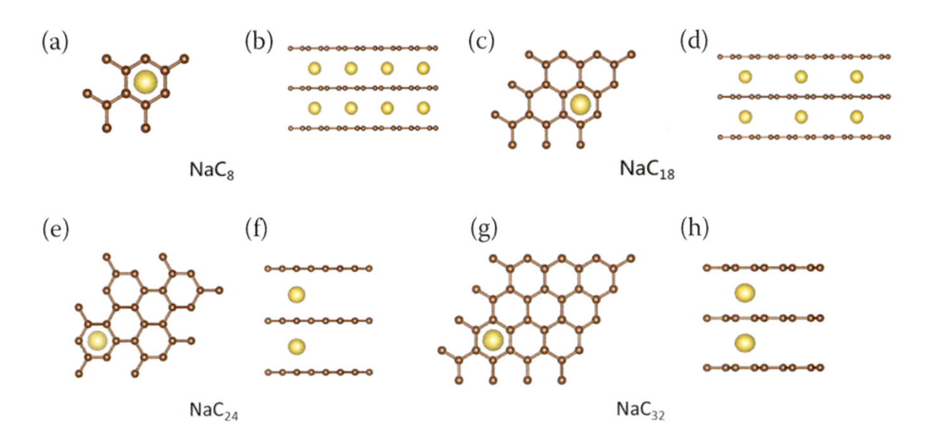

FIGURE 4.1 The optimal crystal structures of sodium-graphite intercalation compounds with the top/side view (the left-/right-hand sides): (a)/(b) NaC_8, (c)/(d) NaC_{18}, (e)/(f) NaC_{24}, (g)/(h) NaC_{32}.

For the stage-1 NaC_x, where x = 8, 18, 24, and 32 (as shown in Figure 4.1), the hollow-site position possesses the lowest ground state energy, that is to say, the hollow-site position is the most stable geometric structure. The interlayer distances of NaC_8, NaC_{18}, NaC_{24}, and NaC_{32} are, respectively, 4.600, 4.511, 4.452, and 4.411 Å; also, the Na-C bond lengths are, respectively, 2.716, 2.668, 2.651, and 2.627 Å. The calculated results reveal that the interlay distances and Na-C bond lengths decrease with the lower concentrations. However, the C-C bond lengths are influenced very slightly (details shown in Table 4.1).

The rich crystal symmetries of 3D graphite intercalation compounds, which are clearly revealed in the graphitic and intercalant layers, as well as, their observable spacing, could be verified by the high-resolution measurements of X-ray diffraction peaks [98–100], reflection electron diffraction patterns (top views) [101,102] and tunneling electronic microscopic spectra (side views) [103], but not those by scanning tunneling microscopy (STM) [104,105]. The former is frequently utilized in these layered materials, e.g., strong evidence of stage-*n* alkali graphite intercalation compounds LiC_{6n} and MC_{8n}, where M = Na, K, Rb, and Cs.

TABLE 4.1
The Calculated Results of NaC_8, NaC_{18}, NaC_{24}, and NaC_{32}

	Ground State Energy (eV)	Intralayer C-C	Na-C	Na-Na	Interlayer Distance	Blue Shift of E_F (eV)
NaC_8	−75.017	1.434	2.716	4.985	4.600	1.48
NaC_{18}	−169.527	1.129	2.688	7.431	4.511	1.17
NaC_{24}	−225.407	1.427	2.651	8.572	4.452	1.02
NaC_{32}	−299.823	1.427	2.627	9.891	4.411	0.92

4.3.2 Band Structure

Pristine graphene, a single carbon-honeycomb crystal, is a well-known semiconductor with zero density of states at the Fermi level [106]. A pair of linear and isotropic valence and conduction subbands intersect there so that a well-defined Dirac-cone structure is initiated from the K/K' valley. This feature is modified by the perpendicular stacking configurations through the weak, but important van der Waals interactions. The linear and/or parabolic energy dispersions come to exist at the K and H valleys, depending on the stacking symmetries. Most importantly, the significant band overlaps, corresponding to the semi-metallic behaviors, are purely induced by the interlayer C-$2p_z$-orbital hybridizations. The AA-/ABC-stacked graphite has the highest/lowest densities of free conduction electrons/valence holes. The semiconductor-semimetal transition would become semimetal-metal after the *n*- or *p*-type dopings.

Band structures of sodium-graphite intercalation compounds, as clearly displayed in Figure 4.2 within a sufficiently wide energy range of -10 eV $\leq E^{c,v} \leq 3$ eV, respectively, are fully supported by the carbon- and intercalant-atom dominances. Their main features cover the number of valence subbands (the active atoms and orbitals in a unit cell) [107,108], the enhanced asymmetry of valence and hole conduction electron energy spectra about the Fermi level (the intercalant-carbon interlayer interactions) [109], the blue shift of E_F (the semimetal-metal transitions due to the charge transfer of metal adatoms) [110], the dependence of E_F on the chemical intercalations (the densities of free conduction electrons under the various intercalation cases), the slight/major modifications of carbon $2p_z$-π & [2s, $2p_x$, $2p_y$]-σ bands along the different wave-vector paths, the subband non-crossing/crossing/anti-crossing behaviors [111], the diverse energy dispersions near the high-symmetry points [112], and their band-edge states (the critical points in energy-wave-vector spaces with zero group velocities) [113]. The Fermi level is mainly determined by

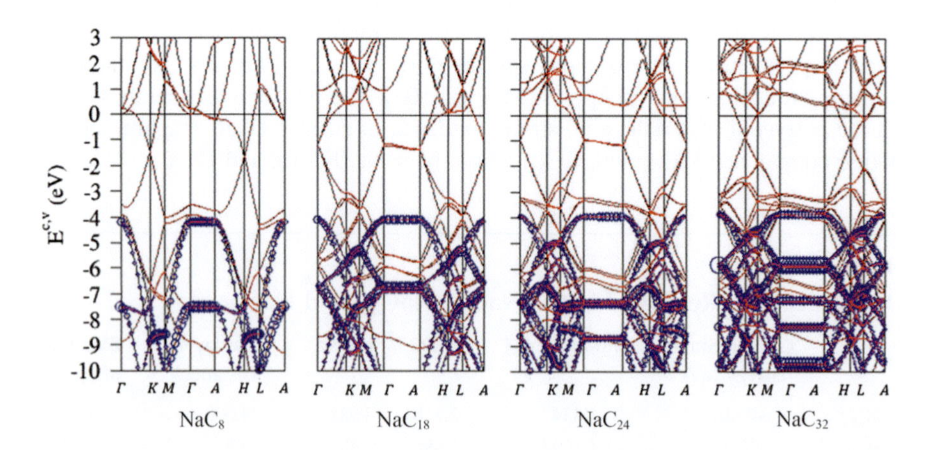

FIGURE 4.2 Band structures of Na-based graphite intercalation compounds within the active valence energy spectra, with atom dominances (red/blue for Na/C): (a) NaC_8, (b) NaC_{18}, (c) NaC_{24}, and (d) NaC_{32}.

the concentration, distribution configuration, and charge transfer of intercalant metal atoms. Its characteristics of the Fermi-Dirac distribution at low temperatures strongly affect the other essential properties, especially for single- and many-particle optical absorption spectra and Coulomb excitations. According to the band structures, the fermi level intersects the conduction bands; that is, the sodium-graphite intercalation compounds exhibit features of *n*-type doping semiconductors.

4.3.3 Density of States

In the density of states (DOS), the critical points presented in the energy-wave-vector space responded to the unusual van Hove singularities. Their special structures are diversified by the different band-edge states of the parabolic, linear, and partially flat energy dispersions. The calculated results are delicately decomposed into the specific contributions of active atoms and orbitals. As a result, significant orbital hybridizations are achieved from the merged unusual structures at the specific energies. These detailed analyses could be generalized to other complicated condensed-matter systems. The current predictions mainly depend on the intercalant concentrations, e.g., the different blue shifts of the Fermi level in sodium-graphite intercalation compounds.

The atom- and orbital-decomposed van Hove singularities are very suitable in determining the active orbital hybridizations of carbon-carbon, intercalant-intercalant, and carbon-intercalant bonds. In general, a Moire superlattice has a lot of carbon atoms with four significant orbitals, compared with a single metal atom with one effective orbital sodium. This leads to the dominating/minor contributions from carbon-honeycomb lattices/the intercalant layers under any intercalation cases (as shown in Figure 4.3). The main features of van Hove singularities mainly arise from the modified π and σ-electronic energy subbands. A local minimum density of states, a dip structure, corresponds to the corrected Dirac-cone structure, and its energy difference with the Fermi level of E_F=0 is an observable blue shift associated with the electron transfer of metal adatoms. The important values (i.e., the blue shift of E_F) of NaC_8, NaC_{18}, NaC_{24}, and NaC_{32} are estimated to be, respectively, 1.48, 1.17, 1.02, and 0.92 eV. It displays that the higher concentrations of Na will provide more electrons transferring from sodium atoms to carbon atoms. Their magnitudes, which correspond to the 3D free electron densities, will have strong effects on the carrier excitations and transports [112,114], e.g., the optical plasmon modes of conduction electrons in the *n*-type graphite intercalation compounds [112,114].

According to four-orbital dominances of carbon atoms, the main features in the density of states could be classified into four specific energy ranges: e.g., for NaC_8: Firstly, the σ-electronic energy spectrum of C-2s orbitals initiated from $E\sim-5.0$ eV; secondly, the σ-valence subbands due to C-[$2p_x$, $2p_y$] orbitals driven from $E\sim-3.5$ eV and dominant within -10.0 eV $\leq E \leq -6.0$ eV; thirdly, the π-electronic valence spectrum of C-$2p_z$ orbitals in -10.0 eV$\leq E \leq -6.0$ eV with dominance above -6.0 eV; fourthly, the π^* conduction spectrum higher than -1.50 eV. There exist three obvious van Hove singularities, corresponding to the dominant carbon-$2p_z$-orbital contributions, respectively, at -3.5 eV, -1.5 eV, and 0.2 eV. Furthermore,

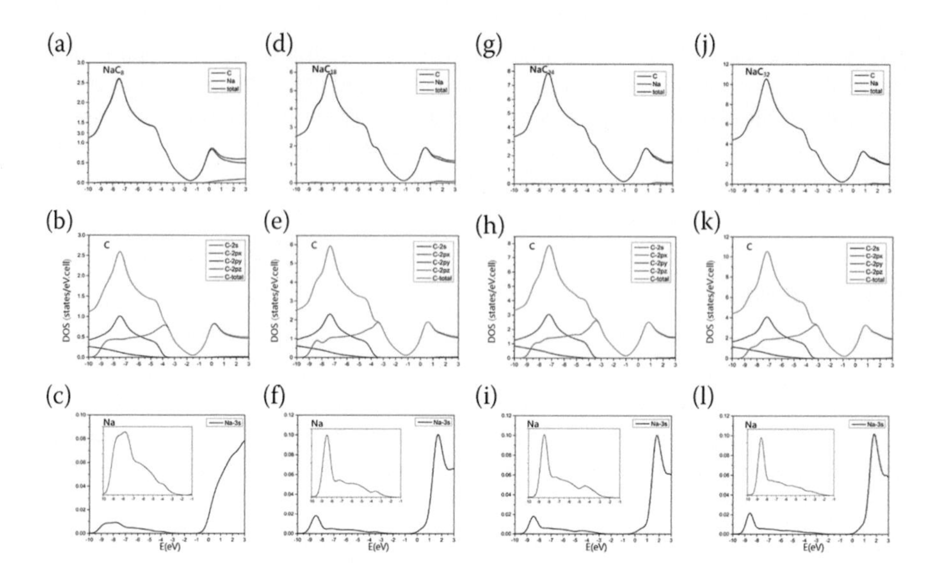

FIGURE 4.3 The density of states in Na-graphite intercalation compounds: (a)/(b)/(c) NaC_8, (d)/(e)/(f) NaC_{18}, (g)/(h)/(i) NaC_{24}, (j)/(k)/(l) NaC_{32}.

the sodium 2s-orbital contributions are merged with them, clearly illustrating the $2p_z$-2s hybridization in each Na-C band. Similar chemical bondings are revealed in NaC_{18}, NC_{24}, and NaC_{32}. The intercalant-decomposed contributions show the reduced amplitudes of van Hove singularities as their concentration decreases. This indicates 3s-3s single-orbital hybridizations for Na-Na bonds. The featured van Hove singularities have successfully identified the active orbital hybridizations of chemical bonds, being consistent with the previous development of the quasiparticle framework [115].

4.3.4 Spatial Charge Distribution

The spatial charge density ($\rho(\mathbf{r})$) and its variation ($\Delta\rho(\mathbf{r})$) after intercalation (as shown in Figure 4.4) are clearly illustrated in a unit cell through the top- and side-views under the different chemical environments. The calculated results could be further utilized to test the X-ray diffraction peak structures [116,117]. By the delicate observations, the critical chemical bondings could be examined from the carbon-honeycomb lattices, the intercalant layers, and their significant spacings. The multi-/single-orbital hybridizations in different chemical bonds are determined by unifying the relevant quantities; that is, the critical quasiparticle pictures are reached through the full cooperation of the atom-dominated band structures, the charge density distributions, and the orbital-decomposed van Hove singularities. These will be very useful in establishing the concise phenomenological models, e.g., the tight-binding model [118–120]/the generalized tight-binding model [114,121] in the absence/presence of a perpendicular magnetic field for the rich magnetic quantization phenomena [112,114,120].

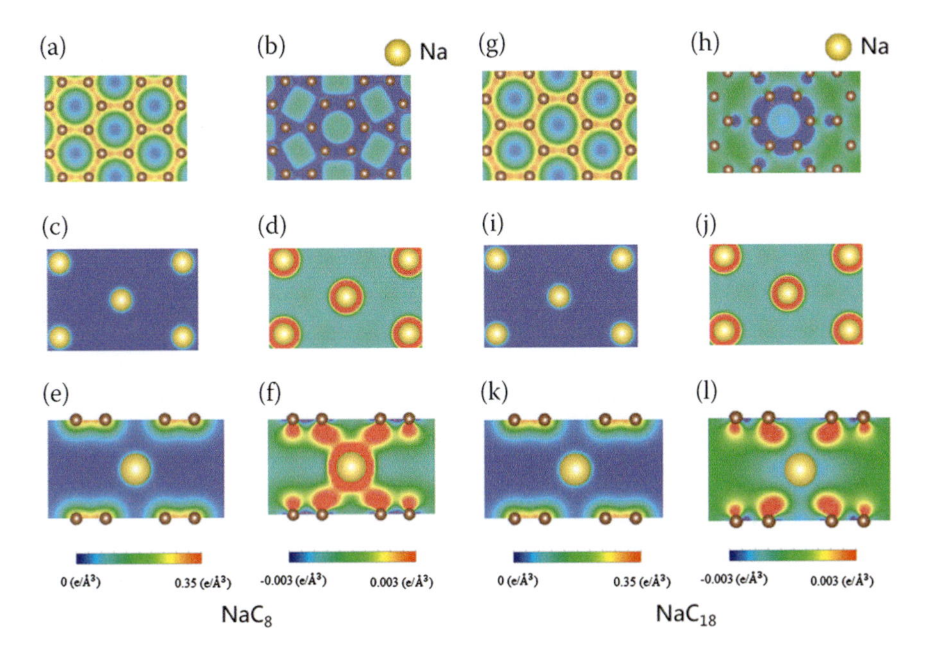

FIGURE 4.4 The spatial charge density distributions ρ(**r**)/the variations before and after the chemical intercalations Δρ(**r**), with the top/side views. The top view on the graphene plane: (a)/(b) NaC_8, (g)/(h) NaC_{18}. The top view on the Na plane: (c)/(d) NaC_8, (i)/(j) NaC_{18}. The side view: (e)/(f) NaC_8, (k)/(l) NaC_{18}.

From carrier distributions about each carbon-honeycomb lattice, the very strong σ bondings and the significant π ones, are respectively, observed in (x, y) and (x, z)/(y, z) planes. For example, ρ(**r**) and Δρ(**r**) of NaC_8 present the high charge density between neighboring carbon atoms (the red color between neighboring carbons) and the peanut-like profile (the light yellow-green color). The former is hardly affected by the sodium-atom intercalations. However, the latter is modified by the interlayer Na-C interactions, as clearly illustrated by the obvious changes of carrier density after interactions (the red color near atoms in Figure 4.4). The detailed analyses further show the specific orbital hybridization of 3s-$2p_z$ orbitals in Na-C bonds. Moreover, 3s-3s mixings in Na-Na bonds are also examined from the intercalant layers, e.g., the significant charge density deformation of 3s-orbital at the height of z = 3.50 Å. Apparently, similar C-C, C-intercalant, and intercalant-intercalant bondings are revealed in the different concentration cases.

On the experimental examinations, the spatial charge density distributions could be directly deduced from the close co-operation of measured X-ray patterns [116,117,122] and elastic scattering predictions associated with atoms and active orbitals in a unit cell [116,117]. For example, a sodium chloride crystal [123], with very strong ionic bonds, is thoroughly investigated and presents an almost isotropic charge distribution near both Na^+-cations and Cl^--anions. Similar analyses are available in exploring those of metal-atom graphite intercalation compounds with concise orbital hybridizations, e.g., the X-ray tests on the theoretical predictions in

Figure 4.4 for Na-graphite intercalation compounds. These investigations will determine the active orbital hybridizations from the intralayer [115,124] and interlayer [115,124] atomic interactions. Such works are very difficult to be done for large-molecule cases. Since the X-ray diffraction spectrum only depends on the whole distribution of charge density, the VASP results should be useful in determining it. How to link them is an open calculation issue.

4.4 OUTLOOK OF NA-ION BATTERIES

Several aspects might be taken into account for the future development of Na-ion batteries, such as low dimensional materials as potential additives, as well as the reuse and recycling of repurposing batteries. The discovery of graphene [125,126] and its incredible essential properties (such as high carrier mobility at room temperature (>200,000 cm^2/Vs) [127–129], superior thermoconductivity (3,000–5,000 W/mK) [128,130], extremely high modulus (1 TPa) [131–133] and tensile strength (130 GPa) [132,134], high transparency to incident light over a broad range of wavelength (97.7%) [135,136], anomalous quantum Hall effect [137–140], and edge-dependent optical selection rules [141,142] and magneto-electronic properties [143–147]) stimulate the development of many low-dimensional materials and researches in various fields for the last decade [148–153]. Up to now, low-dimensional materials with various sizes of bandgap can be synthesized (e.g., gapless graphene, insulating hexagonal boron nitride (h-BN) [154], and semiconducting transition metal dichalcogenides (TMDs) [155] and group III-V compounds [156]), indicating that they can be building blocks similar to bulk materials. Moreover, due to their unique features (e.g., ultrathin, tunneling barrier), low-dimensional materials are expected to be used as next-generation additives in Na-ion batteries beyond the bulk materials.

The successful commercialization and popularization of EVs worldwide [157] cause the widespread and mass production of Li-ion batteries. The retired power batteries have largely increased, causing waste of resources and environmental protection threats. Hence, recycling and utilization of such retired batteries have been promoted [158–160]. Some retired power batteries still possess about 80% initial capacity [161–165]. So they can be repurposed and utilized once again, e.g., serving as the battery modules in the stationary energy storage system [166–169]. Governments in various countries have acknowledged this emergent issue and prepared to launch their policies to deal with the recovery and reuse of repurposing batteries, such as coding principles, traceability management systems, manufacturing factory guidelines, dismantling process guidelines, residual energy measurement, federal and state tax credits, rebates, and other financial support [170–172].

Safety and performance are important in using retired power batteries, i.e., repurposing batteries. Underwriters Laboratories (UL), a global safety certification company established in 1894, published the standard for evaluating the safety and performance of repurposing batteries in 2018 (UL 1974) [173–175]. The document briefly provides a general procedure of the safety operations and performance tests on retired power battery packs, modules, and cells, while the detailed steps and specifics are not given in detail. In real-world applications, the design, form factor, and raw materials of the existing pack/modules/cells often vary greatly from one another, making it difficult to develop a unified technical procedure. Furthermore, information on the detailed technical procedures used is often not easily available in the open literature, except for Schneider *et al.*, who reported the procedure to classify and reuse small cylindrical NiMH battery modules for mobile phones [176,177]; Zhao, who shared the successful experiences of some grid-oriented applications of EV Li-ion batteries in China in a comprehensive technical procedure [178]; and Chung, who announced the procedure described in UL 1974 on LFP repurposing batteries and released the related experimental dataset publicly [179]. Based on the information from these studies, a conceptual procedure for repurposing applications is summarized in five primary steps: (1) judgment of the retired battery system based on historical information, (2) disassembly of retired battery packs/modules, (3) battery performance evaluation (mechanical, electrochemical, and safety), (4) sorting, and (5) developing control and management strategies for repurposing applications. In reality, multiple rounds of inspections and assessments might usually be performed [180,181]. Overall, the technical feasibility of repurposing applications of retired EV power batteries mainly depends on whether these steps could be performed effectively and efficiently. Na-ion batteries as potential candidates for next-generation batteries should consider the possibilities in reuse and recycling of repurposing Na-ion batteries.

ACKNOWLEDGMENTS

The author (H. C. Chung) thanks Prof. Ming-Fa Lin for the book chapter invitation and for inspiring him to study this topic. H. C. Chung would like to thank the contributors to this article for their valuable discussions and recommendations, Jung-Feng Jack Lin, Hsiao-Wen Yang, Yen-Kai Lo, and An-De Andrew Chung. The author (H. C. Chung) thanks Pei-Ju Chien for English discussions and corrections as well as Ming-Hui Chung, Su-Ming Chen, Lien-Kuei Chien, and Mi-Lee Kao for financial support. This work was supported in part by Super Double Power Technology Co., Ltd., Taiwan, under the project "Development of Cloud-native Energy Management Systems for Medium-scale Energy Storage Systems (https://osf.io/7fr9z/)" (Grant number: SDP-RD-PROJ-001–2020). This work was supported in part by the Ministry of Science and Technology (MOST), Taiwan (grant numbers: MOST 108–2112-M-006–016-MY3, MOST 109–2124-M-006-001, and MOST 110–2811-M-006–543).

REFERENCES

1. Chung H-C, Nguyen TDH, Lin S-Y, Li W-B, Tran NTT, Thi Han N, Liu H-Y, Pham HD and Lin M-F 2021 *First-Principles Calculations for Cathode, Electrolyte and Anode Battery Materials*: IOP Publishing. pp. 16-1 to -43. DOI: 10.1088/978-0-7503-4685-6ch16
2. Chung H-C 2020 Engineering integrations, potential applications, and outlooks of Li-ion batteries. *engrXiv*. DOI: 10.31224/osf.io/swcyg

3. Whittingham MS and Gamble FR 1975 The lithium intercalates of the transition metal dichalcogenides. *Mater. Res. Bull.* **10** 363–371. DOI: 10.1016/0025-5408(75)90006-9
4. Whittingham MS 1976 Electrical energy-storage and intercalation chemistry. *Science* **192** 1126. DOI: 10.1126/science.192.4244.1126
5. Yoshino A, Sanechika K and Nakajima T 1985 Secondary battery Japanese patent no. 1989293
6. Yoshino A 2012 The birth of the lithium-ion battery. *Angew. Chem. Int. Ed.* **51** 5798. DOI: 10.1002/anie.201105006
7. Goodenough JB and Kim Y 2010 Challenges for rechargeable Li batteries. *Chem. Mater.* **22** 587–603. DOI: 10.1021/cm901452z
8. Manthiram A 2020 A reflection on lithium-ion battery cathode chemistry. *Nat. Commun.* **11** 1550. DOI: 10.1038/s41467-020-15355-0
9. Mizushima K, Jones PC, Wiseman PJ and Goodenough JB 1980 LixCoO2 (0<x<-1): A new cathode material for batteries of high energy density. *Mater. Res. Bull.* **15** 783. DOI: 10.1016/0025-5408(80)90012-4
10. Thackeray MM, David WIF, Bruce PG and Goodenough JB 1983 Lithium insertion into manganese spinels. *Mater. Res. Bull.* **18** 461. DOI: 10.1016/0025-5408(83)90138-1
11. Manthiram A and Goodenough JB 1989 Lithium insertion into Fe2(SO4)3 frameworks. *J. Power Sources* **26** 403–408. DOI: 10.1016/0378-7753(89)80153-3
12. Manthiram A and Goodenough JB 1987 Lithium insertion into Fe2(MO4)3 frameworks: Comparison of M=W with M=Mo. *J. Solid State Chem.* **71** 349–360. DOI: 10.1016/0022-4596(87)90242-8
13. Masquelier C and Croguennec L 2013 Polyanionic (phosphates, silicates, sulfates) frameworks as electrode materials for rechargeable Li (or Na) batteries. *Chem. Rev.* **113** 6552–6591. DOI: 10.1021/cr3001862
14. Nishi Y 2001 Lithium ion secondary batteries; past 10 years and the future. *J. Power Sources* **100** 101–106. DOI: 10.1016/s0378-7753(01)00887-4
15. Nguyen TDH, Lin S-Y, Chung H-C, Tran NTT and Lin M-F 2021 *First-Principles Calculations for Cathode, Electrolyte and Anode Battery Materials*: IOP Publishing. DOI: 10.1088/978-0-7503-4685-6
16. Du Pasquier A, Plitz I, Menocal S and Amatucci G 2003 A comparative study of Li-ion battery, supercapacitor and nonaqueous asymmetric hybrid devices for automotive applications. *J. Power Sources* **115** 171–178. DOI: 10.1016/s0378-7753(02)00718-8
17. Armstrong AR and Bruce PG 1996 Synthesis of layered LiMnO2 as an electrode for rechargeable lithium batteries. *Nature* **381** 499–500. DOI: 10.1038/381499a0
18. Thackeray MM 1997 Manganese oxides for lithium batteries. *Prog. Solid State Chem.* **25** 1–71. DOI: 10.1016/s0079-6786(97)81003-5
19. Strobel P, Le Cras F and Anne M 1996 Composition-valence diagrams: A new representation of topotactic reactions in ternary transition metal oxide systems. Application to lithium intercalation. *J. Solid State Chem.* **124** 83–94. DOI: 10.1006/jssc.1996.0211
20. Thackeray MM 1999 Spinel electrodes for lithium batteries. *J. Am. Ceram. Soc.* **82** 3347–3354. DOI: 10.1111/j.1151-2916.1999.tb02250.x
21. Thackeray MM, Depicciotto LA, Dekock A, Johnson PJ, Nicholas VA and Adendorff KT 1987 Spinel electrodes for lithium batteries – A review. *J. Power Sources* **21** 1–8. DOI: 10.1016/0378-7753(87)80071-x
22. Doughty DH and Roth EP 2012 A general discussion of Li ion battery safety. *Electrochem. Soc. Interface* **21** 37–44. DOI: 10.1149/2.F03122if

23. Padhi AK, Nanjundaswamy KS, Masquelier C, Okada S and Goodenough JB 1997 Effect of structure on the Fe3+/Fe2+ redox couple in iron phosphates. *J. Electrochem. Soc.* **144** 1609–1613. DOI: 10.1149/1.1837649
24. Padhi AK, Nanjundaswamy KS and Goodenough JB 1997 Phospho-olivines as positive-electrode materials for rechargeable lithium batteries. *J. Electrochem. Soc.* **144** 1188. DOI: 10.1149/1.1837571
25. Yamada A, Chung SC and Hinokuma K 2001 Optimized LiFePO4 for lithium battery cathodes. *J. Electrochem. Soc.* **148** A224–A229. DOI: 10.1149/1.1348257
26. Yabuuchi N and Ohzuku T 2003 Novel lithium insertion material of LiCo1/3Ni1/3Mn1/3O2 for advanced lithium-ion batteries. *J. Power Sources* **119** 171–174. DOI: 10.1016/s0378-7753(03)00173-3
27. Shaju KM and Bruce PG 2006 Macroporous Li(Ni1/3Co1/3Mn1/3)O-2: A high-power and high-energy cathode for rechargeable lithium batteries. *Adv. Mater.* **18** 2330–2334. DOI: 10.1002/adma.200600958
28. Sun YK, Myung ST, Park BC, Prakash J, Belharouak I and Amine K 2009 High-energy cathode material for long-life and safe lithium batteries. *Nat. Mater.* **8** 320–324. DOI: 10.1038/nmat2418
29. Martha SK, Haik O, Zinigrad E, Exnar I, Drezen T, Miners JH and Aurbach D 2011 On the thermal stability of olivine cathode materials for lithium-ion batteries. *J. Electrochem. Soc.* **158** A1115–A1122. DOI: 10.1149/1.3622849
30. Berckmans G, Messagie M, Smekens J, Omar N, Vanhaverbeke L and Van Mierlo J 2017 Cost Projection of state of the art lithium-ion batteries for electric vehicles up to 2030. *Energies* **10** 1314. DOI: 10.3390/en10091314
31. Han CP, He YB, Liu M, Li BH, Yang QH, Wong CP and Kang FY 2017 A review of gassing behavior in Li4Ti5O12-based lithium ion batteries. *J. Mater. Chem. A* **5** 6368–6381. DOI: 10.1039/c7ta00303j
32. Aravindan V, Lee YS and Madhavi S 2015 Research progress on negative electrodes for practical Li-ion batteries: Beyond carbonaceous anodes. *Adv. Energy Mater.* **5** 1402225. DOI: 10.1002/aenm.201402225
33. Sun XC, Radovanovic PV and Cui B 2015 Advances in spinet Li4Ti5O12 anode materials for lithium-ion batteries. *New J. Chem.* **39** 38–63. DOI: 10.1039/c4nj01390e
34. Pohjalainen E, Kallioinen J and Kallio T 2015 Comparative study of carbon free and carbon containing Li4Ti5O12 electrodes. *J. Power Sources* **279** 481–486. DOI: 10.1016/j.jpowsour.2014.12.111
35. Zhao B, Ran R, Liu ML and Shao ZP 2015 A comprehensive review of Li4Ti5O12-based electrodes for lithium-ion batteries: The latest advancements and future perspectives. *Mater. Sci. Eng. R-Rep.* **98** 1–71. DOI: 10.1016/j.mser.2015.10.001
36. Yi TF, Yang SY and Xie Y 2015 Recent advances of Li4Ti5O12 as a promising next generation anode material for high power lithium-ion batteries. *J. Mater. Chem. A* **3** 5750–5777. DOI: 10.1039/c4ta06882c
37. Sandhya CP, John B and Gouri C 2014 Lithium titanate as anode material for lithium-ion cells: A review. *Ionics* **20** 601–620. DOI: 10.1007/s11581-014-1113-4
38. Song MS, Kim RH, Baek SW, Lee KS, Park K and Benayad A 2014 Is Li4Ti5O12 a solid-electrolyte-interphase-free electrode material in Li-ion batteries? Reactivity between the Li4Ti5O12 electrode and electrolyte. *J. Mater. Chem. A* **2** 631–636. DOI: 10.1039/c3ta12728a
39. Reddy MV, Rao GVS and Chowdari BVR 2013 Metal oxides and oxysalts as anode materials for Li ion batteries. *Chem. Rev.* **113** 5364–5457. DOI: 10.1021/cr3001884
40. Chen ZH, Belharouak I, Sun YK and Amine K 2013 Titanium-based anode materials for safe lithium batteries. *Adv. Funct. Mater.* **23** 959–969. DOI: 10.1002/adfm.201200698

41. Kim C, Norberg NS, Alexander CT, Kostecki R and Cabana J 2013 Mechanism of phase propagation during lithiation in carbon-free Li4Ti5O12 battery electrodes. *Adv. Funct. Mater.* **23** 1214–1222. DOI: 10.1002/adfm.201201684
42. Zhu GN, Wang YG and Xia YY 2012 Ti-based compounds as anode materials for Li-ion batteries. *Energy Environ. Sci.* **5** 6652–6667. DOI: 10.1039/c2ee03410g
43. Song MS, Benayad A, Choi YM and Park KS 2012 Does Li4Ti5O12 need carbon in lithium ion batteries? Carbon-free electrode with exceptionally high electrode capacity. *Chem. Commun.* **48** 516–518. DOI: 10.1039/c1cc16462g
44. Yi TF, Jiang LJ, Shu J, Yue CB, Zhu RS and Qiao HB 2010 Recent development and application of Li4Ti5O12 as anode material of lithium ion battery. *J. Phys. Chem. Solids* **71** 1236–1242. DOI: 10.1016/j.jpcs.2010.05.001
45. Guo YG, Hu JS and Wan LJ 2008 Nanostructured materials for electrochemical energy conversion and storage devices. *Adv. Mater.* **20** 2878–2887. DOI: 10.1002/adma.200800627
46. Jiang JW, Chen J and Dahn JR 2004 Comparison of the reactions between Li7/3Ti5/3O4 or LiC6 and nonaqueous solvents or electrolytes using accelerating rate calorimetry. *J. Electrochem. Soc.* **151** A2082–A2087. DOI: 10.1149/1.1817698
47. Battery Pack Prices Fall As Market Ramps Up With Market Average At $156/kWh In 2019 (available at: https://about.bnef.com/blog/battery-pack-prices-fall-as-market-ramps-up-with-market-average-at-156-kwh-in-2019/)
48. Tarascon JM 2010 Is lithium the new gold? *Nat. Chem.* **2** 510-. DOI: 10.1038/nchem.680
49. Kelly JC, Wang M, Dai Q and Winjobi O 2021 Energy, greenhouse gas, and water life cycle analysis of lithium carbonate and lithium hydroxide monohydrate from brine and ore resources and their use in lithium ion battery cathodes and lithium ion batteries. *Res. Conserv. Recycl.* **174** 105762. DOI: 10.1016/j.resconrec.2021.105762
50. Survey U S G 2022 Mineral Commodity Summaries 2022. In: *Mineral Commodity Summaries*,
51. Adelhelm P, Hartmann P, Bender CL, Busche M, Eufinger C and Janek J 2015 From lithium to sodium: Cell chemistry of room temperature sodium-air and sodium-sulfur batteries. *Beilstein J. Nanotechnol.* **6** 1016–1055. DOI: 10.3762/bjnano.6.105
52. Slater MD, Kim D, Lee E and Johnson CS 2013 Sodium-ion batteries. *Adv. Funct. Mater.* **23** 947–958. DOI: 10.1002/adfm.201200691
53. Kim SW, Seo DH, Ma XH, Ceder G and Kang K 2012 Electrode materials for rechargeable sodium-ion batteries: Potential alternatives to current lithium-ion batteries. *Adv. Energy Mater.* **2** 710–721. DOI: 10.1002/aenm.201200026
54. Komaba S, Murata W, Ishikawa T, Yabuuchi N, Ozeki T, Nakayama T, Ogata A, Gotoh K and Fujiwara K 2011 Electrochemical Na insertion and solid electrolyte interphase for hard-carbon electrodes and application to Na-ion batteries. *Adv. Funct. Mater.* **21** 3859–3867. DOI: 10.1002/adfm.201100854
55. Oshima T, Kajita M and Okuno A 2004 Development of sodium-sulfur batteries. *Int. J. Appl. Ceram. Technol.* **1** 269–276. DOI: 10.1111/j.1744-7402.2004.tb00179.x
56. Bones RJ, Teagle DA, Brooker SD and Cullen FL 1989 Development of a Ni, NiCl2 positive electrode for a liquid sodium (ZEBRA) battery cell. *J. Electrochem. Soc.* **136** 1274–1277. DOI: 10.1149/1.2096905
57. Yabuuchi N, Kubota K, Dahbi M and Komaba S 2014 Research development on sodium-ion batteries. *Chem. Rev.* **114** 11636–11682. DOI: 10.1021/cr500192f
58. Delmas C, Braconnier JJ, Fouassier C and Hagenmuller P 1981 Electrochemical intercalation of sodium in NaxCoO2 bronzes. *Solid State Ionics* **3–4** 165–169. DOI: 10.1016/0167-2738(81)90076-x

59. Abraham KM 1982 Intercalation positive electrodes for rechargeable sodium cells. *Solid State Ionics* **7** 199–212. DOI: 10.1016/0167-2738(82)90051-0
60. Delmas C, Braconnier JJ, Maazaz A and Hagenmuller P 1982 Soft chemistry in AxMO2 sheet oxides. *Rev. Chim. Minér.* **19** 343.
61. Liu B, Zhang JG and Xu W 2018 Advancing lithium metal batteries. *Joule* **2** 833–845. DOI: 10.1016/j.joule.2018.03.008
62. Lin DC, Liu YY and Cui Y 2017 Reviving the lithium metal anode for high-energy batteries. *Nat. Nanotechnol.* **12** 194–206. DOI: 10.1038/nnano.2017.16
63. Xu W, Wang JL, Ding F, Chen XL, Nasybutin E, Zhang YH and Zhang JG 2014 Lithium metal anodes for rechargeable batteries. *Energy Environ. Sci.* **7** 513–537. DOI: 10.1039/c3ee40795k
64. Kanno R, Takeda Y, Ichikawa T, Nakanishi K and Yamamoto O 1989 Carbon as negative electrodes in lithium secondary cells. *J. Power Sources* **26** 535–543. DOI: 10.1016/0378-7753(89)80175-2
65. Mohri M, Yanagisawa N, Tajima Y, Tanaka H, Mitate T, Nakajima S, Yoshida M, Yoshimoto Y, Suzuki T and Wada H 1989 Rechargeable lithium battery based on pyrolytic carbon as a negative electrode. *J. Power Sources* **26** 545–551. DOI: 10.1016/0378-7753(89)80176-4
66. Fong R, Vonsacken U and Dahn JR 1990 Studies of lithium intercalation into carbons using nonaqueous electrochemical-cells. *J. Electrochem. Soc.* **137** 2009. DOI: 10.1149/1.2086855
67. Ohzuku T, Iwakoshi Y and Sawai K 1993 Formation of lithium-graphite intercalation compounds in nonaqueous electrolytes and their application as a negative electrode for a lithium ion (shuttlecock) cell. *J. Electrochem. Soc.* **140** 2490–2498. DOI: 10.1149/1.2220849
68. Ge P and Fouletier M 1988 Electrochemical intercalation of sodium in graphite. *Solid State Ionics* **28** 1172–1175. DOI: 10.1016/0167-2738(88)90351-7
69. Barker J, Saidi MY and Swoyer JL 2003 A sodium-ion cell based on the fluorophosphate compound NaVPO4F. *Electrochem. Solid State Lett.* **6** A1–A4. DOI: 10.1149/1.1523691
70. Doeff MM, Ma YP, Visco SJ and Dejonghe LC 1993 Electrochemical insertion of sodium into carbon. *J. Electrochem. Soc.* **140** L169–L170. DOI: 10.1149/1.2221153
71. Thomas P, Ghanbaja J and Billaud D 1999 Electrochemical insertion of sodium in pitch-based carbon fibres in comparison with graphite in NaClO4-ethylene carbonate electrolyte. *Electrochim. Acta* **45** 423–430. DOI: 10.1016/s0013-4686(99)00276-5
72. Jow TR, Shacklette LW, Maxfield M and Vernick D 1987 The role of conductive polymers in alkali-metal secondary electrodes. *J. Electrochem. Soc.* **134** 1730–1733. DOI: 10.1149/1.2100746
73. Ma YP, Doeff MM, Visco SJ and Dejonghe LC 1993 Rechargeable Na / Nax CoO2 and Na15Pb4 / Nax CoO2 polymer electrolyte cells. *J. Electrochem. Soc.* **140** 2726–2733. DOI: 10.1149/1.2220900
74. Stevens DA and Dahn JR 2000 High capacity anode materials for rechargeable sodium-ion batteries. *J. Electrochem. Soc.* **147** 1271–1273. DOI: 10.1149/1.1393348
75. Okada S, Takahashi Y, Kiyabu T, Doi T, Yamaki J-I and Nishida T 2006 Layered Transition Metal Oxides as Cathodes for Sodium Secondary Battery. In: *210th ECS Meeting Abstracts 2006, MA2006-02*, (Cancun, Mexico: Electrochemical Society (ECS)) p. 201.
76. Bauer A, Song J, Vail S, Pan W, Barker J and Lu YH 2018 The scale-up and commercialization of nonaqueous Na-ion battery technologies. *Adv. Energy Mater.* **8** 1702869. DOI: 10.1002/aenm.201702869

77. Vignarooban K, Kushagra R, Elango A, Badami P, Mellander BE, Xu X, Tucker TG, Nam C and Kannan AM 2016 Current trends and future challenges of electrolytes for sodium-ion batteries. *Int. J. Hydrog. Energy* **41** 2829–2846. DOI: 10.1016/j.ijhydene.2015.12.090
78. Deng JQ, Luo WB, Lu X, Yao QR, Wang ZM, Liu HK, Zhou HY and Dou SX 2018 High energy density sodium-ion battery with industrially feasible and air-stable O3-type layered oxide cathode. *Adv. Energy Mater.* **8** 1701610. DOI: 10.1002/aenm.201701610
79. Konarov A, Choi JU, Bakenov Z and Myung ST 2018 Revisit of layered sodium manganese oxides: achievement of high energy by Ni incorporation. *J. Mater. Chem. A* **6** 8558–8567. DOI: 10.1039/c8ta02067a
80. Ortiz-Vitoriano N, Drewett NE, Gonzalo E and Rojo T 2017 High performance manganese-based layered oxide cathodes: Overcoming the challenges of sodium ion batteries. *Energy Environ. Sci.* **10** 1051–1074. DOI: 10.1039/c7ee00566k
81. Liu HD, Xu J, Ma CZ and Meng YS 2015 A new O3-type layered oxide cathode with high energy/power density for rechargeable Na batteries. *Chem. Commun.* **51** 4693–4696. DOI: 10.1039/c4cc09760b
82. Guo SH, Yu HJ, Jian ZL, Liu P, Zhu YB, Guo XW, Chen MW, Ishida M and Zhou HS 2014 A high-capacity, low-cost layered sodium manganese oxide material as cathode for sodium-ion batteries. *ChemSusChem* **7** 2115–2119. DOI: 10.1002/cssc.201402138
83. Chen MZ, Hua WB, Xiao J, Cortiel D, Chen WH, Wang EH, Hu Z, Gu QF, Wang XL, Indris S, Chou SL and Dou SX 2019 NASICON-type air-stable and all-climate cathode for sodium-ion batteries with low cost and high-power density. *Nat. Commun.* **10** 1480. DOI: 10.1038/s41467-019-09170-5
84. Wang BQ, Han Y, Wang X, Bahlawane N, Pan HG, Yan M and Jiang YZ 2018 Prussian blue analogs for rechargeable batteries. *iScience* **3** 110–133. DOI: 10.1016/j.isci.2018.04.008
85. You Y and Manthiram A 2018 Progress in high-voltage cathode materials for rechargeable sodium-ion batteries. *Adv. Energy Mater.* **8** 1701785. DOI: 10.1002/aenm.201701785
86. Delmas C 2018 Sodium and sodium-ion batteries: 50 years of research. *Adv. Energy Mater.* **8** 1703137. DOI: 10.1002/aenm.201703137
87. Zhu XB, Lin TG, Manning E, Zhang YC, Yu MM, Zuo B and Wang LZ 2018 Recent advances on Fe- and Mn-based cathode materials for lithium and sodium ion batteries. *J. Nanopart. Res.* **20** 160. DOI: 10.1007/s11051-018-4235-1
88. Irisarri E, Amini N, Tennison S, Ghimbeu CM, Gorka J, Vix-Guterl C, Ponrouch A and Palacin MR 2018 Optimization of large scale produced hard carbon performance in Na-ion batteries: Effect of precursor, temperature and processing conditions. *J. Electrochem. Soc.* **165** A4058–A4066. DOI: 10.1149/2.1171816jes
89. El Moctar I, Ni Q, Bai Y, Wu F and Wu C 2018 Hard carbon anode materials for sodium-ion batteries. *Funct. Mater. Lett.* **11** 1830003. DOI: 10.1142/s1793604718300037
90. Guo JZ, Wang PF, Wu XL, Zhang XH, Yan QY, Chen H, Zhang JP and Guo YG 2017 High-energy/power and low-temperature cathode for sodium-ion batteries: In situ XRD study and superior full-cell performance. *Adv. Mater.* **29** 1701968. DOI: 10.1002/adma.201701968
91. Kim H, Kwon JE, Lee BN, Hong J, Lee M, Park SY and Kang K 2015 High energy organic cathode for sodium rechargeable batteries. *Chem. Mater.* **27** 7258–7264. DOI: 10.1021/acs.chemmater.5b02569

92. Clement RJ, Billaud J, Armstrong AR, Singh G, Rojo T, Bruce PG and Grey CP 2016 Structurally stable Mg-doped P2-Na2/3Mn1-yMgyO2 sodium-ion battery cathodes with high rate performance: Insights from electrochemical, NMR and diffraction studies. *Energy Environ. Sci.* **9** 3240–3251. DOI: 10.1039/c6ee01750a
93. Kang WP, Zhang ZY, Lee PK, Ng TW, Li WY, Tang YB, Zhang WJ, Lee CS and Yu DYW 2015 Copper substituted P2-type Na0.67CuxMn1-xO2: A stable high-power sodium-ion battery cathode. *J. Mater. Chem. A* **3** 22846–22852. DOI: 10.1039/c5ta06371j
94. Chen MZ, Zhang YY, Xing GC and Tang YX 2020 Building high power density of sodium-ion batteries: Importance of multidimensional diffusion pathways in cathode materials. *Front. Chem.* **8** 152. DOI: 10.3389/fchem.2020.00152
95. Bhauriyal P, Mahata A and Pathak B 2017 The staging mechanism of AlCl4 intercalation in a graphite electrode for an aluminium-ion battery. *Phys. Chem. Chem. Phys.* **19** 7980–7989. DOI: 10.1039/c7cp00453b
96. Howard CA, Dean MPM and Withers F 2011 Phonons in potassium-doped graphene: The effects of electron-phonon interactions, dimensionality, and adatom ordering. *Phys. Rev. B* **84** 241404. DOI: 10.1103/PhysRevB.84.241404
97. Wang QP, Zheng DY, He LX and Ren XG 2019 Cooperative effect in a graphite intercalation compound: Enhanced mobility of AlCl4 in the graphite cathode of aluminum-ion batteries. *Phys. Rev. Appl.* **12** 044060. DOI: 10.1103/PhysRevApplied.12.044060
98. Pramudita JC, Peterson VK, Kimpton JA and Sharma N 2017 Potassium-ion intercalation in graphite within a potassium-ion battery examined using in situ X-ray diffraction. *Powder Diffraction* **32** S43–S48. DOI: 10.1017/s0885715617000902
99. Liang HJ, Hou BH, Li WH, Ning QL, Yang X, Gu ZY, Nie XJ, Wang G and Wu XL 2019 Staging Na/K-ion de-/intercalation of graphite retrieved from spent Li-ion batteries: In operando X-ray diffraction studies and an advanced anode material for Na/K-ion batteries. *Energy Environ. Sci.* **12** 3575–3584. DOI: 10.1039/c9ee02759a
100. Sutto TE, Duncan TT and Wong TC 2009 X-ray diffraction studies of electrochemical graphite intercalation compounds of ionic liquids. *Electrochim. Acta* **54** 5648–5655. DOI: 10.1016/j.electacta.2009.05.026
101. Horio Y, Yamazaki R, Yuhara J, Takakuwa Y and Yoshimura M 2018 Behavior of wave field on graphite surface observed using reflection high-energy electron diffraction technique. *e-J. Surf. Sci. Nanotechnol.* **16** 88. DOI: 10.1380/ejssnt.2018.88
102. Elsayed-Ali HE and Mourou GA 1988 Picosecond reflection high-energy electron diffraction. *Appl. Phys. Lett.* **52** 103–104. DOI: 10.1063/1.99063
103. Ko W, Hus SM, Li XF, Berlijn T, Nguyen GD, Xiao K and Li AP 2018 Tip-induced local strain on MoS2/graphite detected by inelastic electron tunneling spectroscopy. *Phys. Rev. B* **97** 125401. DOI: 10.1103/PhysRevB.97.125401
104. Rong ZY 1994 Extended modifications of electronic structures caused by defects: Scanning tunneling microscopy of graphite. *Phys. Rev. B* **50** 1839–1843. DOI: 10.1103/PhysRevB.50.1839
105. Rong ZY and Kuiper P 1993 Electronic effects in scanning tunneling microscopy: Moiré pattern on a graphite surface. *Phys. Rev. B* **48** 17427–17431. DOI: 10.1103/PhysRevB.48.17427
106. Hashimoto A, Suenaga K, Gloter A, Urita K and Iijima S 2004 Direct evidence for atomic defects in graphene layers. *Nature* **430** 870–873. DOI: 10.1038/nature02817
107. Lin S-Y, Liu H-Y, Nguyen DK, Tran NTT, Pham HD, Chang S-L, Lin C-Y and Lin M-F 2020 *Silicene-Based Layered Materials – Essential properties*: IOP Publishing. DOI: 10.1088/978-0-7503-3299-6

108. Tran NTT, Lin S-Y, Lin C-Y and Lin M-F 2017 *Geometric and Electronic Properties of Graphene-Related Systems – Chemical Bonding Schemes*: CRC Press. DOI: 10.1201/b22450
109. Ho JH, Chang CP and Lin MF 2006 Electronic excitations of the multilayered graphite. *Phys. Lett. A* **352** 446–450. DOI: 10.1016/j.physleta.2005.12.021
110. Ho YH, Wang J, Chiu YH, Lin MF and Su WP 2011 Characterization of Landau subbands in graphite: A tight-binding study. *Phys. Rev. B* **83** 121201. DOI: 10.1103/PhysRevB.83.121201
111. Lin C-Y, Do T-N, Huang Y-K and Lin M-F 2017 *Optical Properties of Graphene in Magnetic and Electric Fields*: IOP Publishing. DOI: 10.1088/978-0-7503-1566-1
112. Lin C-Y, Ho C-H, Wu J-Y, Do T-N, Shih P-H, Lin S-Y and Lin M-F 2019 *Diverse Quantization Phenomena in Layered Materials*: CRC Press. DOI: 10.1201/9781003 004981
113. Lin S-Y, Tran NTT, Chang S-L, Su W-P and Lin M-F eds 2018 *Structure- and Adatom-Enriched Essential Properties of Graphene Nanoribbons*: CRC Press. DOI: 10.1201/9780429400650
114. Chen S-C, Wu J-Y, Lin C-Y and Lin M-F 2017 *Theory of Magnetoelectric Properties of 2D Systems*: IOP Publishing. DOI: 10.1088/978-0-7503-1674-3
115. Li WB, Lin SY, Tran NTT, Lin MF and Lin KI 2020 Essential geometric and electronic properties in stage-ngraphite alkali-metal-intercalation compounds. *RSC Adv.* **10** 23573–23581. DOI: 10.1039/d0ra00639d
116. Li J and Sun JL 2017 Application of X-ray diffraction and electron crystallography for solving complex structure problems. *Acc. Chem. Res.* **50** 2737–2745. DOI: 10.1021/acs.accounts.7b00366
117. Kerber SJ, Barr TL, Mann GP, Brantley WA, Papazoglou E and Mitchell JC 1998 The complementary nature of x-ray photoelectron spectroscopy and angle-resolved x-ray diffraction - Part I: Background and theory. *J. Mater. Eng. Perform.* **7** 329–333. DOI: 10.1361/105994998770347765
118. Matthew W, Foulkes C and Haydock R 1989 Tight-binding models and density-functional theory. *Phys. Rev. B* **39** 12520–12536. DOI: 10.1103/PhysRevB.39.12520
119. Konschuh S, Gmitra M and Fabian J 2010 Tight-binding theory of the spin-orbit coupling in graphene. *Phys. Rev. B* **82** 245412. DOI: 10.1103/PhysRevB.82.245412
120. Lin C-Y, Wu J-Y, Chiu C-W and Lin M-F 2019 *Coulomb Excitations and Decays in Graphene-Related Systems*: CRC Press. DOI: 10.1201/9780429277368
121. Zhang XG, Krstic PS and Butler WH 2003 Generalized tight-binding approach for molecular electronics modeling. *Int. J. Quantum Chem.* **95** 394–403. DOI: 10.1002/qua.10675
122. Baltes H, Yacoby Y, Pindak R, Clarke R, Pfeiffer L and Berman L 1997 Measurement of the x-ray diffraction phase in a 2D crystal. *Phys. Rev. Lett.* **79** 1285–1288. DOI: 10.1103/PhysRevLett.79.1285
123. Tikhomirova KA, Tantardini C, Sukhanova EV, Popov ZI, Evlashin SA, Tarkhov MA, Zhdanov VL, Dudin AA, Oganov AR, Kvashnin DG and Kvashnin AG 2020 Exotic two-dimensional structure: The first case of hexagonal NaCl. *J. Phys. Chem. Lett.* **11** 3821–3827. DOI: 10.1021/acs.jpclett.0c00874
124. Li W-B, Tran NTT, Lin S-Y and Lin M-F 2019 Diverse fundamental properties in stage-n graphite alkali-intercalation compounds: Anode materials of Li+-based batteries. *arXiv e-print* arXiv:2001.02042.
125. Novoselov KS, Jiang D, Schedin F, Booth TJ, Khotkevich VV, Morozov SV and Geim AK 2005 Two-dimensional atomic crystals. *Proc. Natl. Acad. Sci.* **102** 10451. DOI: 10.1073/pnas.0502848102

126. Novoselov KS, Geim AK, Morozov SV, Jiang D, Zhang Y, Dubonos SV, Grigorieva IV and Firsov AA 2004 Electric field effect in atomically thin carbon films. *Science* **306** 666. DOI: 10.1126/science.1102896
127. Bolotin KI, Sikes KJ, Jiang Z, Klima M, Fudenberg G, Hone J, Kim P and Stormer HL 2008 Ultrahigh electron mobility in suspended graphene. *Solid State Commun.* **146** 351. DOI: 10.1016/j.ssc.2008.02.024
128. Morozov SV, Novoselov KS, Katsnelson MI, Schedin F, Elias DC, Jaszczak JA and Geim AK 2008 Giant intrinsic carrier mobilities in graphene and its bilayer. *Phys. Rev. Lett.* **100** 016602. DOI: 10.1103/PhysRevLett.100.016602
129. Berger C, Song ZM, Li XB, Wu XS, Brown N, Naud C, Mayou D, Li TB, Hass J, Marchenkov AN, Conrad EH, First PN and de Heer WA 2006 Electronic confinement and coherence in patterned epitaxial graphene. *Science* **312** 1191. DOI: 10.112 6/science.1125925
130. Balandin AA, Ghosh S, Bao WZ, Calizo I, Teweldebrhan D, Miao F and Lau CN 2008 Superior thermal conductivity of single-layer graphene. *Nano Lett.* **8** 902. DOI: 10.1021/nl0731872
131. Koenig SP, Boddeti NG, Dunn ML and Bunch JS 2011 Ultrastrong adhesion of graphene membranes. *Nat. Nanotechnol.* **6** 543–546. DOI: 10.1038/nnano.2011.123
132. Lee C, Wei XD, Kysar JW and Hone J 2008 Measurement of the elastic properties and intrinsic strength of monolayer graphene. *Science* **321** 385. DOI: 10.1126/science.1157996
133. Bunch JS, Verbridge SS, Alden JS, van der Zande AM, Parpia JM, Craighead HG and McEuen PL 2008 Impermeable atomic membranes from graphene sheets. *Nano Lett.* **8** 2458–2462. DOI: 10.1021/nl801457b
134. Papageorgiou DG, Kinloch IA and Young RJ 2017 Mechanical properties of graphene and graphene-based nanocomposites. *Prog. Mater. Sci.* **90** 75–127. DOI: 10.1016/j.pmatsci.2017.07.004
135. Bae S, Kim H, Lee Y, Xu XF, Park JS, Zheng Y, Balakrishnan J, Lei T, Kim HR, Song YI, Kim YJ, Kim KS, Ozyilmaz B, Ahn JH, Hong BH and Iijima S 2010 Roll-to-roll production of 30-inch graphene films for transparent electrodes. *Nat. Nanotechnol.* **5** 574. DOI: 10.1038/nnano.2010.132
136. Nair RR, Blake P, Grigorenko AN, Novoselov KS, Booth TJ, Stauber T, Peres NMR and Geim AK 2008 Fine structure constant defines visual transparency of graphene. *Science* **320** 1308. DOI: 10.1126/science.1156965
137. Novoselov KS, Jiang Z, Zhang Y, Morozov SV, Stormer HL, Zeitler U, Maan JC, Boebinger GS, Kim P and Geim AK 2007 Room-temperature quantum hall effect in graphene. *Science* **315** 1379. DOI: 10.1126/science.1137201
138. Novoselov KS, McCann E, Morozov SV, Fal'ko VI, Katsnelson MI, Zeitler U, Jiang D, Schedin F and Geim AK 2006 Unconventional quantum Hall effect and Berry's phase of 2 pi in bilayer graphene. *Nat. Phys.* **2** 177. DOI: 10.1038/nphys245
139. Novoselov KS, Geim AK, Morozov SV, Jiang D, Katsnelson MI, Grigorieva IV, Dubonos SV and Firsov AA 2005 Two-dimensional gas of massless Dirac fermions in graphene. *Nature* **438** 197. DOI: 10.1038/nature04233
140. Zhang YB, Tan YW, Stormer HL and Kim P 2005 Experimental observation of the quantum Hall effect and Berry's phase in graphene. *Nature* **438** 201. DOI: 10.1038/nature04235
141. Chung H-C, Chang C-P, Lin C-Y and Lin M-F 2016 Electronic and optical properties of graphene nanoribbons in external fields. *Phys. Chem. Chem. Phys.* **18** 7573. DOI: 10.1039/c5cp06533j
142. Chung HC, Lee MH, Chang CP and Lin MF 2011 Exploration of edge-dependent optical selection rules for graphene nanoribbons. *Opt. Express* **19** 23350–23363. DOI: 10.1364/OE.19.023350

143. Chung H-C, Lin Y-T, Lin S-Y, Ho C-H, Chang C-P and Lin M-F 2016 Magnetoelectronic and optical properties of nonuniform graphene nanoribbons. *Carbon* **109** 883. DOI: 10.1016/j.carbon.2016.08.091
144. Chung H-C, Yang P-H, Li T-S and Lin M-F 2014 Effects of transverse electric fields on Landau subbands in bilayer zigzag graphene nanoribbons. *Philos. Mag.* **94** 1859. DOI: 10.1080/14786435.2014.897009
145. Chung H-C, Su W-P and Lin M-F 2013 Electric-field-induced destruction of quasi-Landau levels in bilayer graphene nanoribbons. *Phys. Chem. Chem. Phys.* **15** 868. DOI: 10.1039/C2CP43631K
146. Chung H-C, Lee M-H, Chang C-P, Huang Y-C and Lin M-F 2011 Effects of transverse electric fields on quasi-Landau levels in zigzag graphene nanoribbons. *J. Phys. Soc. Jpn.* **80** 044602. DOI: 10.1143/jpsj.80.044602
147. Chung HC, Huang YC, Lee MH, Chang CC and Lin MF 2010 Quasi-Landau levels in bilayer zigzag graphene nanoribbons. *Physica E* **42** 711. DOI: 10.1016/j.physe.2009.11.090
148. Kalantar-zadeh K, Ou JZ, Daeneke T, Strano MS, Pumera M and Gras SL 2015 Two-dimensional transition metal dichalcogenides in biosystems. *Adv. Funct. Mater.* **25** 5086–5099. DOI: 10.1002/adfm.201500891
149. Chhowalla M, Shin HS, Eda G, Li LJ, Loh KP and Zhang H 2013 The chemistry of two-dimensional layered transition metal dichalcogenide nanosheets. *Nat. Chem.* **5** 263–275. DOI: 10.1038/nchem.1589
150. Xu MS, Liang T, Shi MM and Chen HZ 2013 Graphene-like two-dimensional materials. *Chem. Rev.* **113** 3766–3798. DOI: 10.1021/cr300263a
151. Osada M and Sasaki T 2012 Two-dimensional dielectric nanosheets: Novel nanoelectronics from nanocrystal building blocks. *Adv. Mater.* **24** 210–228. DOI: 10.1002/adma.201103241
152. Novoselov KS, Fal'ko VI, Colombo L, Gellert PR, Schwab MG and Kim K 2012 A roadmap for graphene. *Nature* **490** 192–200. DOI: 10.1038/nature11458
153. Lin C-Y, Yang C-H, Chiu C-W, Chung H-C, Lin S-Y and Lin M-F 2021 *Rich Quasiparticle Properties of Low Dimensional Systems*: IOP Publishing. DOI: 10.1088/978-0-7503-3783-0
154. Chen TA, Chuu CP, Tseng CC, Wen CK, Wong HSP, Pan SY, Li RT, Chao TA, Chueh WC, Zhang YF, Fu Q, Yakobson BI, Chang WH and Li LJ 2020 Wafer-scale single-crystal hexagonal boron nitride monolayers on Cu (111). *Nature* **579** 219–223. DOI: 10.1038/s41586-020-2009-2
155. Wang QH, Kalantar-Zadeh K, Kis A, Coleman JN and Strano MS 2012 Electronics and optoelectronics of two-dimensional transition metal dichalcogenides. *Nat. Nanotechnol.* **7** 699. DOI: 10.1038/nnano.2012.193
156. Chung H-C, Chiu C-W and Lin M-F 2019 Spin-polarized magneto-electronic properties in buckled monolayer GaAs. *Sci. Rep.* **9** 2332. DOI: 10.1038/s41598-018-36516-8
157. IEA 2020 Global EV Outlook 2020.
158. Martinez-Laserna E, Gandiaga I, Sarasketa-Zabala E, Badeda J, Stroe DI, Swierczynski M and Goikoetxea A 2018 Battery second life: Hype, hope or reality? A critical review of the state of the art. *Renew. Sustain. Energy Rev.* **93** 701–718. DOI: 10.1016/j.rser.2018.04.035
159. Ahmadi L, Yip A, Fowler M, Young SB and Fraser RA 2014 Environmental feasibility of re-use of electric vehicle batteries. *Sustain. Energy Technol. Assess.* **6** 64. DOI: 10.1016/j.seta.2014.01.006
160. Zhu JE, Mathews I, Ren DS, Li W, Cogswell D, Xing BB, Sedlatschek T, Kantareddy SNR, Yi MC, Gao T, Xia Y, Zhou Q, Wierzbicki T and Bazant MZ

2021 End-of-life or second-life options for retired electric vehicle batteries. *Cell Rep. Phys. Sci.* **2** 100537. DOI: 10.1016/j.xcrp.2021.100537

161. Casals LC, Garcia BA and Canal C 2019 Second life batteries lifespan: Rest of useful life and environmental analysis. *J. Environ. Manage.* **232** 354–363. DOI: 10.1016/j.jenvman.2018.11.046
162. Podias A, Pfrang A, Di Persio F, Kriston A, Bobba S, Mathieux F, Messagie M and Boon-Brett L 2018 Sustainability assessment of second use applications of automotive batteries: Ageing of Li-ion battery cells in automotive and grid-scale applications. *World Electr. Veh. J.* **9** 24. DOI: 10.3390/wevj9020024
163. Tong S, Fung T, Klein MP, Weisbach DA and Park JW 2017 Demonstration of reusing electric vehicle battery for solar energy storage and demand side management. *J. Energy Storage* **11** 200–210. DOI: 10.1016/j.est.2017.03.003
164. Wood E, Alexander M and Bradley TH 2011 Investigation of battery end-of-life conditions for plug-in hybrid electric vehicles. *J. Power Sources* **196** 5147–5154. DOI: 10.1016/j.jpowsour.2011.02.025
165. Chung H-C 2018 Failure mode and effects analysis of LFP battery module. *engrXiv*. DOI: 10.31224/osf.io/acxsp
166. Kamath D, Shukla S, Arsenault R, Kim HC and Anctil A 2020 Evaluating the cost and carbon footprint of second-life electric vehicle batteries in residential and utility-level applications. *Waste Manage.* **113** 497–507. DOI: 10.1016/j.wasman.2020.05.034
167. Quinard H, Redondo-Iglesias E, Pelissier S and Venet P 2019 Fast electrical characterizations of high-energy second life lithium-ion batteries for embedded and stationary applications. *Batteries* **5** 33. DOI: 10.3390/batteries5010033
168. Heymans C, Walker SB, Young SB and Fowler M 2014 Economic analysis of second use electric vehicle batteries for residential energy storage and load-levelling. *Energy Policy* **71** 22–30. DOI: 10.1016/j.enpol.2014.04.016
169. Casals LC and Garcia BA 2017 Second-life batteries on a gas turbine power plant to provide area regulation services. *Batteries* **3** 10. DOI: 10.3390/batteries3010010
170. Chung H-C and Cheng Y-C 2019 Action planning and situation analysis of repurposing battery recovery and application in China. *J. Taiwan Energy* **6** 425. DOI: 10.31224/osf.io/nxv7f
171. Hossain E, Murtaugh D, Mody J, Faruque HMR, Sunny MSH and Mohammad N 2019 A comprehensive review on second-life batteries: Current state, manufacturing considerations, applications, impacts, barriers and potential solutions, business strategies, and policies. *IEEE Access* **7** 73215–73252. DOI: 10.1109/access.2019.2917859
172. Gur K, Chatzikyriakou D, Baschet C and Salomon M 2018 The reuse of electrified vehicle batteries as a means of integrating renewable energy into the European electricity grid: A policy and market analysis. *Energy Policy* **113** 535–545. DOI: 10.1016/j.enpol.2017.11.002
173. Laboratories U 2018 *UL 1974 - Standard for evaluation for repurposing batteries*. Underwriters Laboratories.
174. Chung H-C and Cheng Y-C 2020 Summary of safety standards for repurposing batteries. *Monthly J. Taipower's Eng.* **860** 35. DOI: 10.31224/osf.io/d4n3s
175. Chung H-C, Nguyen TDH, Lin S-Y, Li W-B, Tran NTT, Thi Han N, Liu H-Y, Pham HD and Lin M-F 2021 *First-Principles Calculations for Cathode, Electrolyte and Anode Battery Materials*: IOP Publishing. pp. 15–1 to –8 DOI: 10.1088/978-0-7503-4685-6ch15
176. Schneider EL, Oliveira CT, Brito RM and Malfatti CF 2014 Classification of discarded NiMH and Li-Ion batteries and reuse of the cells still in operational conditions in prototypes. *J. Power Sources* **262** 1–9. DOI: 10.1016/j.jpowsour.2014.03.095

177. Schneider EL, Kindlein W, Souza S and Malfatti CF 2009 Assessment and reuse of secondary batteries cells. *J. Power Sources* **189** 1264–1269. DOI: 10.1016/j.jpowsour.2008.12.154
178. Zhao G 2017 *Reuse and Recycling of Lithium-Ion Power Batteries*: John Wiley & Sons, Inc. DOI: 10.1002/9781119321866
179. Chung HC 2021 Charge and discharge profiles of repurposed $LiFePO_4$ batteries based on the UL 1974 standard. *Scientific Data* **8** 165. DOI: 10.1038/s41597-021-00954-3
180. Rallo H, Benveniste G, Gestoso I and Amante B 2020 Economic analysis of the disassembling activities to the reuse of electric vehicles Li-ion batteries. *Resour. Conserv. Recycl.* **159** 104785. DOI: 10.1016/j.resconrec.2020.104785
181. Alfaro-Algaba M and Ramirez FJ 2020 Techno-economic and environmental disassembly planning of lithium-ion electric vehicle battery packs for remanufacturing. *Resour. Conserv. Recycl.* **154** 104461. DOI: 10.1016/j.resconrec.2019.104461

5 Electronic Properties of $LiLaTiO_4$ Compound

Nguyen Thanh Tuan, Thi Dieu Hien Nguyen, and Le Vo Phuong Thuan
Department of Physics, National Cheng Kung University, Tainan City, Taiwan

Ming-Fa Lin
Department of Physics and Hierarchical Green-Energy Material (Hi-GEM) Research Center, National Cheng Kung University, Tainan City, Taiwan

CONTENTS

5.1 QUANTUM QUASIPARTICLES IN $LILATIO_4$ ELECTROLYTE COMPOUND

Solid Electrolytes (SEs) are materials containing charges and ions, which are known as ionic conductions. This feature is different with materials such as semiconductor or metals, where electrons or charges is transported alone [1]. SEs possess potential safety solutions compared to those that result from flammable liquid electrolyte components. Furthermore, significant advantages of SEs such as wide electrochemical windows and flexible temperature ranges make certain attention of researchers. On the other hand, low conductivity in 3D compounds needs to improve further. There are many SEs compounds used as electrolytes in batteries; however, compounds of lithium-ion are the most widely used, especially, compounds with transition metal and rare-earth elements of lithium-ion are new compounds that are researched recently. In which, transition metal and rare earth are groups of elements possessing valence electrons distributed in orbitals d or f. The transition metal is a collection of elements that are arranged mostly right

DOI: 10.1201/9781003367215-5

between the rows on the table, from the group (IIIB) with column 3 on the left side to the group (IIB) with column 12 on the right-side periodic table of elements. These elements have valence electrons in the two outermost shells that can participate in the form of chemical bondings [2]. Besides, the rare-earth elements contain 15 Lanthanide elements in the periodic table, from atomic number Z = 57 (Lanthanum – La) to Z = 71 (Lutetium – Lu) and Scandium – Sc (Z = 21) & Yttrium – Y (Z = 39) in group IIIB, they are called because most of these elements were separated from minerals as oxides around the 18th and 19th centuries. Because of their reactivity, these rare earth elements are difficult to refine into pure metals. Also, separation procedures were not developed until the 20th century due to the similarity in chemical properties of rare-earth elements [3]. With the lithium- and/or oxygen-based main-stream materials such as single-element crystal, binary and ternary compounds, the quaternary $LiLaTiO_4$ compound includes rare-earth element lanthanum (La) and transition metal element titanium (Ti) that have unique electronic properties due to the interaction between the f-orbital of the rare-earth and the d-orbital of the transition metal elements. Therefore, the quaternary $LiLaTiO_4$ compound is a candidate for solid electrolytes materials of lithium-ion batteries. The geometric structure as well as the electron properties of $LiLaTiO_4$ which have been calculated by the first-principles calculation will be presented and discussed in this chapter.

A periodic table of elements, which clearly illustrates the atomic configurations of different subgroups [4], is very useful in understanding the successful syntheses of single element [5] or multi-component crystal structures [6]. In general, all condensed-matte systems could be generated in experimental laboratories according to these kinds of active orbitals: (I) [s, p_x, p_y, p_z] -I/II/III/IV/V/VI/VII/VIII (e.g., group-I alkali atoms), (II) [d_{z^2}, $d_{(x^2-y^2)}$, d_{xy}, d_{yz}, d_{xz}]-transition metal atoms (3d/4d/5d; [7]/[8]/[9]), and (III) [f_{z^3}, f_{xz^2}, f_{yz^2}, f_{xyz}, $f_{z(x^2-y^2)}$, $f_{y(3x^2-y^2)}$, $f_{x(x^2-3y^2)}$]-rare-earth metal atoms (4f/5f; [10]/[11]). Furthermore, ten/fourteen kinds of outmost d/f orbitals are able to generate spin configurations in most main-stream materials, in which their interactions belong to the Hubbard-like ones (the on-site electron-electron Coulomb interactions; [12]). Very interestingly, the multi-orbital hybridizations fully cooperate with spin-dependent interactions [13]; that is, both charge and spin density distributions are mixed together in determining all composite quasiparticles [14], as clearly illustrated in a series of published books (details in [15]). Such research strategies are very suitable for developing an enlarged framework, mainly owing to the simultaneous progress among VASP simulations [16], phenomenological models [17], and experimental examinations [18]. Furthermore, a lot of merits might provide the high potential in generating industrial products (e.g. the functional chips [19]) and the world-class classmate crises. On the other side, their drawbacks, which obviously reveal the giant uncertainties of physical, chemical, and material properties, need to be solved urgently in the near-future science research.

The up-to-date experimental [20] and theoretical studies [21] show that the lithium and/or oxygen-related single-element [22], binary [23], ternary [24], and quaternary materials [25] have become the main-stream materials in ion-based battery applications. For example, $LiFeO_2$/$LiFePO_4$/$LiCoO_2$/$LiNiO_2$ [26]/[27]/[28]/[29], $LiSiO_2$/LiO_2 [30]/[31], and $LiScO_2$/$LiTiO_2$ [32]/[33], respectively, frequently serve as cathode, electrolyte, and anode materials in lithium-ion-based batteries [34]. Their basic and applied sciences are expected to display rich and unique composite quasiparticles because of various oxygen-created bonds [35]. The various crystal phases have been successfully synthesized by the different ways of chemical engineering and delicately examined under high-resolution X-ray diffraction patterns [36]. Very interesting, the diversified compound materials could be generated under the rather efficient methods through the significant variations about (1) the different atoms [37], (2) the relative atom concentrations [38], (3) the same one, but the different arrangements [39], (4) the intercalation/de-intercalation processes [38], (5) the creation of defects/vacancies [40], (6) the necessary heterojunctions [41], and (6) the lower dimensionalities [2D/1D/0D; [42]/[43]/[44]. First, the crystal growth processes could be fully understood from the strong cooperation of the first-principles methods [45] and molecular dynamics [46]. And then, the various crystal symmetries are initiated from many isomer structures of Li- and O-related materials. The highly non-uniform and anisotropic environments are very sensitive to the above-mentioned critical factors. Obviously, the atom- and orbital-dependent chemical bondings [47] and spin configurations [48] will be responsible for all intrinsic interactions (e.g., the Hubbard-like model, [49]) and thus the composite quasiparticle phenomena [50]. Specifically, ternary $LiFeO_2$ and quaternary $LiFePO_4$ are outstanding candidates in the quasiparticle developments [51] and commercial products [52], especially for the cathode materials of the latter [53]. Such samples could be generated in a large-scale manner in industrial factories at the lowest cost. It is well known that the transition-metal iron atoms. are able to generate the unusual magnetic configuration due to five kinds of 3d orbitals [54]. Their quasiparticle behaviors will establish the well-characterized partnerships between the atom- and orbital-dependent charge (the exchange and correlation energies [55]) and spin interactions (the on-site electron-electron Coulomb interactions [56]) How to distinguish the significant differences of ternary and quaternary compounds from the quasiparticle viewpoints are an urgent and interesting research topic. A similar research strategy could be further generalized to ternary $LiTiO_2$ [57] and quaternary $LiLaTiO_4$ compounds [58], where they can serve as the anode and electrolyte components of lithium-ion-based batteries [59]. Compared with the former, the latter presents more rich environments in terms of different atoms [60], active orbitals [61], chemical bondings [62], and spin configurations [63]. Most of the experimental measurements and theoretical predictions are focused on growth- and geometry-related properties. The other quasiparticle phenomena are worthy of systematic explorations, such as the quasi-Moire superlattice with a lot of active orbitals [64], the atom- and spin-dominated band structures and wave functions, the high-resolution measurements of angle-resolved photoemission spectroscopy (ARPES [65]) about valence energy subbands, and the atom-, orbital- and spin-decomposed contributions on charge density distributions and Van Hove singularities. Such necessary research will open a wider quasiparticle framework, following a series of books with the full research strategies.

The sodium compounds, which contain $NaLiTiO_4$ and $NaEuTiO_4$ are used to synthesize the $LiLaTiO_4$ compounds by the ion-exchange reaction. Because of the evaporation of the sodium component, an excess amount of sodium carbonate was added to compensate. The mixture of sodium carbonate, rare earth oxide, and titanium oxide was used as the parent materials compounds. The mixture was fired at a temperature of 900–1,000°C in the air for 30 minutes. The compounds include $LiEuTiO_4$ and $LiLaTiO_4$, which were prepared by the ion-exchange reaction from the precursor compounds. The ion-exchange reaction of interlayer sodium ions in these compounds with lithium-ion was carried out in molten $LiNO_3$ at 300–310°C for 12 hours. The compounds after synthesis were washed distilled by water and air-dried at room temperature. Besides, the $LiLaTiO_4$ compounds can be synthesized. The lattice parameters are a = b = 3.772 Å, c = 12.083 Å calculated from the X-ray diffraction pattern. These results are similar to the lattice parameter that is calculated by first-principle a = b = 3.81 Å, c = 12.23 Å. The space group of both researches is P4/nmm [58]. Besides the ion-exchange reaction, the hydrothermal method is used for synthesizing $LiLaTiO_4$ compounds and this method is more advantageous than the ion-exchange reaction. Because the ion-exchange reaction requires a high temperature for reaction, thus, the products are often contaminated by high-temperature decomposition, and it is not easy to synthesize some of these compounds by solid-state reaction. Additionally, the difficulty in using ion-exchange reaction for the synthesis of some layered compounds is that reaction depends on the characteristics of the parent-layered compound and exchanged ions [66]. With hydrothermal method, the lattice parameters of $LiLaTiO_4$ are a = b = 3.773 Å, c = 12.080 Å. This result is quite similar to the previous method, ion-exchange reaction, and close to the result based on first-principles calculation.

Only the first-principles methods [67], but not the phenomenological models can fully explore the unusual science phenomena in 3D quaternary $LiLaTiO_4$ compound. This lithium oxide system consists of rare-earth- and transition-metal atoms. Both charge- and spin-dependent interactions should play critical roles in composite quasiparticles, where the former will be very sensitive to the variations of atoms, active orbitals, and their occupied/unoccupied statuses. The concise pictures of chemistry, physics, and material engineering are expected to be unique through the delicate VASP calculations and analyses. The very complicated multi-orbital chemical bondings should be closely related to the active orbitals about Li-2s, La-[6s, $5d_{z^2}$, $5d_{(x^2-y^2)}$, $5d_{xy}$, $5d_{yz}$, $5d_{xz}$], Ti-[4s, $3d_{z^2}$, $3d_{(x^2-y^2)}$, $3d_{xy}$, $3d_{yz}$, $3d_{xz}$] and O-[2s, $2p_x$, $2p_y$, $2p_z$]. Their significant features are able to determine the various quasiparticles and are thus very useful in enlarging a unified framework in the previously published [68] and current books. All calculated results agree with one another through the optimal crystal symmetries with a quasi-Moire superlattice, the atom- and orbital-dependent magnetic moments, the ferromagnetic energy spectra and wave functions, the highly non-uniform and anisotropic orbital mixings, the unusual spin density distribution, and the atom-, orbital and spin-decomposed density of states. However, most of experimental examinations are absent except for the X-ray diffraction patterns. How to create the simultaneous progresses among the VASP simulations, phenomenological models and experimental observations will be discussed in detail.

In this book, the first-principles calculation method was performed using density functional theory (DFT) under Vienna ab initio Simulate Package (VASP). The generalized approximation of Perdew-Burke-Ernzerhof (PBE) is chosen for model studies. The cutoff energy for plane-wave expansion was 600 eV. For the $LiLaTiO_4$, the first Brillouin zone was performed using Γ-centered 4 × 4 × 1 for geometry optimization and electronic structure calculations.

The lithium- and/or oxygen-based main-stream materials, which cover single-element crystals [69], binary [70], ternary [71], and quaternary compounds [72], will clearly illustrate the diverse quasiparticle phenomena in terms of sample growths, various crystal phases, electronic properties, optical transitions, Coulomb excitations, quasiparticle decay rates, temperature-dependent transport properties, phonon spectra, and mechanical deformations. Similar studies have been successfully generalized to the lithium- and sulfur-related unusual crystal symmetries, as shown in the previously proposed book. In this work. Both quaternary $Li_4La_4Ti_4O_8$ and ternary $Li_4Ti_5O_{12}$ compounds are chosen for a model study because of the rich chemical bondings, mainly owing due to the rare-earth- and transition-metal atoms, respectively, with seven and five kinds of f- and d- orbital in a periodic table. The charge- and spin-created intrinsic interactions [the single-particle and many-body ones] can greatly diversify the fundamental properties under the highly non-uniform and anisotropic environments.

Very interesting, the quaternary compound of $LiLaTiO_4$ exhibits unusual crystal symmetries after the delicate VASP simulations, as clearly shown in Figures 5.1(a)–(d) along the distinct [x, y, z]-directions. There exist 22 chemical bonds in a primitive unit cell covering Li-O (4), La-O (8), Ti-O (8), and La-Ti ones (2). Their bonds lengths are, respectively, observable within 2.0602 Å, 2.3496–2.7371 Å, 1.8219–2.5781 Å, and 3.4043 Å. The second and third bonds show drastic changes and directly reflect the sensitive f- and d-orbital-dominated intrinsic many-body

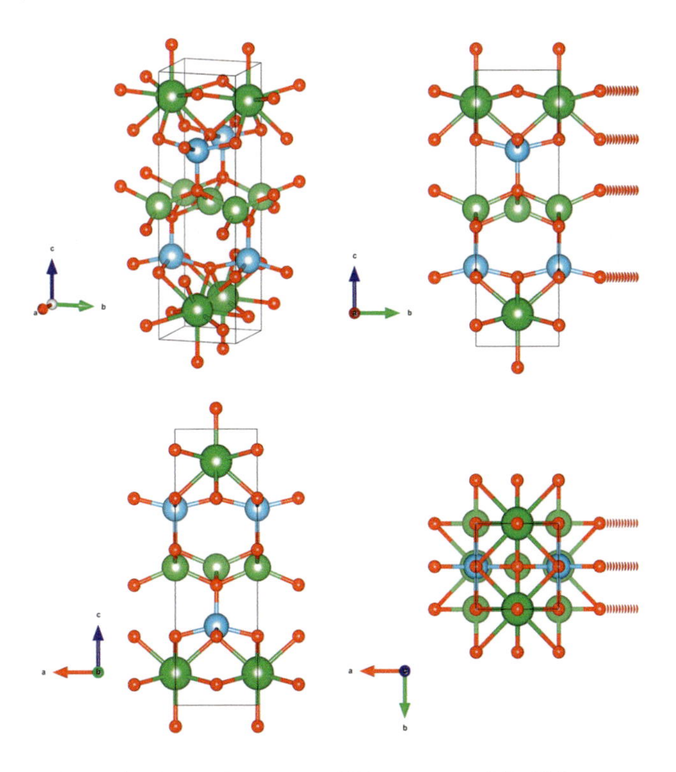

FIGURE 5.1 The optimal crystal structure of 3D quaternary $LiLaTiO_4$ compound after the VASP simulations (a)/(b)/(c)/(d) along the different [x, y, z] directions.

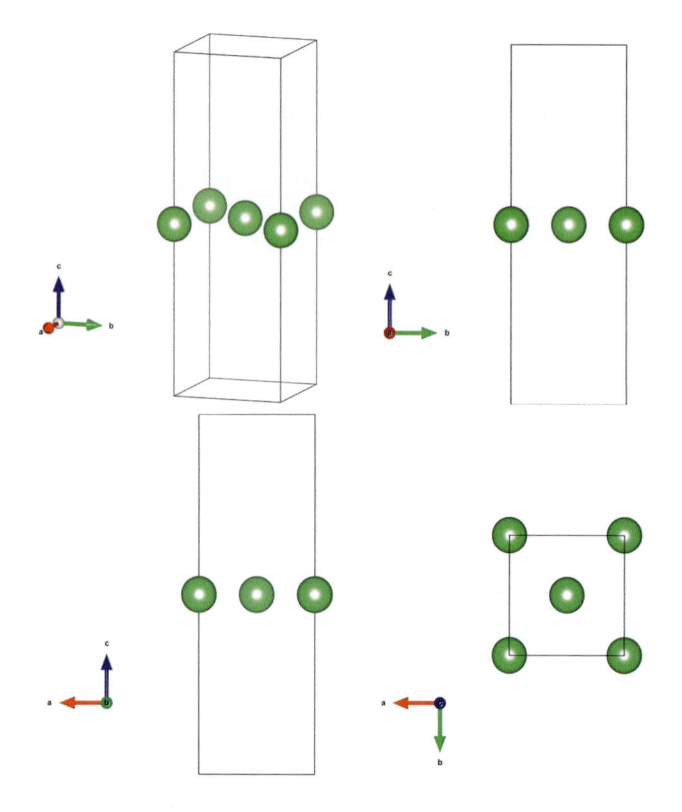

FIGURE 5.2 Similar plot as Fig. 1 but shown for a 3D bcc lithium crystal.

interactions. A quasi-Moire superlattice will be responsible for composite quasi-particle behaviors, as it includes 14 different atoms, and four types of [s, p, d, f]-orbitals, as well as spin-up and spin-down configurations (Figure 5.2).

5.2 RICH ENERGY SPECTRA AND WAVE FUNCTIONS WITH FERROMAGNETIC CONFIGURATIONS

The VASP calculations of band structures and atom-dominated wave functions can reveal the diverse phenomena of electronic properties, being clearly shown along high-symmetry points of the first Brillouin zone. Their main features can cover the band property across the Femi level (e.g., the wide-/middle-/narrow-/ or zero-gap semi-conductors, semimetals, and metals [73]/[74]/[75]/[76]/[77]/[78]), the number of valence and conduction energy subbands, their asymmetry/symmetry about $E_F = 0$, the various energy dispersions (the parabolic, linear, oscillatory and partially flat ones [79]), the band-edge states with vanishing or abnormal group velocities (the creation of van Hove singularities [80]), and atom- and orbital-dominated band widths. The theoretical VASP predictions, which need to be done for ternary [81] and quaternary [82] lithium oxide compounds, cannot be verified from the high-resolution angle-resolved photo-emission spectroscopy. In addition, ARPES is available in determining the quasiparticle decay rates [83]. The intrinsic limits mainly arise from the non-conservation of the normal momentum transfers (the surface confinements effects [84]) and too many valence subbands below E_F (proportional to the number of active atoms and orbitals). On the contrary, the quasiparticle energy spectra of certain bulk materials could be well examined from ARPES, e.g., those of layered graphite, and diamond-like silicon and germanium. How to greatly the strong partnerships between theoretical and experimental research should be the next-step research focuses.

Band structure of 3D quaternary $LiLaTiO_4$ compound is clearly shown in Figures 5.3(a)–(d) along the high-symmetry points about the first Brillouin zone of the tetragonal crystal. The unusual quasiparticle behaviors lie in a wide band gap, the specific energy ranges of atom dependences, and super-large energy spacing of certain valence states. A direct gap, in which the highest occupied valence states and the lowest unoccupied conduction ones appear at the Γ point, is about 2.10 eV. This indicates the sufficiently strong covalent bondings so that free carriers cannot survive in lithium oxide compounds. Furthermore, the rich semiconducting behaviors are capable of driving the excitonic effects of optical transitions (the many-body excitation phenomena [85]). The perturbations of electromagnetic waves are under current investigations. According to the optimal hole and electron spectra, respectively, above and below the Fermi level, there exist three subgroups: (I) $E^v < -12$eV, (II) -6 eV $< E^v < 0$ eV, and (III) 0 eV $< E^c < 6.0$ eV. A very large energy spacing between (I) and (II) subgroups, mainly owing to the important differences of ionization energies between [2s,4s,6s] and [2p,3d,4f] orbitals. Whether band gap and very large valence energy spacing could be examined from the high-resolution ARPES measurements is urgently requested in the strong cooperation of theoretical and experimental research. Specifically, a lot of valence

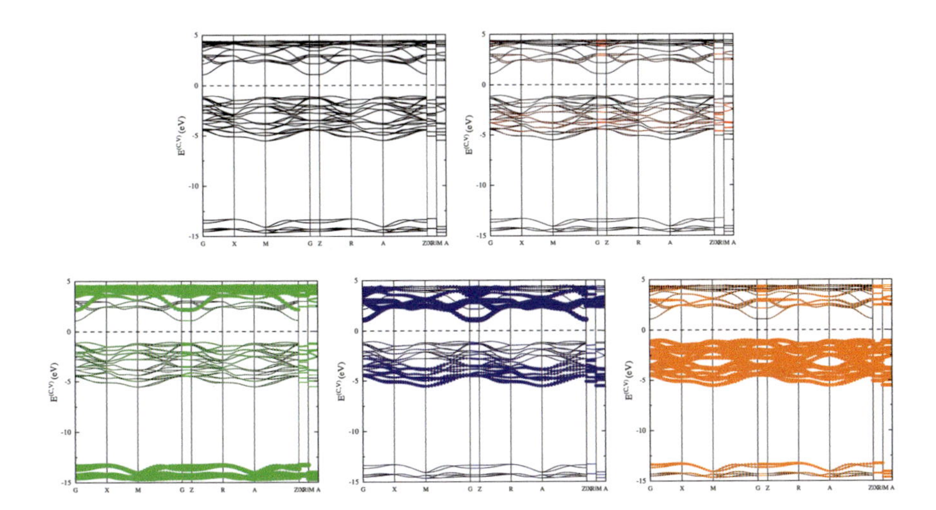

FIGURE 5.3 The band structure for LiLaTiO4 compound along the high-symmetry pints of the first Brillouin zone, as clearly illustrated by the solid and dashed black curves in (a), respectively for the spin-up and spin-down configurations. Those atom-projected contributions are shown by the (b) red, (c) green, (d) blue, and (e) orange balls.

states in (II) which determine all essential quasiparticle properties, are co-dominated by [Li, La, Ti, O] red, blue, green, and orange balls especially for the third ones with the higher weight. The atom-dependent contributions are expected to have strong effects on the intrinsic response abilities, e.g., the single-particle absorption peaks and the collective, plasmon modes. Similar quasiparticle behaviors are revealed in ternary $Li_4Ti_5O_{12}$, while certain important differences include band gap, as well as, the number, energy dispersions, and band-edge states f valence energy subbands. The diversified quantities can serve as the critical roles of the modulation strategies.

How to develop the phenomenological models and achieve concise pictures should obtain the well-fitting of the low-lying valence and conduction energy subbands with the VASP results (Figures 5.3(a)–(e)). This strategy would have giant difficulties under too many occupied states within the energy range of 6 eV about the Fermi level. For example, the suitable parameters, which might cover the orbital-dependent hopping integrals and on-site Coulomb potential energies, as well as the spin-induced electron-electron interactions [51], will determine the reliable tight-binding model. As for the quaternary $LiLaTiO_4$ compound [86], the active orbitals cover Li-2s, La-[6s, $5d_{z^2}$, $5d_{(x^2-y^2)}$, $5d_{xy}$, $5d_{yz}$, $5d_{xz}$], Ti-[4s, $3d_{z^2}$, $3d_{(x^2-y^2)}$, $3d_{xy}$, $3d_{yz}$, $3d_{xz}$] and O-[2s, $2p_x$, $2p_y$, $2p_z$], in which the Li-O, La-O, Ti-O, and La-Ti bonds need to be taken into consideration [Table 5.1]. Similar results appear in ternary compounds of $Li_4Ti_5O_{12}$. Most of the low-lying energy subbands are due to the prominent chemical bondings of three-p, five-d, and seven-f orbitals (discussed later in density of states; details in Figure 5.6(a)–(e)). Furthermore, all [2s,4s,6s] orbitals make main contributions at the deeper energies of −12 ~ −20 eV. Such complicated multi-orbital hybridizations are able to provide an outstanding platform for developing a grand quasiparticle framework.

TABLE 5.1
The Optimal Geometric Parameters for an Optimal Quaternary $LiLaTiO_4$ Compound

Lattice Parameter	a = 3.81 Å; b = 3.81 Å; c = 12.23 Å.		
	α = 90.000°; β = 90.000°; γ = 90.000°.		
Unit-cell volume	177.29 $Å^3$		
Space Group	P4/nmm		
Crystal System	Tetragonal		
Chemical Bondings	22 bondings		
Bond Lengths	Bonds	Min	Max
	La – O (8 bondings)	2.3496 Å	2.7371 Å
	Ti – O (8 bondings)	1.8219 Å	2.5781 Å
	La –Ti (2 bondings)	3.4043 Å	
	Li – O (4 bondings)	2.0602 Å	
Atoms	14 atoms/unit cell (2 Li-atoms, 2 La-atoms, 2 Ti-atoms & 8 O-atoms)		

5.3 COMPLICATED CHARGE AND SPIN DENSITY DISTRIBUTIONS

The chemical bondings associated with single- and multi-orbital hybridizations will exhibit the dramatic transformations after the formation of crystal (the various growth methods), the chemical modifications (substitutions/absorptions/intercalations/decorations). Concerning the lithium- and oxygen-based compounds, there exist rich charge density distributions in a primitive unit cell along the distinct axial directions. It is well known that each isolated atom has a spherical charge distribution, being sensitive to the [s, p, d, f]-orbital probability cases. Within the chemical bonds of neighboring atoms, their active orbitals are able to create significant overlaps so that the covalent bonding charge distribution appears in a quasi-Moire superlattice. The highly non-uniform and anisotropic environments of chemistry, physics, and material engineering are clearly revealed in the drastic changes in the spatial charge distribution (details in Figures 5.4(a)–(d)). Obviously, the covalent crystal bondings are quite different from the weak van der Waals (e.g., the interlayer C-$2p_z$ hybridizations in a layered graphite [87]), metallic (a bcc lithium [88]) and ionic ones (binary LiCl compound [89]). According to the up-to-date high-resolution X-ray patterns, only the ionic crystals, with the almost isotropic

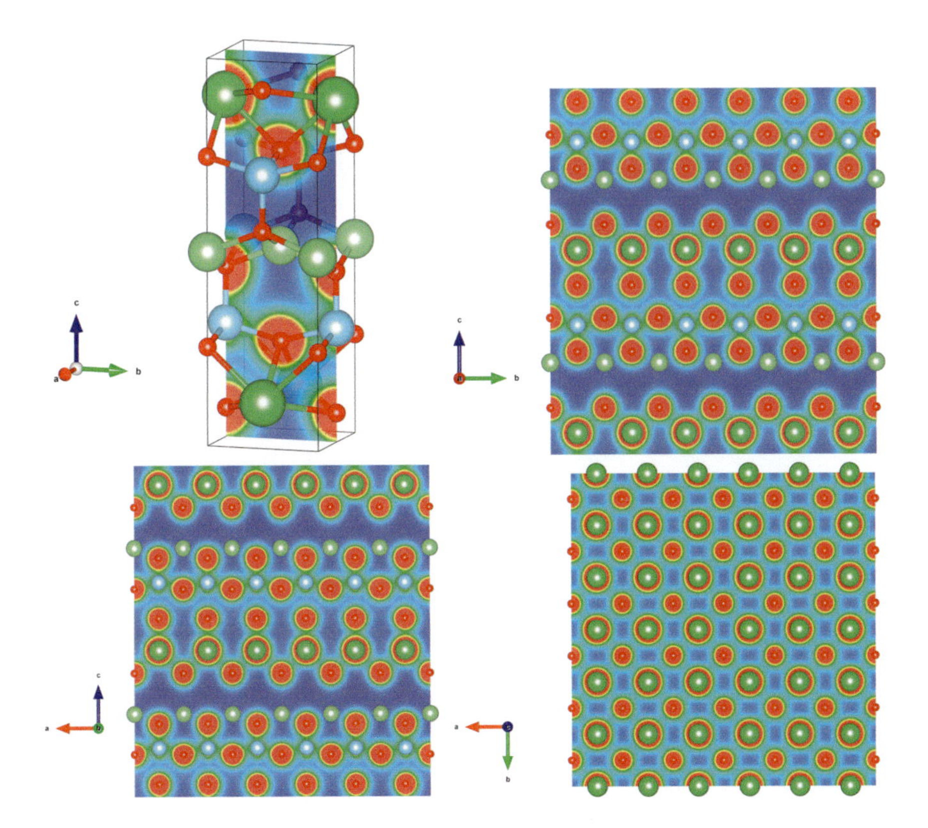

FIGURE 5.4 The spatial charge density distribution of quaternary $LiLaTiO_4$ compound along the distinct [x, y, z] directions in (a), (b), (c), and (d).

charge distributions, could be estimated from the measured data, e.g., the weak distortion of probability distribution only close to the middle of sodium and chloride ions. Direct examinations on the orbital hybridizations are absent for covalent crystals up to now.

The quaternary $LiLaTiO_4$ clearly illustrates the unusual charge distribution in the complicated covalent bonding, as shown in Figures 5.5(a)–(d). When Li, La, Ti, and O atoms survive in the isolated case, their spherical charge densities, respectively, appear in the orderings of 2s, 6s-5d, 4s-3d, and 2s-2p with distinct colors. For example, as to oxygen atom, 2s and [$2p_x$, $2p_y$, $2p_z$] come to exist in the red and yellow/green regions (the inner and outer regions). When the critical chemical bonds are generated by the successful syntheses [90], the orbital hybridizations of the nearest-neighboring atoms dramatically change the whole charge distribution. The main features cover the observable orbital mixings about

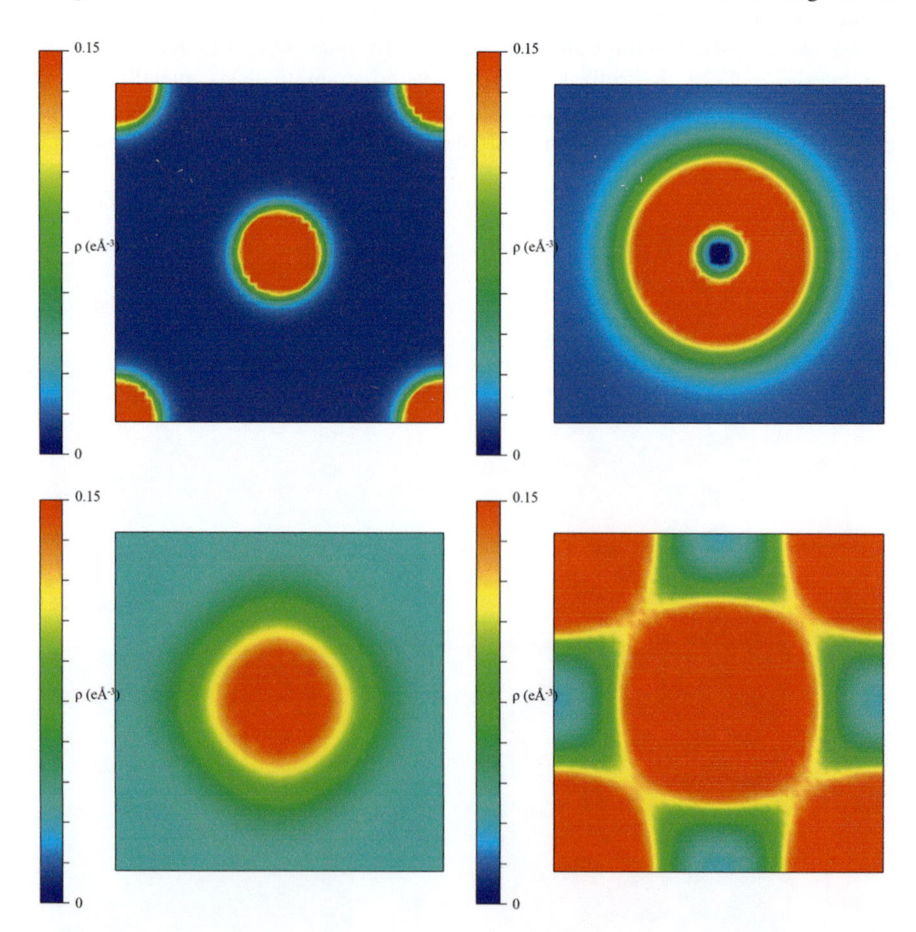

FIGURE 5.5 The complicated spin configuration of ternary $LiLaTiO_4$ compound near each atom in a primitive unit cell, being due to 10-5d and 10-3d orbitals of La and Ti atoms, respectively. The sensitive physical environment is clearly observed for the different [x, y, z] directions within (a), (b), (c), and (d).

Li-O/La-O/Ti-O/La-Ti bonds, the high distortions of La-d, Ti-d, and O-p orbitals, as well as the weak, but significant deformations of Li-2s, La-6s, Ti-4s, and O-2s. According to the delicate VASP simulations and analyses, the critical mechanisms are deduced to be 2s-[2s, $2p_x$, $2p_y$, $2p_z$], [6s, $5d_{z^2}$, $5d_{(x^2-y^2)}$, $5d_{xy}$, $5d_{yz}$, $5d_{xz}$]-[2s, $2p_x$, $2p_y$, $2p_z$], [4s, $3d_{z^2}$, $3d_{(x^2-y^2)}$, $3d_{xy}$, $3d_{yz}$, $3d_{xz}$]-[2s, $2p_x$, $2p_y$, $2p_z$] and [6s, $5d_{z^2}$, $5d_{(x^2-y^2)}$, $5d_{xy}$, $5d_{yz}$, $5d_{xz}$]-[4s, $3d_{z^2}$, $3d_{(x^2-y^2)}$, $3d_{xy}$, $3d_{yz}$, $3d_{xz}$], respectively, corresponding to Li-O, La-O, Ti-O, and La-Ti bonds. This conclusion agrees with the crystal symmetries (Figure 5.1(a)–(d)), band structure (Figure 5.3(a)–(e)), and van Hove singularities (Figure 5.6(a)–(e)). In short, the lithium- and oxygen-based main-stream materials exhibit the rich charge density distributions with the prominent covalent bondings, being responsible for the composite quasiparticles.

Very interestingly, only the significant orbital hybridizations dominate the ground state energies of quaternary $LiLaTiO_4$ and ternary $Li_4Ti_5O_{12}$, while the spin-dependent interactions are negligible in the intrinsic Hamiltonian. Their 5d and 3d orbital do not create the ferromagnetic configurations. This indicates strong competition between the charge- ad spin-induced interactions, in which the former and the latter, respectively, include the kinetic/exchange/correlation energies and the on-site electron-electron interactions (the Hubbard-like ones [91]). Furthermore, the covalent bondings fully suppress the ferromagnetic configurations. On the other hand, certain lithium oxide compounds are able to present the ferromagnetic spin configurations, e.g., the spin density distributions of ternary $LiFeO_2$/$LiCoO_2$/$LiNiO_2$ materials [92]/[93]/[94]. The existence or absence of the spin-dependent configurations requires systematic VASP investigations in the near-future research.

5.4 [S, P, D, F]- AND SPIN-INDUCED MERGED VAN HOVE SINGULARITIES

The energy-dependent 3D/2D/1D density of states is characterized by the specific integration of the inverse of group velocity along a constant-energy closed surfaces/loops/fixed point. Its special structures are created by the band-edge states (the extreme/saddle/irregular points in the energy-wave-vector space) with zero or abnormal group velocities. Such van Hove singularities are very sensitive to dimensionalities. The analytic forms are summarized as follows: (1) delta-function-like structures for dispersion-less states, (2) $\sqrt{E}$ for 3D parabolic extreme points, (3) plateaus/shoulders for 2D parabolic ones, (4) the logarithmic divergences for 2D saddle points, (5) the square-root divergences for 2D constant-energy loops, and (6) the similar ones, but for 1D parabolic maxima/minima points. Only 2D, 1D, and 0D features could be directly observed from the high-precision measurements of scanning tunneling spectroscopy (STS [95]), mainly owing to the intrinsic limits of very weak quantum currents. Very interesting, van Hove singularities, which represent the available channel under the various excitations, will have strong effects on any response abilities. Their strong relations with electronic [96], optical [97], magnetic [98], transport [99], thermal [100], and mechanical properties [101], are worthy of systematic investigations.

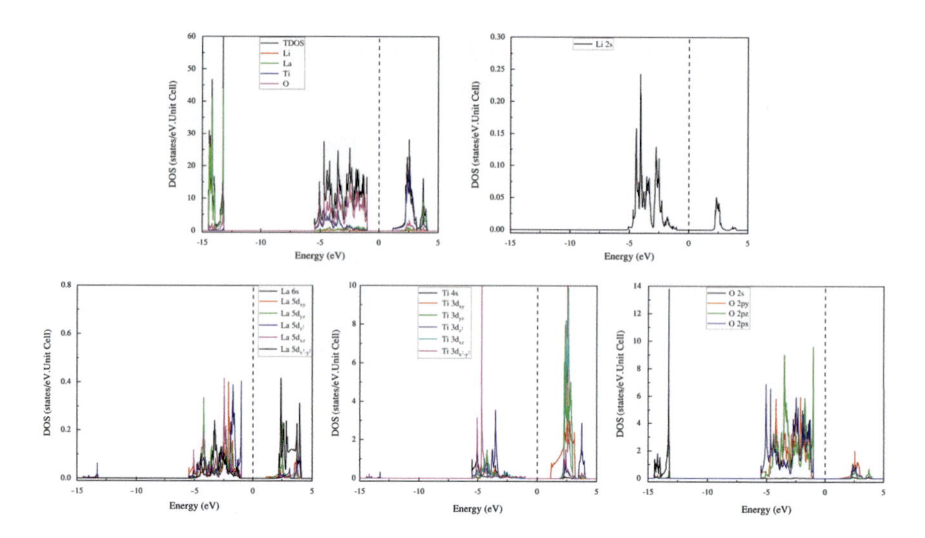

FIGURE 5.6 The energy- and spin-dependent density of states about quaternary $LiLaTiO_4$ compound under the various cases: (a) The total and atom-decomposed contributions in the presence of spin splitting. (b) The lithium-, 2s-orbital- and spin-decomposed contributions. (c) The, La-[6s, $5d_{z^2}$, $5d_{(x^2-y^2)}$, $5d_{xy}$, $5d_{yz}$, $5d_{xz}$] and spin-projected weights. (d) The Ti-[$3d_{z^2}$, $3d_{(x^2-y^2)}$, $3d_{xy}$, $3d_{yz}$, $3d_{xz}$] and spin-decomposed probabilities. (e) The O-[2s, $2p_x$, $2p_y$, $2p_z$] and spin-split dominances.

The quaternary $LiLaTiO_4$ compound, as shown Figures 5.3(a)–(e), clearly illustrates the rich and unique van Hove singularities because of a lot of band-edge states about valence and conduction energy subbands, The main features cover the atom- and orbital-decomposed compositions (the various colored curves), but without spin splitting, an obviously semiconducting behavior with zero contribution within an energy gap centered about the Fermi level (the black curve in Figure 5.3(a)), many merged van Hove singularities in the non-well-characterized forms, the high asymmetry of hole and electron relative to $E_F = 0$, three specific energy ranges due to the significant multi-orbital hybridizations in various chemical bonds. Three energy categories are divided according to the sufficient wide valence energy spacing and band gap: (I) -16 eV < E^v < -12 eV] dominated by O-2s orbitals [the deep on-site ionization energy], (II) -5.5 eV < E^v < -1.3 eV associated all active orbitals and (III) E^c > 1.3 eV for the similar ones. By the delicate analyses, specially chosen for the (II) region, the concise chemical pictures in chemical bonds present the following results: (1) Li-O with 2s-[2s, $2p_x$, $2p_y$, $2p_z$], (2) La-O under [6s, $5d_{z^2}$, $5d_{(x^2-y^2)}$, $5d_{xy}$, $5d_{yz}$, $5d_{xz}$]-[2s, $2p_x$, $2p_y$, $2p_z$], (3) Ti-O through [4s, $3d_{z^2}$, $3d_{(x^2-y^2)}$, $3d_{xy}$, $3d_{yz}$, $3d_{xz}$]-[2s, $2p_x$, $2p_y$, $2p_z$], and (4) La-Ti using [6s, $5d_{z^2}$, $5d_{(x^2-y^2)}$, $5d_{xy}$, $5d_{yz}$, $5d_{xz}$]-[2s, $2p_x$, $2p_y$, $2p_z$], [4s, $3d_{z^2}$, $3d_{(x^2-y^2)}$, $3d_{xy}$, $3d_{yz}$, $3d_{xz}$]. Of course, the fourth type of La-Ti bonds disappear in the ternary $Li_4Ti_5O_{12}$ compound. The similar ways could be generalized to any condensed-matter systems, such as graphene-related emergent materials [102], semiconductor compounds [103], and green energies [104]. The various chemical bondings and spin configurations will greatly diversify the composite quasiparticles and be useful in finishing a series of high-performance books [105].

In general, the joint density of states, which is associated with the initial unoccupied and final occupied states due to the external perturbations [106], will dominate the available excitation channels. For example, the electromagnetic waves can create vertical transitions because of the almost vanishing photon momenta. That is, only the energy transfers occur during the photon absorptions by electrons. The significant quasiparticles of photon-electron composites come to exist at any time in the forms of single-particle and collective excitations (electron-hole pairs and plasmon modes [107]). The former and the latter are, respectively, characterized by the prominent structures in the imaginary part of the transverse dielectric function and the strong peaks in the energy loss function (the screened response function [108]). By the delicate VASP simulations, the optical transitions could be further examined by the atom-. orbital- and spin-decomposed contributions, since their related van Hove singularities show very strong evidence in the joint density of states, this research strategy is very successful in fully exploring the enlarged quasiparticle framework about all group-IV pristine (graphene/silicene/germanene/tinene/plumbene [109]/[110]/[111]/[112]/[113]) and hydrogenated systems (double- and single-side full hydrogenations [114]/[115]). The featured optical excitations cover the threshold absorption frequency [116], the excitonic peaks in semiconductors [117], the orbital-related plasmons at distinct frequencies [118], and each strong absorption peak due to the specific orbital hybridizations of the initial and final states (the composite quasiparticles [119]). As for quaternary $LiLaTiO_4$ and ternary $Li_4Ti_5O_{12}$, the main features, the first one excepted, are very difficult to be verified from the reflectance/absorption/transmission/photoluminescence spectroscopies [120]/[121]/[122]/[123], The very complicated chemical bondings in a quasi-Moire superlattice are responsible for the diversified behaviors. Similar research barriers would frequently appear in the other physical perturbations, e.g., the coherent electron/atom oscillations under the time-dependent Coulomb ionic fields [124]/[125]. Such open issues would become the very strong motivations of the next-step framework developments.

5.5 MULTI-ATOM, ACTIVE-ORBITAL-, AND SPIN-CREATED DIVERSE QUASIPARTICLES

Very interestingly, the up-to-date experimental syntheses and theoretical predictions have successfully created the various crystal phases based on the lithium and oxygen atoms, e.g., the diverse crystal symmetries in $LiLaTiO_4$ [58], $Li_4Ti_5O_{12}$ [126], $LiFePO_4$/$LiMnPO_4$ [127]/[128], and $LiFeO_2$/$LiCoO_2$/$LiNiO_2$/$LiMnO_2$ [129]/[130]/[131]/[132]. The first-principles methods should be directly combined with the molecular dynamic dynamics for thoroughly exploring the real chemical reaction processes. This is very useful in understanding the time-dependent growths by the different chemical ways and the charging/discharging operations of ion-based batteries. The almost continuous intermediate configurations will clearly illustrate the giant Moire superlattices. How to observe the giant unit cells by the time-resolved X-ray patterns is a very interesting topic. The strong relations between the theoretical and experimental research need to be established under the current investigations. The crystal structures could serve as the critical modulation factor in observing the enlarged composite quasiparticles, since they are very sensitive to the following mechanisms: (1) the relative atom concentrations, (2) the same weight, but distinct distribution, (3) the distinct atoms, the various chemical modifications (adsorptions/substitutions/hetero junctions/decorations/intercalations [133]/134]/[135]/[136]/[137]), (4) the physical perturbations (gate voltages/magnetic fields/Coulomb fields/electromagnetic waves/pressures/thermal excitations/mechanical strains [138]/[139]/[140]/[139]/[141]/[142]/[143]), and (5) the quantum confinements in 2D/1D/0D materials [144]/[145]/[146].

In addition to Li- and O-related materials, the Li-, P-, and S-related materials (e.g., various Li-P/P-S/Li-S/Li-P-S compounds [147]/[148]/[149]/[150]), as well as the semiconductor compounds of III-VI/III-V/II-VI/II-V/IV-IV systems [151]/[152]/[153]/[154]/[155]. By the delicate VASP simulations for all calculated results and experimental observations, the concise multi-orbital hybridizations clearly illustrate the diverse quasiparticle couplings: (1) 2s-[2s, $2p_x$, $2p_y$, $2p_z$] of Li-O bonds [156], (2) [6s, $5d_{z^2}$, $5d_{(x^2-y^2)}$, $5d_{xy}$, $5d_{yz}$, $5d_{xz}$]-[2s, $2p_x$, $2p_y$, $2p_z$] through La-O bonds [157], [4s, $3d_{z^2}$, $3d_{(x^2-y^2)}$, $3d_{xy}$, $3d_{yz}$, $3d_{xz}$]- [2s, $2p_x$, $2p_y$, $2p_z$] by Ti-O/Fe-O/Co-O/Ni-O/Mn-O bonds [158]/[159]/[160]/[161]/[162], (4) [6s, $5d_{z^2}$, $5d_{(x^2-y^2)}$, $5d_{xy}$, $5d_{yz}$, $5d_{xz}$]-[2s, $2p_x$, $2p_y$, $2p_z$]-[4s, $3d_{z^2}$, $3d_{(x^2-y^2)}$, $3d_{xy}$, $3d_{yz}$, $3d_{xz}$] under La-Ti bond, (5) [3s, $3p_x$, $3p_y$, $3p_z$]-[3s, $3p_x$, $3p_y$, $3p_z$] by P-S bonds [163], and (6) [3s, $3p_x$, $3p_y$, $3p_z$]-[3s, $3p_x$, $3p_y$, $3p_z$] available from Ga-S bonds [164].

The chemical bondings are expected to be greatly diversified by the various crystal structures, such as the significant hydrogenations about graphene/silicene/germanene/tinene/plumbene [165]/[166]/[167]/[168]/[169], the dimensionality crossover about the layered systems (GaS/GaSe/GaTe/InSe [170]/[171]/[172]/[173]), the interlayer/intramolecular/intermolecular orbital hybridizations in graphite intercalation compounds [174]/[175]/[176], and the enriched environments in curved/scrolled/folded/buckled structures [177]/[178]/[179]/[180]. The whole pictures of orbital hybridizations, which correspond to the well-defined chemical bonds, need to be thoroughly established for the basic and applied sciences. This very interesting topic will be done in a series of published books.

Certain transition-metal lithium oxides exhibit spin-induced magnetic properties. For example, the $LiFeO_2$/$LiFePO_4$/$LiCoO_2$/$LiNiO_2$/$LiMnO_4$ [181]/[182]/[183]/[184]/[185] clearly exhibit the spin-split valence and conduction subbands, the atom- and orbital-decomposed magnetic moments, the spin-up and spin-down density distributions, spin-decomposed van Hove singularities, and spin-dominated optical gaps. However, the ferromagnetic configuration might disappear in other rare-earth- and transition-metal-related lithium oxides, e.g., $LiLaTiO_4$ and $Li_4Ti_5O_{12}$. The existence/absence of spin configuration should be attributed to the very competition between the many-particle Coulomb interactions, respectively, associated with charges (exchange and correlation energies) and spins (Hubbard-like on-site Coulomb interactions). This is able to generate the atom-, orbital-, and spin- composite quasiparticles in terms of essential properties, e.g., the well characterizations of band gap, threshold absorption frequency, stable excitons, prominent electron-hole responses, and energy-related plasmon modes. Whether the phenomenological models could be well built from the VASP simulations are worthy of systematic investigations. Moreover, how to directly measure the spatial spin density distributions and the magnetic moments, respectively, requires the high-precision measurements of STS (scanning tunneling spectroscopy [186]) and NMR (nuclear magnetic resonance [187]) The close partnerships between charges and spins, which need the delicate works, will be very helpful for enlarging the quasiparticle framework in near-future books.

5.6 CONCISE CONCLUSIONS: CHARGE- AND SPIN-DOMINATED COMPOSITE QUASIPARTICLES

The lithium- and oxygen-related main-stream materials, which cover the single-element, binary, ternary, and quaternary crystals, are outstanding candidates for fully studying the diverse quasiparticle phenomena under an enlarged framework. Such work is very useful in promoting the simultaneous progresses among the first-principles simulations, the phenomenological models, and the experimental observations. In this chapter, the quaternary $LiLaTiO_4$ compounds are thoroughly identified to exhibit the rich chemical bondings mainly from the active orbitals of one-s, three-p, five-d, and seven-f types [188]. The fourth-type orbitals are absent in $Li_4Ti_5O_{12}$, as clearly revealed in the previous study [37]. The critical mechanisms, being closely related to charge- and spin-induced single-particle and many-body interactions, The non-magnetic configurations indicate that the former completely suppress the latter through a strong competition (the negligible Hubbard-like on-site spin-dependent Coulomb interactions [189]). In addition to geometric and electronic properties, the VASP calculations are available to explore the other fundamental properties, e.g., the optical transitions with/without the excitonic effects [190], and the Coulomb excitations [191]. and the quantum phonon spectra. The delicate analyses are required to achieve concise pictures because of too many low-lying valence and conduction energy subbands within a quasi-Moire superlattice (Figures 5.3(a)–(e)). How to utilize these VASP electronic states to get the reliable parameters about the tight-binding model/the generalized tight-biding model [192]/[193] is the nest-step studying focus. On the experimental side, the high-resolution measurements, the X-ray prominent patterns excepted, cannot be accurately analyzed under the intrinsic physics limits, The strong partnerships with the theoretical predictions might be one of the most efficient ways.

The delicate VASP simulations and analyses have been successfully done for lithium oxide compounds [51], as clearly illustrated in lithium-sulfur ones. The crystal asymmetries present the position-dependent chemical bonds, e.g., the Li-O, La-O, Ti-O, and La-Ti bonds, The highly non-uniform and isotropic environment, which is characteristic of sensitive covalent bondings, can greatly create the strong modifications of chemical reactions (substitutions/adsorptions/intercalations/decorations) and physical perturbations (gate voltages/magnetic fields/electromagnetic waves/Coulomb fields/thermal excitations/mechanical strains). The featured band structures include a sufficiently large gap of more than 2 eV, a large asymmetry of the hole and electron energy spectrum about the Fermi level, a lot of valence energy subbands within $E^v > -6$ eV, many weak, but irregular energy dispersions, ambiguous band-edge states, a forbidden energy spacing between -12 eV $< E^v < -6$ eV (the separation of s- and [p, d, f]-dominated chemical bondings), and the distinct atom dominances at the specific energy ranges [118]. The charge density distributions in unit cells clearly show the main features of covalent bondings, the diverse orbital distortions near the distinct atoms. The qualitative pictures of Li-O, La-O, Ti-O, and La-Ti bonds, being examined from the various projection examinations, in chemical bonds, are as follows, respectively: (1) 2s-[2s, $2p_x$, $2p_y$, $2p_z$], (2) [6s, $5d_{z^2}$, $5d_{(x^2-y^2)}$, $5d_{xy}$, $5d_{yz}$, $5d_{xz}$]-[2s, $2p_x$, $2p_y$, $2p_z$], [4s, $3d_{z^2}$, $3d_{(x^2-y^2)}$, $3d_{xy}$, $3d_{yz}$, $3d_{xz}$]-[2s, $2p_x$, $2p_y$, $2p_z$], and (4) [6s, $5d_{z^2}$, $5d_{(x^2-y^2)}$, $5d_{xy}$, $5d_{yz}$, $5d_{xz}$]-[2s, $2p_x$, $2p_y$, $2p_z$], [4s, $3d_{z^2}$, $3d_{(x^2-y^2)}$, $3d_{xy}$, $3d_{yz}$, $3d_{xz}$]. The significant orbital mixings are further revealed as the merged van Hove singularities. Since the direct combination of too many special structures appears in the initial/most important valence states, this will induce much trouble for the experimental verifications and theoretical identifications. The strong relations among the three approaches need to be established in the further book proposals.

REFERENCES

1. G.-y. Adachi, N. Imanaka, and S. Tamura, "Ionic conducting lanthanide oxides," *Chemical Reviews*, vol. 102, no. 6, pp. 2405–2430, 2002.
2. W. B. Jensen, "The place of zinc, cadmium, and mercury in the periodic table," *Journal of Chemical Education*, vol. 80, no. 8, p. 952, 2003.
3. S. B. Castor and J. B. Hedrick, "Rare earth elements," *Industrial Minerals and Rocks*, pp. 769–792, 2006.
4. W. M. Latimer and W. H. Rodebush, "Polarity and ionization from the standpoint of the Lewis theory of valence," *Journal of the American Chemical Society*, vol. 42, no. 7, pp. 1419–1433, 1920.
5. S. Ji, Y. Chen, X. Wang, Z. Zhang, D. Wang, and Y. Li, "Chemical synthesis of single atomic site catalysts," *Chemical Reviews*, vol. 120, no. 21, pp. 11900–11955, 2020.
6. D. Yan *et al.*, "Modification of luminescent properties of a coumarin derivative by formation of multi-component crystals," *CrystEngComm*, vol. 14, no. 16, pp. 5121–5123, 2012.
7. H. Valencia, A. Gil, and G. Frapper, "Trends in the adsorption of 3d transition metal atoms onto graphene and nanotube surfaces: a DFT study and molecular orbital analysis," *The Journal of Physical Chemistry C*, vol. 114, no. 33, pp. 14141–14153, 2010.
8. A. Cox, J. Louderback, S. Apsel, and L. Bloomfield, "Magnetism in 4d-transition metal clusters," *Physical Review B*, vol. 49, no. 17, p. 12295, 1994.
9. F. Cyrot-Lackmann and F. Ducastelle, "Binding energies of transition-metal atoms adsorbed on a transition metal," *Physical Review B*, vol. 4, no. 8, p. 2406, 1971.
10. A. El Hachimi, H. Zaari, A. Benyoussef, M. El Yadari, and A. El Kenz, "First-principles prediction of the magnetism of 4f rare-earth-metal-doped wurtzite zinc oxide," *Journal of Rare Earths*, vol. 32, no. 8, pp. 715–721, 2014.
11. P. G. Huray, S. Nave, and R. Haire, "Magnetism of the heavy 5f elements," *Journal of the Less Common Metals*, vol. 93, no. 2, pp. 293–300, 1983.
12. J. Hirsch, "Effect of coulomb interactions on the peierls instability," *Physical Review Letters*, vol. 51, no. 4, p. 296, 1983.
13. S.-C. Chen, J.-Y. Wu, and M.-F. Lin, "Feature-rich magneto-electronic properties of bismuthene," *New Journal of Physics*, vol. 20, no. 6, p. 062001, 2018.

14. J. Dong *et al.*, "Competing orders and spin-density-wave instability in La (O1− xFx) FeAs," *EPL (Europhysics Letters)*, vol. 83, no. 2, p. 27006, 2008.
15. S.-Y. Lin, H.-Y. Liu, S.-J. Tsai, and M.-F. Lin, "7 Geometric and electronic properties of Li battery cathode," *Lithium-Ion Batteries and Solar Cells: Physical, Chemical, and Materials Properties*, p. 117, 2021.
16. J. Hafner, "Materials simulations using VASP—a quantum perspective to materials science," *Computer Physics Communications*, vol. 177, no. 1–2, pp. 6–13, 2007.
17. N. Plyusnin, "Phenomenological models of nucleation and growth of metal on a semiconductor," *Physics of the Solid State*, vol. 61, no. 12, pp. 2431–2433, 2019.
18. K. Buschow and J. Fast, "Crystal structure and magnetic properties of some rare earth germanides," *Physica Status Solidi (b)*, vol. 21, no. 2, pp. 593–600, 1967.
19. Y. Huang *et al.*, "Printable functional chips based on nanoparticle assembly," *Small*, vol. 13, no. 4, p. 1503339, 2017.
20. A.-M. Sapse and P. v. R. Schleyer, "Lithium chemistry: a theoretical and experimental overview," 1995.
21. X. Yuan, H. Liu, and J. Zhang, *Lithium-ion batteries: advanced materials and technologies*. CRC Press, 2011.
22. Y. Ji, H. Dong, C. Liu, and Y. Li, "The progress of metal-free catalysts for the oxygen reduction reaction based on theoretical simulations," *Journal of Materials Chemistry A*, vol. 6, no. 28, pp. 13489–13508, 2018.
23. A. E. Reed, F. Weinhold, L. A. Curtiss, and D. J. Pochatko, "Natural bond orbital analysis of molecular interactions: theoretical studies of binary complexes of HF, H2O, NH3, N2, O2, F2, CO, and CO2 with HF, H2O, and NH3," *The Journal of Chemical Physics*, vol. 84, no. 10, pp. 5687–5705, 1986.
24. Z. Feng *et al.*, "Raman, infrared, photoluminescence and theoretical studies of the II-VI-VI ternary CdSeTe," *Journal of Crystal Growth*, vol. 138, no. 1–4, pp. 239–243, 1994.
25. A. J. Mackus, J. R. Schneider, C. MacIsaac, J. G. Baker, and S. F. Bent, "Synthesis of doped, ternary, and quaternary materials by atomic layer deposition: a review," *Chemistry of Materials*, vol. 31, no. 4, pp. 1142–1183, 2018.
26. K.-S. Park, D. Im, A. Benayad, A. Dylla, K. J. Stevenson, and J. B. Goodenough, "$LiFeO_2$-incorporated Li_2MoO_3 as a cathode additive for lithium-ion battery safety," *Chemistry of Materials*, vol. 24, no. 14, pp. 2673–2683, 2012.
27. C. Zhao, H. Yin, and C. Ma, "Quantitative evaluation of LiFePO $ _4 $ battery cycle life improvement using ultracapacitors," *IEEE Transactions on Power Electronics*, vol. 31, no. 6, pp. 3989–3993, 2015.
28. X. Wang, Q. Qu, Y. Hou, F. Wang, and Y. Wu, "An aqueous rechargeable lithium battery of high energy density based on coated Li metal and $LiCoO_2$," *Chemical Communications*, vol. 49, no. 55, pp. 6179–6181, 2013.
29. W. Ebner, D. Fouchard, and L. Xie, "The $LiNiO_2$/carbon lithium-ion battery," *Solid State Ionics*, vol. 69, no. 3-4, pp. 238–256, 1994.
30. E. Sivonxay, M. Aykol, and K. A. Persson, "The lithiation process and Li diffusion in amorphous SiO_2 and Si from first-principles," *Electrochimica Acta*, vol. 331, p. 135344, 2020.
31. L. Wang, Y. Zhang, Z. Liu, L. Guo, and Z. Peng, "Understanding oxygen electrochemistry in aprotic LiO_2 batteries," *Green Energy & Environment*, vol. 2, no. 3, pp. 186–203, 2017.
32. Z. Liu, H. Deng, S. Zhang, W. Hu, and F. Gao, "Theoretical prediction of $LiScO_2$ nanosheets as a cathode material for Li–O 2 batteries," *Physical Chemistry Chemical Physics*, vol. 20, no. 34, pp. 22351–22358, 2018.

33. W. Borghols, D. Lützenkirchen-Hecht, U. Haake, E. Van Eck, F. Mulder, and M. Wagemaker, "The electronic structure and ionic diffusion of nanoscale $LiTiO_2$ anatase," *Physical Chemistry Chemical Physics*, vol. 11, no. 27, pp. 5742–5748, 2009.
34. Y. E. Durmus *et al.*, "Side by side battery technologies with lithium-ion based batteries," *Advanced Energy Materials*, vol. 10, no. 24, p. 2000089, 2020.
35. C. R. Ryder, J. D. Wood, S. A. Wells, and M. C. Hersam, "Chemically tailoring semiconducting two-dimensional transition metal dichalcogenides and black phosphorus," *ACS Nano*, vol. 10, no. 4, pp. 3900–3917, 2016.
36. J. Miao, T. Ishikawa, B. Johnson, E. H. Anderson, B. Lai, and K. O. Hodgson, "High resolution 3D x-ray diffraction microscopy," *Physical Review Letters*, vol. 89, no. 8, p. 088303, 2002.
37. T. D. H. Nguyen, H. D. Pham, S.-Y. Lin, and M.-F. Lin, "Featured properties of Li+-based battery anode: $Li_4Ti_5O_{12}$," *RSC Advances*, vol. 10, no. 24, pp. 14071–14079, 2020.
38. W.-b. Li, N. T. T. Tran, S.-y. Lin, and M.-F. Lin, "Diverse fundamental properties in stage-n graphite alkali-intercalation compounds: anode materials of Li+-based batteries," *arXiv preprint arXiv:2001.02042*, 2019.
39. G. Centi and S. Perathoner, "Carbon nanotubes for sustainable energy applications," *ChemSusChem*, vol. 4, no. 7, pp. 913–925, 2011.
40. I. N. Martyanov, S. Uma, S. Rodrigues, and K. J. Klabunde, "Structural defects cause TiO_2-based photocatalysts to be active in visible light," *Chemical Communications*, no. 21, pp. 2476–2477, 2004.
41. S. Guan, Q. Fan, Z. Shen, Y. Zhao, Y. Sun, and Z. Shi, "Heterojunction TiO_2@ $TiOF_2$ nanosheets as superior anode materials for sodium-ion batteries," *Journal of Materials Chemistry A*, vol. 9, no. 9, pp. 5720–5729, 2021.
42. R. Rojaee and R. Shahbazian-Yassar, "Two-dimensional materials to address the lithium battery challenges," *ACS Nano*, vol. 14, no. 3, pp. 2628–2658, 2020.
43. C. Lv *et al.*, "1D Nb-doped $LiNi_1/3Co_1/3Mn_1/3O_2$ nanostructures as excellent cathodes for Li-ion battery," *Electrochimica Acta*, vol. 297, pp. 258–266, 2019.
44. J. Zheng, Y. Wu, Y. Sun, J. Rong, H. Li, and L. Niu, "Advanced anode materials of potassium ion batteries: from zero dimension to three dimensions," *Nano-Micro Letters*, vol. 13, no. 1, pp. 1–37, 2021.
45. Y. Wang, J. Li, Z. Zhu, Z. You, J. Xu, and C. Tu, "Bulk crystal growth, first-principles calculation, and optical properties of pure and Er3+-doped $SrLaGa_3O_7$ single crystals," *Crystal Growth & Design*, vol. 16, no. 4, pp. 2289–2294, 2016.
46. G. C. Sosso *et al.*, "Crystal nucleation in liquids: open questions and future challenges in molecular dynamics simulations," *Chemical Reviews*, vol. 116, no. 12, pp. 7078–7116, 2016.
47. A. Majid, A. Batool, S. U. D. Khan, and A. Ahmad, "First-principles study of f-orbital-dependent band topology of topological rare earth hexaborides," *International Journal of Quantum Chemistry*, vol. 121, no. 4, p. e26452, 2021.
48. E. Bertaut, "Spin configurations of ionic structures: theory and practice," *Spin Arrangements and Crystal Structure, Domains, and Micromagnetics* (eds Rado, G. T. & Suhl, H.), pp. 149–209, 1963.
49. M. Martins and P. Ramos, "The quantum inverse scattering method for Hubbard-like models," *Nuclear Physics B*, vol. 522, no. 3, pp. 413–470, 1998.
50. G. S. Jeon and J. K. Jain, "Nature of quasiparticle excitations in the fractional quantum Hall effect," *Physical Review B*, vol. 68, no. 16, p. 165346, 2003.
51. T. D. H. Nguyen, N. T. T. Tran, and M.-F. Lin, "Open Issues and Potential Applications," in *Lithium-Ion Batteries and Solar Cells*: CRC Press, 2021, pp. 261–277.

52. T. D. H. Nguyen, S.-Y. Lin, H.-C. Chung, N. T. T. Tran, and M.-F. Lin, "First-Principles Calculations for Cathode, Electrolyte and Anode Battery Materials."
53. S.-Y. Lin *et al.*, "Diversified quasiparticle properties of four-component $LixFePO_4$ cathode materials in Li-based batteries: charges and spins."
54. T. P. Kaloni, "Tuning the structural, electronic, and magnetic properties of germanene by the adsorption of 3d transition metal atoms," *The Journal of Physical Chemistry C*, vol. 118, no. 43, pp. 25200–25208, 2014.
55. Q. Zhao, R. C. Morrison, and R. G. Parr, "From electron densities to Kohn-Sham kinetic energies, orbital energies, exchange-correlation potentials, and exchange-correlation energies," *Physical Review A*, vol. 50, no. 3, p. 2138, 1994.
56. S. Hesselmann, T. C. Lang, M. Schuler, S. Wessel, and A. M. Läuchli, "Comment on "The role of electron-electron interactions in two-dimensional Dirac fermions"," *Science*, vol. 366, no. 6470, p. eaav6869, 2019.
57. D. Murphy, R. Cava, S. Zahurak, and A. Santoro, "Ternary $LixTiO_2$ phases from insertion reactions," *Solid State Ionics*, vol. 9, pp. 413–417, 1983.
58. K. Toda, S. Kurita, and M. Sato, "New layered perovskite compounds, $LiLaTiO_4$ and $LiEuTiO_4$," *Journal of the Ceramic Society of Japan*, vol. 104, no. 1206, pp. 140–142, 1996.
59. C. Daniel, "Materials and processing for lithium-ion batteries," *JOM*, vol. 60, no. 9, pp. 43–48, 2008.
60. W. Lee *et al.*, "Advances in the cathode materials for lithium rechargeable batteries," *Angewandte Chemie International Edition*, vol. 59, no. 7, pp. 2578–2605, 2020.
61. Y. Shu, E. G. Hohenstein, and B. G. Levine, "Configuration interaction singles natural orbitals: An orbital basis for an efficient and size intensive multireference description of electronic excited states," *The Journal of Chemical Physics*, vol. 142, no. 2, p. 024102, 2015.
62. A. Janotti and C. G. Van de Walle, "Hydrogen multicentre bonds," *Nature Materials*, vol. 6, no. 1, pp. 44–47, 2007.
63. B. Cordero *et al.*, "Covalent radii revisited," *Dalton Transactions*, no. 21, pp. 2832–2838, 2008.
64. W.-B. Li *et al.*, "Diversified phenomena in sodium-, potassium-and magnesium-related graphite intercalation compounds."
65. K. Sugawara, T. Sato, S. Souma, T. Takahashi, and H. Suematsu, "Fermi surface and edge-localized states in graphite studied by high-resolution angle-resolved photoemission spectroscopy," *Physical Review B*, vol. 73, no. 4, p. 045124, 2006.
66. D. Chen, X. Jiao, and R. Xu, "Hydrothermal synthesis and characterization of the layered titanates MLaTiO4 (M= Li, Na, K) powders," *Materials Research Bulletin*, vol. 34, no. 5, pp. 685–691, 1999.
67. G. Ceder, G. Hautier, A. Jain, and S. P. Ong, "Recharging lithium battery research with first-principles methods," *MRS Bulletin*, vol. 36, no. 3, pp. 185–191, 2011.
68. C.-Y. Lin, C.-H. Yang, C.-W. Chiu, H.-C. Chung, S.-Y. Lin, and M.-F. Lin, *Rich Quasiparticle Properties of Low Dimensional Systems*. IOP Publishing, 2021.
69. Y. Wang, E. Wang, X. Zhang, and H. Yu, "High-voltage "single-crystal" cathode materials for lithium-ion batteries," *Energy & Fuels*, vol. 35, no. 3, pp. 1918–1932, 2021.
70. M. Yu, C. Zhang, and C. B. Kah, "Computational discovery of 2D materials: a fundament study of boron sheets and phosphide binary compounds," *Video Proceedings of Advanced Materials*, vol. 1, no. 1, pp. 1–2, 2020.
71. J. Shu, "RETRACTED: Li–Ti–O compounds and carbon-coated Li–Ti–O compounds as anode materials for lithium ion batteries," ed: Elsevier, 2009.

72. P. W. Jaschin, Y. Gao, Y. Li, and S.-H. Bo, "A materials perspective on magnesium-ion-based solid-state electrolytes," *Journal of Materials Chemistry A*, vol. 8, no. 6, pp. 2875–2897, 2020.
73. S. Pearton *et al.*, "Wide band gap ferromagnetic semiconductors and oxides," *Journal of Applied Physics*, vol. 93, no. 1, pp. 1–13, 2003.
74. V. V. Fedorov, A. Gallian, I. Moskalev, and S. B. Mirov, "En route to electrically pumped broadly tunable middle infrared lasers based on transition metal doped II–VI semiconductors," *Journal of Luminescence*, vol. 125, no. 1–2, pp. 184–195, 2007.
75. R. Dornhaus, G. Nimtz, and B. Schlicht, *Narrow-gap semiconductors*. Springer, 2006.
76. A. Abrikosov, "Calculation of critical indices for zero-gap semiconductors," *Journal of Experimental and Theoretical Physics*, vol. 66, pp. 1443–1460, 1974.
77. P. Hosur and X. Qi, "Recent developments in transport phenomena in Weyl semimetals," *Comptes Rendus Physique*, vol. 14, no. 9–10, pp. 857–870, 2013.
78. C. J. Smithells, *Metals reference book*. Elsevier, 2013.
79. C.-Y. Lin, S.-C. Chen, J.-Y. Wu, and M.-F. Lin, "Curvature effects on magnetoelectronic properties of nanographene ribbons," *Journal of the Physical Society of Japan*, vol. 81, no. 6, p. 064719, 2012.
80. S. Xu *et al.*, "Tunable van Hove singularities and correlated states in twisted monolayer–bilayer graphene," *Nature Physics*, vol. 17, no. 5, pp. 619–626, 2021.
81. O. Reckeweg, B. Blaschkowski, and T. Schleid, "Li5OCl_3 and Li_3OCl: two remarkably different lithium oxide chlorides," *Zeitschrift für Anorganische und Allgemeine Chemie*, vol. 638, no. 12–13, pp. 2081–2086, 2012.
82. F. Kong, X. Xu, and J.-G. Mao, "A series of new ternary and quaternary compounds in the LiI– GaIII– TeIV– O system," *Inorganic chemistry*, vol. 49, no. 24, pp. 11573–11580, 2010.
83. C.-W. Chiu, Y.-L. Chung, C.-H. Yang, C.-T. Liu, and C.-Y. Lin, "Coulomb decay rates in monolayer doped graphene," *RSC Advances*, vol. 10, no. 4, pp. 2337–2346, 2020.
84. M. Alcoutlabi and G. B. McKenna, "Effects of confinement on material behaviour at the nanometre size scale," *Journal of Physics: Condensed Matter*, vol. 17, no. 15, p. R461, 2005.
85. C. Hofmann *et al.*, "An experimental approach for investigating many-body phenomena in Rydberg-interacting quantum systems," *Frontiers of Physics*, vol. 9, no. 5, pp. 571–586, 2014.
86. R. Jalem, Y. Tateyama, K. Takada, and M. Nakayama, "First-principles DFT study on inverse Ruddlesden–Popper tetragonal compounds as solid electrolytes for all-solid-state Li+-ion batteries," *Chemistry of Materials*, vol. 33, no. 15, pp. 5859–5871, 2021.
87. A. Grüneis and D. V. Vyalikh, "Tunable hybridization between electronic states of graphene and a metal surface," *Physical Review B*, vol. 77, no. 19, p. 193401, 2008.
88. G. Ernst, C. Artner, O. Blaschko, and G. Krexner, "Low-temperature martensitic phase transition of bcc lithium," *Physical Review B*, vol. 33, no. 9, p. 6465, 1986.
89. A. Krasovskiy and P. Knochel, "A LiCl-mediated Br/Mg exchange reaction for the preparation of functionalized aryl-and heteroarylmagnesium compounds from organic bromides," *Angewandte Chemie International Edition*, vol. 43, no. 25, pp. 3333–3336, 2004.
90. T. W. Yip, E. J. Cussen, and D. A. MacLaren, "Synthesis of $H_xLi_{1-x}LaTiO_4$ from quantitative solid-state reactions at room temperature," *Chemical Communications*, vol. 46, no. 5, pp. 698–700, 2010.

91. A. Taheridehkordi, S. Curnoe, and J. LeBlanc, "Algorithmic Matsubara integration for Hubbard-like models," *Physical Review B*, vol. 99, no. 3, p. 035120, 2019.
92. W. Gu, Y. Guo, Q. Li, Y. Tian, and K. Chu, "Lithium iron oxide ($LiFeO_2$) for electroreduction of dinitrogen to ammonia," *ACS Applied Materials & Interfaces*, vol. 12, no. 33, pp. 37258–37264, 2020.
93. J. Van Elp *et al.*, "Electronic structure of CoO, Li-doped CoO, and $LiCoO_2$," *Physical Review B*, vol. 44, no. 12, p. 6090, 1991.
94. S. Laubach *et al.*, "Changes in the crystal and electronic structure of $LiCoO_2$ and $LiNiO_2$ upon Li intercalation and de-intercalation," *Physical Chemistry Chemical Physics*, vol. 11, no. 17, pp. 3278–3289, 2009.
95. K. Hipps, "Scanning tunneling spectroscopy (STS)," in *Handbook of applied solid state spectroscopy*: Springer, 2006, pp. 305–350.
96. P. Kim, T. W. Odom, J.-L. Huang, and C. M. Lieber, "Electronic density of states of atomically resolved single-walled carbon nanotubes: Van Hove singularities and end states," *Physical Review Letters*, vol. 82, no. 6, p. 1225, 1999.
97. R. W. Havener, Y. Liang, L. Brown, L. Yang, and J. Park, "Van Hove singularities and excitonic effects in the optical conductivity of twisted bilayer graphene," *Nano Letters*, vol. 14, no. 6, pp. 3353–3357, 2014.
98. P. Igoshev, A. Katanin, and V. Y. Irkhin, "Magnetic fluctuations and itinerant ferromagnetism in two-dimensional systems with Van Hove singularities," *Journal of Experimental and Theoretical Physics*, vol. 105, no. 5, pp. 1043–1056, 2007.
99. Y. Shi *et al.*, "Tunable van Hove singularities and correlated states in twisted trilayer graphene," *arXiv preprint arXiv:2004.12414*, 2020.
100. Y. Quan and W. E. Pickett, "Van Hove singularities and spectral smearing in high-temperature superconducting H3S," *Physical Review B*, vol. 93, no. 10, p. 104526, 2016.
101. G. Li *et al.*, "Observation of Van Hove singularities in twisted graphene layers," *Nature Physics*, vol. 6, no. 2, pp. 109–113, 2010.
102. N. T. T. Tran, S.-Y. Lin, C.-Y. Lin, and M.-F. Lin, *Geometric and electronic properties of graphene-related systems: Chemical bonding schemes*. CRC Press, 2017.
103. E. Monaico, I. Tiginyanu, and V. Ursaki, "Porous semiconductor compounds," *Semiconductor Science and Technology*, vol. 35, no. 10, p. 103001, 2020.
104. V. Flexer, C. F. Baspineiro, and C. I. Galli, "Lithium recovery from brines: a vital raw material for green energies with a potential environmental impact in its mining and processing," *Science of the Total Environment*, vol. 639, pp. 1188–1204, 2018.
105. T. D. H. Nguyen *et al.*, "Theoretical frameworks on geometric, electronic, magnetic, and optical properties."
106. W. Heisenberg and H. Euler, "Consequences of dirac theory of the positron," *arXiv preprint physics/0605038*, 2006.
107. C. Tegenkamp, H. Pfnür, T. Langer, J. Baringhaus, and H. Schumacher, "Plasmon electron–hole resonance in epitaxial graphene," *Journal of Physics: Condensed Matter*, vol. 23, no. 1, p. 012001, 2010.
108. P. Littlewood, "Screened dielectric response of sliding charge-density waves," *Physical Review B*, vol. 36, no. 6, p. 3108, 1987.
109. L. Matthes, O. Pulci, and F. Bechstedt, "Massive Dirac quasiparticles in the optical absorbance of graphene, silicene, germanene, and tinene," *Journal of Physics: Condensed Matter*, vol. 25, no. 39, p. 395305, 2013.
110. P. De Padova *et al.*, "The quasiparticle band dispersion in epitaxial multilayer silicene," *Journal of Physics: Condensed Matter*, vol. 25, no. 38, p. 382202, 2013.

111. H. Shu, Y. Li, S. Wang, and J. Wang, "Quasi-particle energies and optical excitations of hydrogenated and fluorinated germanene," *Physical Chemistry Chemical Physics*, vol. 17, no. 6, pp. 4542–4550, 2015.
112. S.-Y. Lin *et al.*, "Battery-related problems."
113. F. Bechstedt, P. Gori, and O. Pulci, "Beyond graphene: clean, hydrogenated and halogenated silicene, germanene, stanene, and plumbene," *Progress in Surface Science*, vol. 96, no. 3, p. 100615, 2021.
114. C. Reddy and Y.-W. Zhang, "Structure manipulation of graphene by hydrogenation," *Carbon*, vol. 69, pp. 86–91, 2014.
115. B. S. Pujari, S. Gusarov, M. Brett, and A. Kovalenko, "Single-side-hydrogenated graphene: density functional theory predictions," *Physical Review B*, vol. 84, no. 4, p. 041402, 2011.
116. V. K. Dien *et al.*, "Orbital-hybridization-created optical excitations in Li_2GeO_3," *Scientific Reports*, vol. 11, no. 1, pp. 1–10, 2021.
117. A. Klots *et al.*, "Probing excitonic states in suspended two-dimensional semiconductors by photocurrent spectroscopy," *Scientific Reports*, vol. 4, no. 1, pp. 1–7, 2014.
118. T. D. H. Nguyen, K. D. Vo, H. D. Pham, T. M. D. Huynh, and M.-F. Lin, "Electronic and optical excitation properties of vanadium pentoxide V_2O_5," *Computational Materials Science*, vol. 198, p. 110675, 2021.
119. S. Weinberg, "Quasiparticles and the Born series," *Physical Review*, vol. 131, no. 1, p. 440, 1963.
120. G. Kortüm, *Reflectance spectroscopy: principles, methods, applications*. Springer Science & Business Media, 2012.
121. H. Wende, "Recent advances in x-ray absorption spectroscopy," *Reports on Progress in Physics*, vol. 67, no. 12, p. 2105, 2004.
122. L. Sanche and G. Schulz, "Electron transmission spectroscopy: rare gases," *Physical Review A*, vol. 5, no. 4, p. 1672, 1972.
123. G. Gilliland, "Photoluminescence spectroscopy of crystalline semiconductors," *Materials Science and Engineering: R: Reports*, vol. 18, no. 3–6, pp. 99–399, 1997.
124. V. N. Litvinenko and Y. S. Derbenev, "Coherent electron cooling," *Physical Review Letters*, vol. 102, no. 11, p. 114801, 2009.
125. M. Sabbar *et al.*, "State-resolved attosecond reversible and irreversible dynamics in strong optical fields," *Nature Physics*, vol. 13, no. 5, pp. 472–478, 2017.
126. B. Zhao, R. Ran, M. Liu, and Z. Shao, "A comprehensive review of $Li_4Ti_5O_{12}$-based electrodes for lithium-ion batteries: the latest advancements and future perspectives," *Materials Science and Engineering: R: Reports*, vol. 98, pp. 1–71, 2015.
127. W.-J. Zhang, "Structure and performance of $LiFePO_4$ cathode materials: a review," *Journal of Power Sources*, vol. 196, no. 6, pp. 2962–2970, 2011.
128. T. Drezen, N.-H. Kwon, P. Bowen, I. Teerlinck, M. Isono, and I. Exnar, "Effect of particle size on $LiMnPO_4$ cathodes," *Journal of Power Sources*, vol. 174, no. 2, pp. 949–953, 2007.
129. R. Kanno *et al.*, "Synthesis, structure, and electrochemical properties of a new lithium iron oxide, $LiFeO_2$, with a corrugated layer structure," *Journal of the Electrochemical Society*, vol. 143, no. 8, p. 2435, 1996.
130. S. Kikkawa, S. Miyazaki, and M. Koizumi, "Deintercalated $NaCoO_2$ and $LiCoO_2$," *Journal of Solid State Chemistry*, vol. 62, no. 1, pp. 35–39, 1986.
131. J.-H. Lim, H. Bang, K.-S. Lee, K. Amine, and Y.-K. Sun, "Electrochemical characterization of Li_2MnO_3–Li [Ni1/3Co1/3Mn1/3] O2–$LiNiO_2$ cathode synthesized via co-precipitation for lithium secondary batteries," *Journal of Power Sources*, vol. 189, no. 1, pp. 571–575, 2009.

132. A. R. Armstrong and P. G. Bruce, "Synthesis of layered $LiMnO_2$ as an electrode for rechargeable lithium batteries," *Nature*, vol. 381, no. 6582, pp. 499–500, 1996.
133. C. Wei, J. Fan, and H. Gong, "Tungsten adsorption on La_2O_3 (001) surfaces," *Materials Letters*, vol. 161, pp. 313–316, 2015.
134. M. Nocuń and M. Handke, "Identification of Li–O absorption bands based on lithium isotope substitutions," *Journal of Molecular Structure*, vol. 596, no. 1–3, pp. 145–149, 2001.
135. A. G. Milnes, *Heterojunctions and metal semiconductor junctions*. Elsevier, 2012.
136. L. Mao, X.-Y. Cai, and M.-S. Zhu, "Hierarchically 1D CdS decorated on 2D perovskite-type $La_2Ti_2O_7$ nanosheet hybrids with enhanced photocatalytic performance," *Rare Metals*, vol. 40, no. 5, pp. 1067–1076, 2021.
137. B. C. Viana, O. P. Ferreira, A. G. Souza Filho, A. A. Hidalgo, J. Mendes Filho, and O. L. Alves, "Alkali metal intercalated titanate nanotubes: a vibrational spectroscopy study," *Vibrational Spectroscopy*, vol. 55, no. 2, pp. 183–187, 2011.
138. A. Liu *et al.*, "Solution-processed alkaline lithium oxide dielectrics for applications in n-and p-type thin-film transistors," *Advanced Electronic Materials*, vol. 2, no. 9, p. 1600140, 2016.
139. J. Baker, A. Jenkins, and R. Ward, "Electron magnetic resonance in lithium oxide from a centre containing Fe3+," *Journal of Physics: Condensed Matter*, vol. 3, no. 43, p. 8467, 1991.
140. X. Dong *et al.*, "Predicted lithium oxide compounds and superconducting low-pressure LiO 4," *Physical Review B*, vol. 100, no. 14, p. 144104, 2019.
141. E. V. Castro, M. P. López-Sancho, and M. A. Vozmediano, "Effect of pressure on the magnetism of bilayer graphene," *Physical Review B*, vol. 84, no. 7, p. 075432, 2011.
142. M. Obrovac, R. Dunlap, R. Sanderson, and J. Dahn, "The electrochemical displacement reaction of lithium with metal oxides," *Journal of the Electrochemical Society*, vol. 148, no. 6, p. A576, 2001.
143. D. Varshney and S. Shriya, "Elastic, mechanical, and thermodynamical properties of superionic lithium oxide for high pressures," *Physics and Chemistry of Minerals*, vol. 40, no. 6, pp. 521–530, 2013.
144. X. Liu and M. C. Hersam, "2D materials for quantum information science," *Nature Reviews Materials*, vol. 4, no. 10, pp. 669–684, 2019.
145. A. K. Roy, "Quantum confinement in 1D systems through an imaginary-time evolution method," *Modern Physics Letters A*, vol. 30, no. 37, p. 1550176, 2015.
146. H. Gotoh and H. Ando, "Excitonic quantum confinement effects and exciton electroabsorption in semiconductor thin quantum boxes," *Journal of Applied Physics*, vol. 82, no. 4, pp. 1667–1677, 1997.
147. N.-S. Roh, S.-D. Lee, and H.-S. Kwon, "Effects of deposition condition on the ionic conductivity and structure of amorphous lithium phosphorus oxynitrate thin film," *Scripta Materialia*, vol. 42, no. 1, 1999.
148. K. S. Siow, L. Britcher, S. Kumar, and H. J. Griesser, "XPS study of sulfur and phosphorus compounds with different oxidation states," *Sains Malaysiana*, vol. 47, no. 8, pp. 1913–1922, 2018.
149. X. Zhang *et al.*, "Structure-related electrochemical performance of organosulfur compounds for lithium–sulfur batteries," *Energy & Environmental Science*, vol. 13, no. 4, pp. 1076–1095, 2020.
150. Z.-L. Xu *et al.*, "Exceptional catalytic effects of black phosphorus quantum dots in shuttling-free lithium sulfur batteries," *Nature Communications*, vol. 9, no. 1, pp. 1–11, 2018.
151. Y. Depeursinge, "Electronic properties of the layer III–VI semiconductors. A comparative study," *Il Nuovo Cimento B (1971-1996)*, vol. 64, no. 1, pp. 111–150, 1981.

152. C. Duke, "Structure and bonding of tetrahedrally coordinated compound semiconductor cleavage faces," *Journal of Vacuum Science & Technology A: Vacuum, Surfaces, and Films*, vol. 10, no. 4, pp. 2032–2040, 1992.
153. M. C. Tamargo, *II-VI semiconductor materials and their applications*. CRC Press, 2002.
154. W. Turner, A. Fischler, and W. Reese, "Electrical and optical properties of the II–V compounds," *Journal of Applied Physics*, vol. 32, no. 10, pp. 2241–2245, 1961.
155. F. Meyer *et al.*, "Schottky barrier heights on IV-IV compound semiconductors," *Journal of Electronic Materials*, vol. 25, no. 11, pp. 1748–1753, 1996.
156. D. Xue and S. Zhang, "The role of Li-O bonds in calculations of nonlinear optical coefficients of $LiXO_3$-type complex crystals," *Philosophical Magazine B*, vol. 78, no. 1, pp. 29–36, 1998.
157. F. Zocchi, "Accurate bond valence parameters for M–O bonds (M= C, N, La, Mo, V)," *Chemical Physics Letters*, vol. 421, no. 1-3, pp. 277–280, 2006.
158. A. Gansäuer *et al.*, "Catalysis via homolytic substitutions with C– O and Ti– O bonds: oxidative additions and reductive eliminations in single electron steps," *Journal of the American Chemical Society*, vol. 131, no. 46, pp. 16989–16999, 2009.
159. D. Krishnamurthy *et al.*, "A low-spin alkylperoxo– Iron (III) complex with weak Fe–O and O–O bonds: implications for the mechanism of superoxide reductase," *Journal of the American Chemical Society*, vol. 128, no. 44, pp. 14222–14223, 2006.
160. R. M. Wood and G. J. Palenik, "Bond valence sums in coordination chemistry. A simple method for calculating the oxidation state of cobalt in complexes containing only Co– O bonds," *Inorganic Chemistry*, vol. 37, no. 16, pp. 4149–4151, 1998.
161. J. Liu *et al.*, "Ni–O Cooperation versus nickel (II) hydride in catalytic hydroboration of N-heteroarenes," *ACS Catalysis*, vol. 9, no. 5, pp. 3849–3857, 2019.
162. G. J. Palenik, "Bond valence sums in coordination chemistry using oxidation state independent R 0 values. A simple method for calculating the oxidation state of manganese in complexes containing only Mn–O bonds," *Inorganic Chemistry*, vol. 36, no. 21, pp. 4888–4890, 1997.
163. D. J. Jones, E. M. O'Leary, and T. P. O'Sullivan, "Modern synthetic approaches to phosphorus-sulfur bond formation in organophosphorus compounds," *Advanced Synthesis & Catalysis*, vol. 362, no. 14, pp. 2801–2846, 2020.
164. V. Atuchin, L. Isaenko, V. Kesler, and S. Lobanov, "Core level photoelectron spectroscopy of $LiGaS_2$ and Ga–S bonding in complex sulfides," *Journal of Alloys and Compounds*, vol. 497, no. 1–2, pp. 244–248, 2010.
165. Z. Luo *et al.*, "Modulating the electronic structures of graphene by controllable hydrogenation," *Applied Physics Letters*, vol. 97, no. 23, p. 233111, 2010.
166. P. Zhang, X. Li, C. Hu, S. Wu, and Z. Zhu, "First-principles studies of the hydrogenation effects in silicene sheets," *Physics Letters A*, vol. 376, no. 14, pp. 1230–1233, 2012.
167. A. Nijamudheen, R. Bhattacharjee, S. Choudhury, and A. Datta, "Electronic and chemical properties of germanene: the crucial role of buckling," *The Journal of Physical Chemistry C*, vol. 119, no. 7, pp. 3802–3809, 2015.
168. B. Cai, S. Zhang, Z. Hu, Y. Hu, Y. Zou, and H. Zeng, "Tinene: a two-dimensional Dirac material with a 72 meV band gap," *Physical Chemistry Chemical Physics*, vol. 17, no. 19, pp. 12634–12638, 2015.
169. H. Zhao *et al.*, "Unexpected giant-gap quantum spin Hall insulator in chemically decorated plumbene monolayer," *Scientific Reports*, vol. 6, no. 1, pp. 1–8, 2016.
170. M. Mosaferi, I. A. Sarsari, and M. Alaei, "Band structure engineering in gallium sulfide nanostructures," *Applied Physics A*, vol. 127, no. 2, pp. 1–9, 2021.
171. H. Arora and A. Erbe, "Recent progress in contact, mobility, and encapsulation engineering of InSe and GaSe," *InfoMat*, vol. 3, no. 6, pp. 662–693, 2021.

172. X. Xia, X. Li, and H. Wang, "Metal–insulator transition in few-layered GaTe transistors," *Journal of Semiconductors*, vol. 41, no. 7, p. 072902, 2020.
173. W. Li, S. Poncé, and F. Giustino, "Dimensional crossover in the carrier mobility of two-dimensional semiconductors: the case of InSe," *Nano Letters*, vol. 19, no. 3, pp. 1774–1781, 2019.
174. M. Posternak, A. Baldereschi, A. Freeman, E. Wimmer, and M. Weinert, "Prediction of electronic interlayer states in graphite and reinterpretation of alkali bands in graphite intercalation compounds," *Physical Review Letters*, vol. 50, no. 10, p. 761, 1983.
175. A. Panich, "Nuclear magnetic resonance study of fluorine–graphite intercalation compounds and graphite fluorides," *Synthetic Metals*, vol. 100, no. 2, pp. 169–185, 1999.
176. J. Purewal, *Hydrogen adsorption by alkali metal graphite intercalation compounds*. California Institute of Technology, 2010.
177. S.-Y. Lin, N. T. T. Tran, S.-L. Chang, W.-P. Su, and M.-F. Lin, *Structure- and adatom-enriched essential properties of graphene nanoribbons*. CRC Press, 2018.
178. S.-Y. Lin, S.-L. Chang, C.-R. Chiang, W.-B. Li, H.-Y. Liu, and M.-F. Lin, "Feature-rich geometric and electronic properties of carbon nanoscrolls," *Nanomaterials*, vol. 11, no. 6, p. 1372, 2021.
179. N. Budisa, V. Kubyshkin, and D. Schulze-Makuch, "Fluorine-rich planetary environments as possible habitats for life," *Life*, vol. 4, no. 3, pp. 374–385, 2014.
180. Y. Kiani, "Buckling of functionally graded graphene reinforced conical shells under external pressure in thermal environment," *Composites Part B: Engineering*, vol. 156, pp. 128–137, 2019.
181. R. Makkus, K. Hemmes, and J. De Wit, "A comparative study of NiO (Li), $LiFeO_2$, and $LiCoO_2$ porous cathodes for molten carbonate fuel cells," *Journal of the Electrochemical Society*, vol. 141, no. 12, p. 3429, 1994.
182. A. Yamada, S.-C. Chung, and K. Hinokuma, "Optimized LiFePO4 for lithium battery cathodes," *Journal of the Electrochemical Society*, vol. 148, no. 3, p. A224, 2001.
183. V. Galakhov, V. Karelina, D. Kellerman, V. Gorshkov, N. Ovechkina, and M. Neumann, "Electronic structure, x-ray spectra, and magnetic properties of the LiCoO2− δ and NaxCoO2 nonstoichiometric oxides," *Physics of the Solid State*, vol. 44, no. 2, pp. 266–273, 2002.
184. V. Bianchi *et al.*, "Synthesis, structural characterization and magnetic properties of quasistoichiometric LiNiO2," *Solid State Ionics*, vol. 140, no. 1–2, pp. 1–17, 2001.
185. C. M. Julien, A. Ait-Salah, A. Mauger, and F. Gendron, "Magnetic properties of lithium intercalation compounds," *Ionics*, vol. 12, no. 1, pp. 21–32, 2006.
186. R. M. Feenstra, "Scanning tunneling spectroscopy," *Surface Science*, vol. 299, pp. 965–979, 1994.
187. P. J. Hore, *Nuclear magnetic resonance*. Oxford University Press, USA, 2015.
188. I. Kruk, *Environmental toxicology and chemistry of oxygen species*. Springer Science & Business Media, 1997.
189. D. Rai, A. Shankar, M. Ghimire, and R. Thapa, "A comparative study of a Heusler alloy Co2FeGe using LSDA and LSDA+ U," *Physica B: Condensed Matter*, vol. 407, no. 18, pp. 3689–3693, 2012.
190. L. Yang, M. L. Cohen, and S. G. Louie, "Excitonic effects in the optical spectra of graphene nanoribbons," *Nano letters*, vol. 7, no. 10, pp. 3112–3115, 2007.

191. J.-H. Ho, C. Lu, C. Hwang, C. Chang, and M.-F. Lin, "Coulomb excitations in AA-and AB-stacked bilayer graphites," *Physical Review B*, vol. 74, no. 8, p. 085406, 2006.
192. D. Thouless, "Bandwidths for a quasiperiodic tight-binding model," *Physical Review B*, vol. 28, no. 8, p. 4272, 1983.
193. Y. Hancock, A. Uppstu, K. Saloriutta, A. Harju, and M. J. Puska, "Generalized tight-binding transport model for graphene nanoribbon-based systems," *Physical Review B*, vol. 81, no. 24, p. 245402, 2010.

6 Electronic Properties of Li_2S-Si Heterojunction

Nguyen Thi Han and Vo Khuong Dien
Department of Physics, National Cheng Kung University, Tainan City, Taiwan

Ming-Fa Lin
Department of Physics and Hierarchical Green-Energy Material (Hi-GEM) Research Center, National Cheng Kung University, Tainan City, Taiwan

CONTENTS

6.1 INTRODUCTION AND MOTIVATION

Rechargeable batteries have attracted a great deal of attention presently due to their application in portable devices, electric vehicles, and electricity storing from solar and wind energy[1–3]. Generally, commercial rechargeable batteries include the positive electrode (anode) and the negative electrode (cathode) and are separated by an electrolyte to prevent the inner circuits[3–5]. Wherein each component should have good physical, chemical, and material properties, especially for the contact electrode/electrolyte region to ensure the rapid lithium-ion transmission[6–8]. Apparently, a drastic change of geometric structures is revealed in the cathode/electrolyte/anode materials[9,10]. The structural transformations between two-metastable configurations during the battery operation are rather complex and strongly related to the lithium transportations[11–13].

In general, the cathode and anode systems of Li^+-based batteries belong to a class of solid-state materials[14,15], such as the three-dimensional (3D) ternary transition metal oxides $LiCoO_2$/graphite[16] and $Li_4Ti_5O_{12}$/graphite[17] compounds, respectively. On the other hand, the candidate for the solid electrolytes could be

DOI: 10.1201/9781003367215-6

named as Li_3OCl, Li_2SiO_3, Li_2GeO_3, and Li_2S[15,18–26]. Among them, the latter one has attracted significant research interest due to the following reasons: such nanoparticles present very high discharge capacities up to 1,360 $mAhg^{-1}$ at a 0.1C rate[27], showing an economically viable and scalable reaction routine for future lithium-sulfur batteries. Furthermore, lithium sulfide is also can be used as cathode material since it can be coupled to Li-free anodes, such as graphite, Si, or Sn[28].

The interfacial resistance problems originated from the cathode-/anode-electrolyte interfaces[29–32]. Such resistances create some crucial difficulties for practical applications as the low current drains and the low power density[33,34]. Previous theoretical predictions have been successfully investigating the geometric, electronic, and transport properties of cathode-/anode-electrolyte interfaces by first-principles calculations[35–37]. However, the critical mechanisms are absent; that is to say, the physical, chemical, and material pictures have been not achieved yet. For example, systematic investigations into the interaction of the chemical bonds of the interfaces, especially, the multi-orbital hybridization that is related to the essential properties of the anode-/cathode-/electrolyte- compound[38] is absent in investigations up to now.

The previous publication states that the combinations of mesoporous carbon cathode[39–41], the Li_2S electrolyte[42], and silicon nanowire anode[43–45] is an ideal for the batteries systems due to their high capacity, low reaction potential, and long circle lifetime[46–48]. Inspired by this research, in this work, the geometric symmetries and electronic properties of the 3D heterojunction Si-Li_2S compound (the anode-electrolyte composite of Li+-based batteries) are systematically investigated. The state-of-the-art analysis conducted on the various chemical bonds in a large Moire cell, the band structure with atomic domination, the atom-/orbital projected density of states (DOS) and the spatial charge density is capable of providing the critical multi-orbital hybridizations. The spin density distribution, the spin-degenerate/spin-split energy bands around the low-energy regions, and the net magnetic moment also be examined in detail as to whether the magnetism could exist in this compound. The theoretical predictions on the relaxation structure, the valence states, the whole energy spectrum, and the band gap could be examined via Powder X-ray Diffraction[49] (PXRD)/Tunneling Electron Microscopy (TEM)[50]/Scanning Electron Microscopy (SEM)[51]/Scanning Tunneling Microscopy (STM)[52], Angle-Resolved Photo Emission Spectroscopy (ARPECT)[53], Scanning Tunneling Spectroscopy (STS)[54] and optical absorption spectra, respectively. The present work provides more perceptive insights into the understanding of the diversified chemical bonding, as well as the electronic properties of Li_2S-Si heterostructure for the future promising anode-electrolytes of LIBs.

6.2 COMPUTATIONAL DETAILS

We used the density functional theory method via the Vienna Ab-initio Simmulation[55,56] Package (VASP) to perform the optimization of crystal structures and the calculation of the electronic properties. The Perdew-Burke-Ernzerhof (PBE) generalized gradient approximation was used for the exchange-correlation functional[57]. The interaction between the valence electrons and ions core was evaluated by the projector augmented wave (PAW) method[58]. The cutoff energy for the expansion of the plane wave basis set is 500 eV for all calculations[59].

The Brillouin zone was integrated with a special k-point mesh of 9 × 9 × 9 and 21 × 21 × 21 in the Monkhorst-Pack sampling technique for the geometric relaxation and electronic calculation, respectively. The convergence condition of the ground state is set to be 10^{-6} eV between two consecutive simulation steps, and all atoms were allowed to fully relax during the geometry optimization until the Hellmann-Feynman force acting on each atom was smaller than 0.01 eV/Å[60]. Spin-polarized calculations were performed for the geometry optimization and the calculation of the band structure. In the k-point sample, the cutoff energy has been checked for convergence of the calculations.

6.3 RESULTS AND DISCUSSIONS

6.3.1 Geometric Structure

In this work, we have chosen meta-stable configurations, which are the cubic Li_2S (Figure 6.1(a)) and cubic Silicon (Figure 6.1(b)) to form the heterostructure. The geometrics of isolated structures before constructing the heterojunction are shown in Figure 6.1(c). After an optimal relaxation of the first-principles simulations, 3D binary lithium-sulfur compound presents the cubic phase with unusual crystal symmetry. The unit cell of this cubic system presents identical lengths of three perpendicular lattice vectors along x-, y-, and z-directions. There is only one kind of Li-S chemical bonding that survives in the unit cell with identical lengths of 2.46 Å. The current calculation presents the lattice constant a=b=c of 5.710 Å, being fully consistent with the high-precision X-ray measurements of 5.72 Å and the previous theoretical predictions. Similarly, the silicon also presents a diamond structure; the optimal geometric structure presents a lattice constant of 5.47 Å and a Si-Si

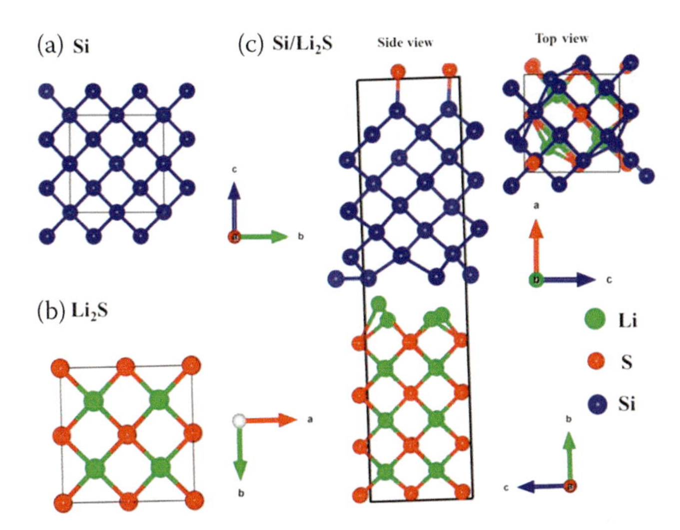

FIGURE 6.1 The optimal lattice structure: (a) Si 3D compound with cubic symmetry with 8 atoms in a unit cell (b) Li_2S 3D compound with cubic symmetry with 12 atoms in a unit cell (c) Li_2S-Si heterojunction with 40 atoms in unit cell.

chemical bond length of 2.37 Å. The lattice mismatch of Silicon and Li_2S is about 0.95 at the (100) surface and thus, easy to form the heterojunctions. By comparing their lattice mismatch along different crystallographic orientations, we found that the mismatch is quite small between the Li_2S (100) surface and the Si (100) surface so the interface models are constructed between these two surfaces.

After contact to form the heterostructure, the geometric structure of Li_2S-Silicon compound presents a very unusual geometric structure, a large Moire super-lattice with a lattice constant of a=c= 5.58 Å and b= 22.38 Å. This large unit cell contains 40 atoms (12- Li, 8- S, and 16- Si) (Table 6.2). A non-uniform environment includes many chemical bonds in the unit cell. Apart from the isolated Li_2S and Si systems, there are service three kinds of chemical bonds in the heterostructure, Si-Si, Li-S, and Si-S at the anode and the electrolyte and at the interface (Table 6.1).

TABLE 6.1
Chemical Bonding in Li_2S-Si Heterojunction

Number	Atom1	Atom2	Length
1	Li1	S1	2.4163
2	Li1	S6	2.3506
3	Li1	S3	2.4157
4	Li1	S7	2.4107
5	Li3	S1	2.4257
6	Li3	S3	2.395
7	Li3	S6	2.395
8	Li3	S7	2.3745
9	Li4	S1	2.3911
10	Li4	S2	2.3686
11	Li4	S5	2.3686
12	Li4	S7	2.4755
13	Li6	S1	2.3809
14	Li6	S2	2.4374
15	Li6	S5	2.4439
16	Li6	S7	2.3589
17	Li7	S3	2.452
18	Li7	S4	2.2998
19	Li7	S6	2.372
20	Li7	S8	2.3009
21	Li8	S1	2.4163
22	Li8	S3	2.3506
23	Li8	S6	2.4157
24	Li8	S7	2.4107
25	Li9	S2	2.2887
26	Li9	Si12	2.4719
27	Li9	S5	2.3324

(Continued)

TABLE 6.1 (Continued)
Chemical Bonding in Li_2S-Si Heterojunction

Number	Atom1	Atom2	Length
28	Li10	S1	2.386
29	Li10	S3	2.391
30	Li10	S6	2.391
31	Li10	S7	2.4289
32	Li11	S1	2.4208
33	Li11	S2	2.4166
34	Li11	S5	2.4166
35	Li11	S7	2.3776
36	Li12	S3	2.4177
37	Li12	S6	2.4177
38	Li12	S4	2.3114
39	Li12	S8	2.2798
40	Li13	S1	2.3809
41	Li13	S5	2.4374
42	Li13	S2	2.4439
43	Li13	S7	2.3589
44	Li14	S6	2.452
45	Li14	S3	2.372
46	Li14	S4	2.2998
47	Li14	S8	2.3009
48	Li15	S5	2.2887
49	Li15	Si4	2.4719
50	Li15	S2	2.3324
51	Li16	S2	2.4584
52	Li16	S5	2.4584
53	Li16	Si4	2.6233
54	Li16	Si12	2.6233
55	S2	Li4	2.3686
56	S2	Li13	2.4439
57	S2	Li15	2.3324
58	S3	Li5	2.4174
59	S3	Li1	2.4157
60	S3	Li3	2.395
61	S3	Li14	2.372
62	S4	Si5	2.0958
63	S4	Li7	2.2998
64	S4	Li12	2.3114
65	S4	Li14	2.2998
66	S4	Li5	2.285
67	S5	Li4	2.3686
68	S5	Li6	2.4439

(Continued)

TABLE 6.1 (Continued)
Chemical Bonding in Li_2S-Si Heterojunction

Number	Atom1	Atom2	Length
69	S5	Li9	2.3324
70	S6	Li3	2.395
71	S6	Li7	2.372
72	S6	Li8	2.4157
73	S6	Li5	2.4174
74	S7	Li1	2.4107
75	S7	Li3	2.3745
76	S7	Li4	2.4755
77	S7	Li6	2.3589
78	S7	Li8	2.4107
79	S7	Li13	2.3589
80	S8	Si13	2.1181
81	S8	Li7	2.3009
82	S8	Li12	2.2798
83	S8	Li14	2.3009
84	S8	Li5	2.3077
85	Si1	Si11	2.3649
86	Si1	Si16	2.3731
87	Si1	Si3	2.3633
88	Si1	Si7	2.3728
89	Si2	Si12	2.3891
90	Si2	Si15	2.3609
91	Si2	Si4	2.404
92	Si2	Si8	2.3588
93	Si3	Si6	2.3406
94	Si3	Si9	2.3649
95	Si3	Si1	2.3633
96	Si3	Si14	2.3753
97	Si4	Si10	2.3891
98	Si4	Li16	2.6233
99	Si4	Si2	2.404
100	Si4	Si12	2.5436
101	Si5	Si7	2.3083
102	Si5	Si16	2.2963
103	Si6	Si8	2.3177
104	Si6	Si11	2.3406
105	Si6	Si15	2.3284
106	Si7	Si1	2.3728
107	Si7	Si9	2.3728
108	Si7	Si13	2.2923
109	Si8	Si2	2.3588

(Continued)

TABLE 6.1 (Continued)
Chemical Bonding in Li_2S-Si Heterojunction

Number	Atom1	Atom2	Length
110	Si8	Si10	2.3588
111	Si8	Si14	2.3915
112	Si9	Si16	2.3731
113	Si9	Si7	2.3728
114	Si9	Si11	2.3633
115	Si10	Si15	2.3609
116	Si10	Si8	2.3588
117	Si10	Si12	2.404
118	Si11	Si9	2.3633
119	Si11	Si14	2.3753
120	Si12	Li16	2.6233
121	Si12	Si4	2.5436
122	Si12	Si10	2.404
123	Si13	Si16	2.3059
124	Si13	Si7	2.2923
125	Si14	Si15	2.3835
126	Si14	Si3	2.3753
127	Si14	Si8	2.3915
128	Si14	Si11	2.3753
129	Li5	S4	2.285
130	Li5	S6	2.4174
131	Li5	S8	2.3077
132	Li3	S7	2.3745
133	Li3	S6	2.395
134	Li3	S7	2.3745
135	Li4	S7	2.4755
136	Li4	S7	2.4755
137	Li7	S8	2.3009
138	Li9	Si12	2.4719
139	Li14	S4	2.2998
140	Li14	S8	2.3009
141	Li15	Si4	2.4719
142	Li16	S2	2.4584
143	Li16	Si12	2.6233
144	Li16	S5	2.4584
145	Li16	Si4	2.6233
146	S8	Li5	2.3077
147	Si4	Si12	2.5436
148	Si8	Si14	2.3915
149	Si8	Si14	2.3915

TABLE 6.2
No. of Bonds, Bond Length, Lattice Constant of Li_2S-Si Heterojunction

Systems (Cubic)	No. of Bonds	Unit Cell	Bond Length ()	Lattice Constant ()
Si	Si-Si (25)	8 atoms	2.37	a=b=c=5.47
Li_2S	Li-S (62)	12 atoms	2.48	a=b=c5.72
Si/Li2S	Si-Si	40 atoms	2.29–2.54	a=c= 5.58; b= 22.38
	Li-S		2.28–2.42	
	Si-S		2.09–2.11	

The chemical fluctuations of the first, the second, and the last one are in the range of 2.29 Å –2.54 Å, 2.28 Å –2.42 Å, and 2.09 Å –2.11 Å, respectively. Many chemical bonds with serious fluctuations are originated from multi-orbital hybridizations, which are directly reflect to the unusual electronic band structure, the spatial charge density and the merge of decompound density of states (discussed later). Most importantly, the strong fluctuation of bond lengths at the interface, which generates an extremely non-uniform environment, is expected to create interfacial resistance and thus, strongly related to the battery's efficiency.

The theoretical predictions in this chapter, including the optimal geometric of the isolated structures and the chemical modifications of the materials in the battery heterojunctions, are vital important under the current research. On the experimental expects, the X-ray Powder Diffractions (XRD) is a powerful technique to measure the lattice constant, phase, and position-dependent chemical interactions in three-dimensional materials, and thus, suitable to verify the geometric structure of the Li_2S-Si heterostructure. The delicate combinations of experimental measurements and theoretical predictions are very useful in elucidating the critical mechanisms of ion migrations in the battery. In addition, high-tech measurements such as STM and TEM[61,62] are useful in such respect.

6.3.2 Electronic Properties

The Li_2S-Si compound, as shown in Figure 6.2(c), presents unusual band structures. The rich and unique electronic properties of pure Silicon (Figure 6.2(a)) cover an indirect band gap of 0.68 eV with the highest valence band located at a difference wavevector with the lowest conduction band, the high asymmetry of the occupied hole and unoccupied electron spectra about the Fermi level [indication of multi-orbital chemical bonds], the anisotropic and oscillatory dependences[63,64], the parabolic, linear and partial flat energy dispersions near the high-symmetry points, a lot of band-edge states, the critical points in energy-wave-vector space, the crossings/non-crossings/anti-crossing behaviors, and the multi-degenerate states at the high-symmetry points. The purely Li_2S compound (Figure 6.2(b)) also presents the diamond structure and thus, processes the same the first Brillouin zone. The energy spectra present at least nine energy sub-bands with corresponding to the numbers of

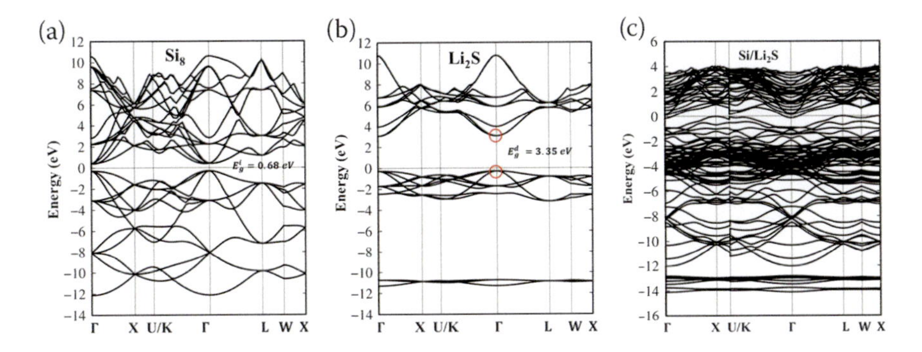

FIGURE 6.2 (a)/(b) The significant valence and conduction band of Si_8 and Li_2S, respectively along the high-symmetry points within the first Brillouin zone for the energy range (−14.0 eV, $E^{c,v}$, 12.0 eV), (c) The significant valence and conduction band of Li_2S-Si heterojunction along the high-symmetry points within the first Brillouin zone for the energy range (−16.0 eV, $E^{c,v}$, 6.0 eV).

Li, S atoms and their orbitals. The energy-dependent wave vector space mostly presents dispersionless dispersions due to the weak and important Li-S orbital hybridizations. The presence of a large direct band gap of 3.35 eV at the center of the first Brillouin zone, and thus, an insulator Li_2S can be used as a candidate electrolyte material.

The electronic properties of Li_2S and Si structure are changed dramatically through the extremely strong chemical modification, as shown in Figure 6.2(a)-(b), compared to Silicon and Li_2S compounds, the Li_2S-Si heterostructure presents more valence and conduction energy subbands because of the many Li, S and Si atoms in the Moiré super-lattice. That is to say, there exists a very complicated electronic structure. The asymmetric spectra of the electron and hole energy subbands around the Fermi level are enhanced greatly by the chemical modification at the Si, Li_2S, and the interface region. Many band-edge states, e.g., the extreme and saddle points in energy–wave-vector space, are created in the presence of heterojunctions. Very interestingly, for the low-lying energy spectrum, the unoccupied states, which exhibit the non-uniform energy dispersions along the high-symmetry directions, cross E_F. Such an unusual result indicates obvious metallic properties in the presence of n-type free carrier density.

In addition to the electronic energy band structure, the electronic wave functions and atom distributions, as present in Figure 6.3, can provide partial information about the multi-orbital hybridizations in the heterojunctions. The domination of the Li_2S portion is denoted by the blue circles while the red one belongs to Si's distribution. In general, the contribution to the electronic wave functions relates to the number of atoms and orbitals, the electronic weight of both Li_2 S and Si are the same. Very interestingly, the distribution of the former and the latter are almost merged with each other, illustrated by the complicated multi-orbital hybridizations. The full information about the orbital hybridizations, the physical/chemical mechanism will be elucidated in the spatial charge density distributions and the atom-/orbital-projected density of states. Wave-vector dependence of the occupied

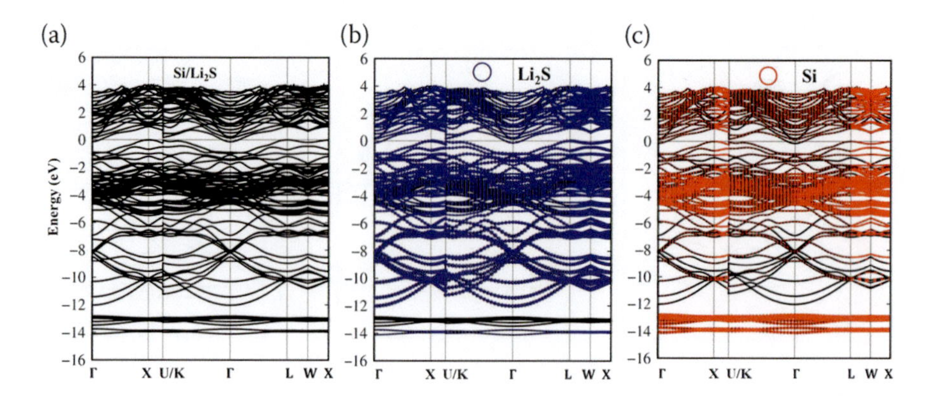

FIGURE 6.3 (a) The significant valence and conduction band of the Li_2S-Si heterojunction along the high-symmetry points within the first Brillouin zone for the energy range (−16.0 eV, $E^{c,v}$, 6.0 eV), with the specific (b) Li_2S and (c) Si dominances (blue, and red balls, respectively).

electronic states in Li_2S-Si heterostructure could be exanimated by angle-resolved photoemission spectroscopy [ARPES]. Such measurements are very appreciated for 2D layered materials. For example, the linear and isotropic Dirac cone in graphene monolayer[65]. The Mexican hat is sharp in monolayer group III-VI semiconductor compounds[66–68]. The observed change in the location of the VBM in monolayer MoS_2 provides support for the indirect-to-direct band gap transition in going from few-layer/bulk to monolayer MoS_2[69,70]. However, the ARPES measurement survives some intrinsic limits due to the non-conservation of the perpendicular momentum transfers during the photo-electron scattering process. How to overcome this measurement uncertainty has become an open issue, especially for the mainstream materials with Moire superlattices as battery materials.

The active orbital hybridizations could be comprehended through the charge density difference and the spatial charge distributions. The former could be achieved via the formula: $\Delta\rho_{hetero} = \rho_{hetero} - \rho_{Li2S} - \rho_{Si}$. In which, ρ_{hetero}, ρ_{Li2S}, and ρ_{Si} are the charge density of the optimized Li_2S/Si heterostructure, the isolated Li_2S system, and the isolated Si system, respectively. Figure 6.4 shows that the charge depletion is mainly distributed around the Li_2S crystal, while the accumulation is mostly dispersed on the Si system. The heterogeneous charge density difference reveals that the Si system interacts with the Li_2S system via charge redistribution. The spatial charge density could provide a partly information about the orbital hybridization in the chemical bonding. As for Si-Si and Li-S chemical bonding, they present a bit change after the formation of the heterojunction in Figure 6.5. The orbital hybridizations of the former and the latter mainly come from the (3s, $3p_x$, $3p_y$, $3p_z$) and (2s) – (3s, $3p_x$, $3p_y$, $3p_z$) orbitals, respectively. By the shape contract, Si-S chemical bonds exhibit a large fluctuation and are consistent with the atom- and orbital-decomposed van Hove singularities (see Figures 6.6 and 6.7).

The electronic density of states (DOS) is defined as the electronic states in the ω to $\omega + d\omega$ frequency range. The density of states of 3D materials, on the other hand, is calculated as the integration of inverse group velocity on the surface. The van

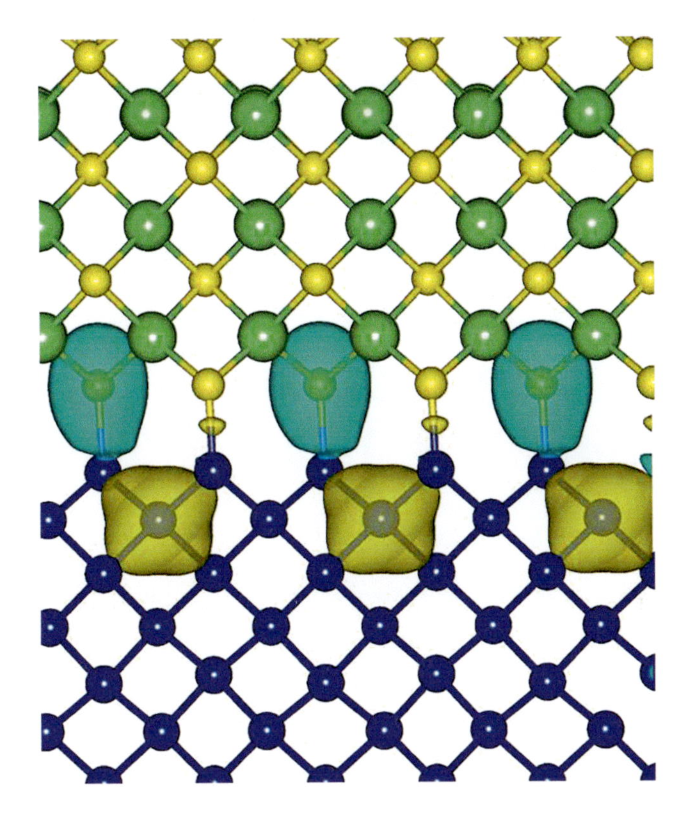

FIGURE 6.4 The charge density diference $\Delta\rho(r)$ at an isosurface value of 2×10^{-4} eÅ^{-3}. Cyan and yellow colors denote charge depletion and accumulation, respectively.

Hove singularities, which are strong/prominent peaks in the density of states, are closely related to energy dispersions and material dimensions. As shown in Figure 6.6, the atom- and orbital-projected density of states are capable of detecting multi-orbital hybridizations in Si-Si and Li-S bonds of Silicon and Li_2S crystals. The Si and Li, S atoms contribute significantly to the density of states for the whole energy spectrum [Figure 6.7]. This further reflects the fact that the Si-Si and Li-S bonds survive in the unit cell of Silicon and Li_2S. Concerning the Li-, S- orbital-projected density of states [Figures 6.7], the significant structures are clearly attributable to the merged van Hove singularities at the distinct ranges, which are associated with the on-site Coulomb potential energies [atomic orbital ionization energies]. The Li-2s and S-(3s, 3px, 3py, 3pz)/Si-(3s, $3p_x$, $3p_y$, $3p_z$), respectively, dominate in the whole energy ranges and thus, evident sp^3 hybridizations in Li_2S and Silicon crystals.

The density of states of Li_2S/Silicon heterostructure, on the other hand, presents a significant change. The presence of many van Hove singularities with a variety of shapes since they originated from a lot of energy sub-bands with various energy dispersions, the non-vanishing of the electronic states around the Fermi level further support the n-type metallic properties. The merge of van Hove singularities of

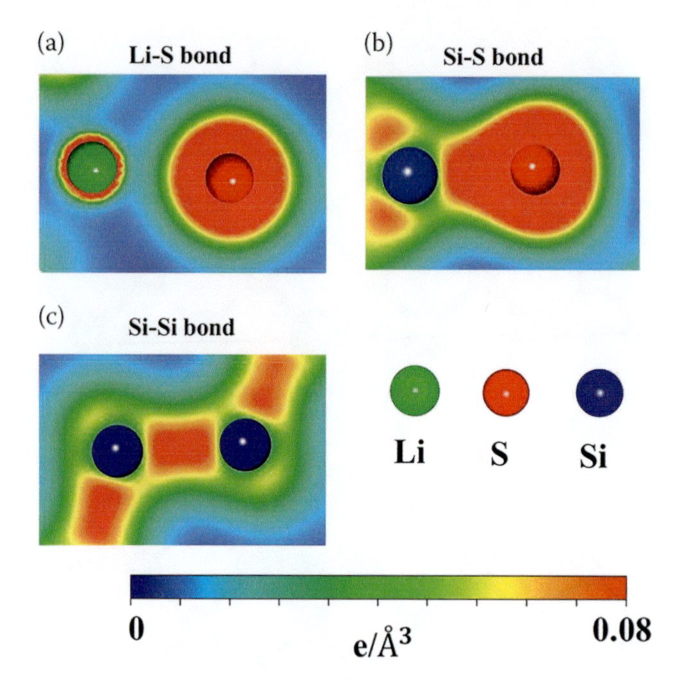

FIGURE 6.5 The spatial charge density distribution for (a) Li–S, (b) Si–S, and (c) Si-Si bonds.

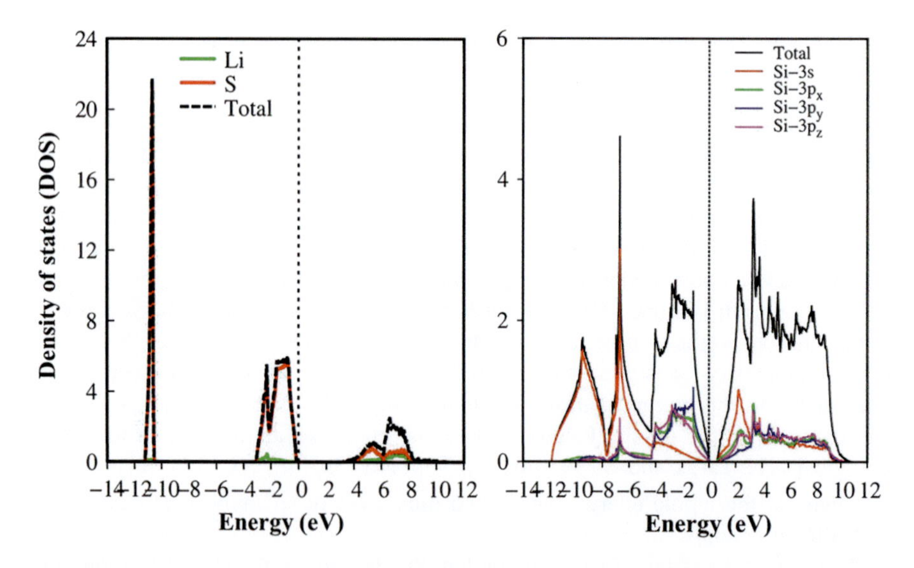

FIGURE 6.6 The density of states: those coming from (a) Li, Si atoms (green and red curves), (b) Si-(3s, $3p_x$, $3p_y$, $3p_z$) orbitals (red, green, blue and pink curves).

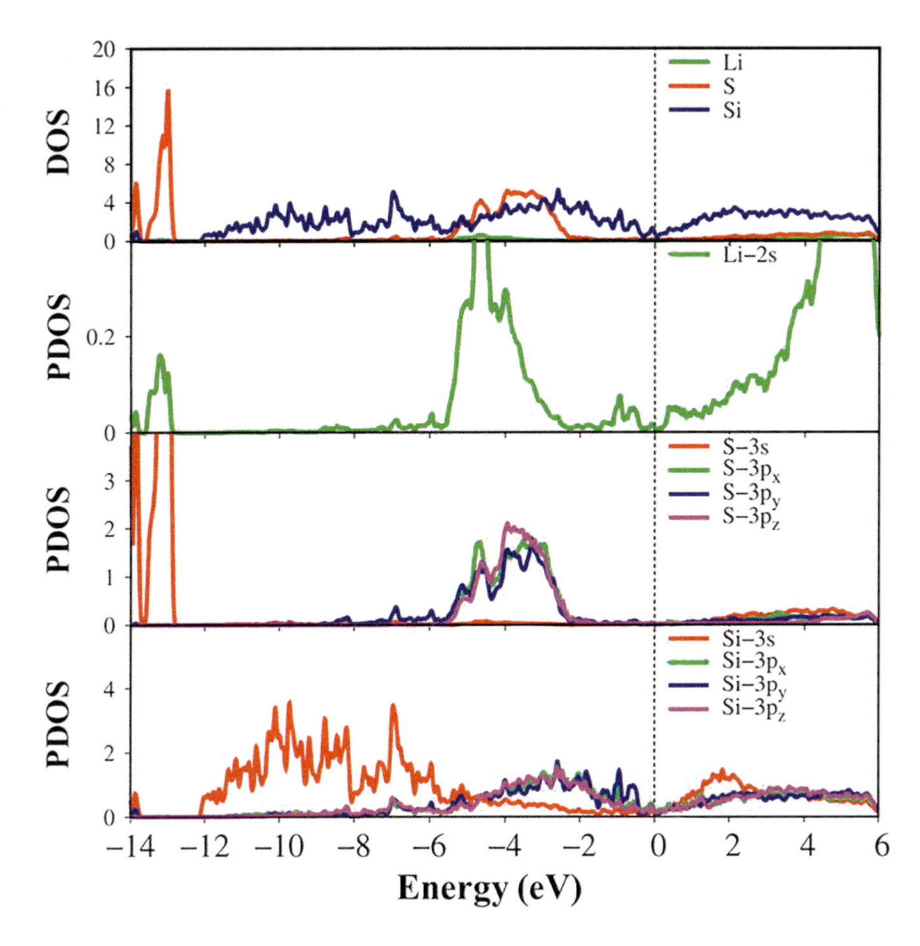

FIGURE 6.7 The atom- and orbital-projected density of states: those coming from (a) Li, S, and Si atoms (green, red and blue curves), (b) Li-2s orbitals (green curve), (c) S-(3s, $3p_x$, $3p_y$, $3p_z$) orbitals (red, green, blue and pink curves), and (d) Si-(3s, $3p_x$, $3p_y$, $3p_z$) orbitals (red, green, blue and pink curves).

different atoms and orbitals is evident in the survival of Li-S, Si-Si, and S-Si chemical bondings in the heterostructure. These critical quasiparticle behaviors are expected to determine the concise physical pictures of the other essential properties, e.g., the threshold transition frequency and prominent absorption structures closely related to the specific orbital hybridizations. Direct studies of several van Hove singularities for low-dimensional materials could be carried out using high-resolution measurements. Scanning tunneling spectroscopy, an extension of scanning tunneling microscopy[71,72], could be used to detect the form, intensity, and energy position of special van Hove singularities. Up-to-now, this high-tech measurement can be well applied for low-dimensional carbon-related systems. For example, 1D carbon nanotubes and GNRs present the square-root relation van Hove singularities[73,74], asymmetric V-shape structure for monolayer graphene[75] and finite

DOS at Femi level in the AB graphite stacking[76] have all been investigated. The main characteristics of the density of states of Li_2S-Si as the prominence van Hove singularities, the non-vanishing of the density of states around the Fermi level, the asymmetry of the occupied and unoccupied states, and so on have not been carried out yet due to the intrinsic limitations.

6.4 CONCLUSIONS AND PERSPECTIVES

In this chapter, using first-principles calculations, the optimal geometric structure and the electronic properties of 3D Li_2S-Si heterostructure have been achieved. The physical and chemical pictures are created on the various chemical bonds in a huge Moire superlattice, the band structure with complicated energy dispersions and its wave functions, the van Hove singularity in density of states (DOS), and the spatial charge density is competent of confirming the critical multi-orbital hybridizations. The intrinsic Li_2S and Silicon present a direct band gap of 3.35 eV and an indirect band gap of 0.68 eV, respectively. However, the Li_2S-Si heterostructure exhibits the n-type metallic behavior. This property mainly originated from the strong chemical modifications at the electrode-electrolyte interfaces. The rich and diversified orbital hybridizations in the chemical bonds are achieved from the spatial charge density fluctuations/charge density differences and atoms-/orbitals-decompound density of states. Our work provides new combinations which are useful in lithium batteries as well as theoretical basic sciences.

ACKNOWLEDGMENTS

This work was financially supported by the Hierarchical Green-Energy Materials (Hi-GEM) Research Center, from The Featured Areas Research Center Program within the framework of the Higher Education Sprout Project by the Ministry of Education (MOE) and the Ministry of Science and Technology (MOST 110–2634-F-006 –017) in Taiwan.

Disclosures: The authors declare no conflicts of interest.

REFERENCES

1. Abruna, H. D., Kiya, Y., Henderson, J. C., Batteries and electrochemical capacitors. *Physics Today* 2008, *61* (12), 43–47.
2. Choi, N. S., Chen, Z., Freunberger, S. A., Ji, X., Sun, Y. K., Amine, K., Yushin, G., Nazar, L. F., Cho, J., Bruce, P. G., Challenges facing lithium batteries and electrical double-layer capacitors. *Angewandte Chemie International Edition* 2012, *51* (40), 9994–10024.
3. Cheng, F., Liang, J., Tao, Z., Chen, J., Functional materials for rechargeable batteries. *Advanced Materials* 2011, *23* (15), 1695–1715.
4. Luo, W., Cheng, S., Wu, M., Zhang, X., Yang, D., Rui, X., A review of advanced separators for rechargeable batteries. *Journal of Power Sources* 2021, *509*, 230372.
5. Wen, J., Yu, Y., Chen, C., A review on lithium-ion batteries safety issues: existing problems and possible solutions. *Materials Express* 2012, *2* (3), 197–212.

6. Zhang, X., Ju, Z., Zhu, Y., Takeuchi, K. J., Takeuchi, E. S., Marschilok, A. C., Yu, G., Multiscale understanding and architecture design of high energy/power lithium-ion battery electrodes. *Advanced Energy Materials* 2021, *11* (2), 2000808.
7. Zhang, T., Ran, F., Design Strategies of 3D carbon-based electrodes for charge/ion transport in lithium ion battery and sodium ion battery. *Advanced Functional Materials* 2021, *31* (17), 2010041.
8. Shi, Y., Zhou, X., Yu, G., Material and structural design of novel binder systems for high-energy, high-power lithium-ion batteries. *Accounts of Chemical Research* 2017, *50* (11), 2642–2652.
9. Liu, C., Li, H., Kong, X., Zhao, J., Modeling analysis of the effect of battery design on internal short circuit hazard in $LiNi_{0.8}Co_{0.1}Mn_{0.1}O_2$/SiOx-graphite lithium ion batteries. *International Journal of Heat and Mass Transfer* 2020, *153*, 119590.
10. Nowroozi, M. A., Wissel, K., Donzelli, M., Hosseinpourkahvaz, N., Plana-Ruiz, S., Kolb, U., Schoch, R., Bauer, M., Malik, A. M., Rohrer, J., High cycle life all-solid-state fluoride ion battery with La_2NiO_{4+d} high voltage cathode. *Communications Materials* 2020, *1* (1), 1–16.
11. Chernova, N. A., Roppolo, M., Dillon, A. C., Whittingham, M. S., Layered vanadium and molybdenum oxides: batteries and electrochromics. *Journal of Materials Chemistry* 2009, *19* (17), 2526–2552.
12. BO, D., High Performance Nanostructured Phospho-olivine Cathodes for Lithium-ion Batteries. 2014.
13. Horrocks, G. A. Phase Transitions in Binary and Ternary Vanadium Oxides: Implications for Thermochromic and Intercalation Batteries. 2017.
14. Lin, M.-F., Hsu, W.-D., Huang, J.-L., *Lithium-ion Batteries and Solar Cells: Physical, Chemical, and Materials Properties*. CRC Press: 2021.
15. Khuong Dien, V., Thi Han, N., Nguyen, T. D. H., Huynh, T. M. D., Pham, H. D., Lin, M.-F., Geometric and electronic properties of Li_2GeO_3. *Frontiers in Materials* 2020, 288.
16. Diao, W., Saxena, S., Pecht, M., Accelerated cycle life testing and capacity degradation modeling of $LiCoO_2$-graphite cells. *Journal of Power Sources* 2019, *435*, 226830.
17. He, Z.-Q., Xiong, L.-Z., Liang, K., Wet method preparation and electrochemical characterization of $Li_4Ti_5O_{12}$/graphite anode material. *Journal of Jishou University (Natural Sciences Edition)* 2010, *31* (1), 91.
18. Han, N. T., Dien, V. K., Tran, N. T. T., Nguyen, D. K., Su, W.-P., Lin, M.-F., First-principles studies of electronic properties in lithium metasilicate (Li_2SiO_3). *RSC Advances* 2020, *10* (41), 24721–24729.
19. Han, N. T., Dien, V. K., Lin, M.-F., Excitonic effects in the optical spectra of lithium metasilicate (Li_2SiO_3). *arXiv preprint arXiv:2010.11621* 2020.
20. Thi, H. N., Tran, N. T. T., Dien, V. K., Nguyen, D. K., Lin, M.-F., Diversified Properties in 3D Ternary Oxide Compound: Li_2SiO_3. In *Lithium-Ion Batteries and Solar Cells*, CRC Press: 2021; pp. 79–101.
21. Cai, K., Song, M.-K., Cairns, E. J., Zhang, Y., Nanostructured Li_2S–C composites as cathode material for high-energy lithium/sulfur batteries. *Nano Letters* 2012, *12* (12), 6474–6479.
22. Kondo, S., Takada, K., Yamamura, Y., New lithium ion conductors based on Li_2S-SiS_2 system. *Solid State Ionics* 1992, *53*, 1183–1186.
23. Lu, Z., Chen, C., Baiyee, Z. M., Chen, X., Niu, C., Ciucci, F., Defect chemistry and lithium transport in Li_3 OCl anti-perovskite superionic conductors. *Physical Chemistry Chemical Physics* 2015, *17* (48), 32547–32555.
24. Chen, M.-H., Emly, A., Van der Ven, A., Anharmonicity and phase stability of antiperovskite Li_3 OCl. *Physical Review B* 2015, *91* (21), 214306.

25. Dien, V. K., Pham, H. D., Tran, N. T. T., Han, N. T., Huynh, T. M. D., Nguyen, T. D. H., Fa-Lin, M., Orbital-hybridization-created optical excitations in Li_2GeO_3. *Scientific Reports* 2021, *11* (1), 1–10.
26. Han, N. T., Dien, V. K., Lin, S.-Y., Chung, H.-C., Li, W.-B., Tran, N. T. T., Liu, H.-Y., Pham, H. D., Lin, M.-F., Comprehensive understanding of electronic and optical properties of Li_2SiO_3 compound. 1998.
27. Kohl, M., Brückner, J., Bauer, I., Althues, H., Kaskel, S., Synthesis of highly electrochemically active Li 2 S nanoparticles for lithium–sulfur-batteries. *Journal of Materials Chemistry A* 2015, *3* (31), 16307–16312.
28. Kaiser, M. R., Han, Z., Liang, J., Dou, S.-X., Wang, J., Lithium sulfide-based cathode for lithium-ion/sulfur battery: recent progress and challenges. *Energy Storage Materials* 2019, *19*, 1–15.
29. Maleki Kheimeh Sari, H., Li, X., Controllable cathode–electrolyte interface of Li $[Ni_{0.8}Co_{0.1}Mn_{0.1}]$ O_2 for lithium ion batteries: a review. *Advanced Energy Materials* 2019, *9* (39), 1901597.
30. Yan, C., Xu, R., Xiao, Y., Ding, J. F., Xu, L., Li, B. Q., Huang, J. Q., Toward critical electrode/electrolyte interfaces in rechargeable batteries. *Advanced Functional Materials* 2020, *30* (23), 1909887.
31. Hu, D., Chen, G., Tian, J., Li, N., Chen, L., Su, Y., Song, T., Lu, Y., Cao, D., Chen, S., Unrevealing the effects of low temperature on cycling life of 21700-type cylindrical Li-ion batteries. *Journal of Energy Chemistry* 2021, *60*, 104–110.
32. Wang, M., Yang, H., Wang, K., Chen, S., Ci, H., Shi, L., Shan, J., Xu, S., Wu, Q., Wang, C., Quantitative analyses of the interfacial properties of current collectors at the mesoscopic level in lithium ion batteries by using hierarchical graphene. *Nano Letters* 2020, *20* (3), 2175–2182.
33. Fei, C., Lee, F. C., Li, Q., High-efficiency high-power-density LLC converter with an integrated planar matrix transformer for high-output current applications. *IEEE Transactions on Industrial Electronics* 2017, *64* (11), 9072–9082.
34. Chen, K. J., Häberlen, O., Lidow, A., Lin Tsai, C., Ueda, T., Uemoto, Y., Wu, Y., GaN-on-Si power technology: Devices and applications. *IEEE Transactions on Electron Devices* 2017, *64* (3), 779–795.
35. Zhang, P. *First-Principles Study on the Mechanical Properties of Lithiated Sn Anode Materials for Li-Ion Batteries*. Curtin University, 2019.
36. Lin, S.-Y., Liu, H.-Y., Tsai, S.-J., Lin, M.-F., 7 Geometric and Electronic Properties of Li Battery Cathode. *Lithium-Ion Batteries and Solar Cells: Physical, Chemical, and Materials Properties* 2021, 117.
37. Ling, C., A review of the recent progress in battery informatics. *npj Computational Materials* 2022, *8* (1), 1–22.
38. Li, X., Liu, J., He, J., Wang, H., Qi, S., Wu, D., Huang, J., Li, F., Hu, W., Ma, J., Hexafluoroisopropyl trifluoromethanesulfonate-driven easily Li+ desolvated electrolyte to afford Li|| NCM811 cells with efficient anode/cathode electrolyte interphases. *Advanced Functional Materials* 2021, *31* (37), 2104395.
39. Wang, H., Zhang, C., Chen, Z., Liu, H. K., Guo, Z., Large-scale synthesis of ordered mesoporous carbon fiber and its application as cathode material for lithium–sulfur batteries. *Carbon* 2015, *81*, 782–787.
40. Chen, S.-R., Zhai, Y.-P., Xu, G.-L., Jiang, Y.-X., Zhao, D.-Y., Li, J.-T., Huang, L., Sun, S.-G., Ordered mesoporous carbon/sulfur nanocomposite of high performances as cathode for lithium–sulfur battery. *Electrochimica Acta* 2011, *56* (26), 9549–9555.
41. Xie, J., Yao, X., Cheng, Q., Madden, I. P., Dornath, P., Chang, C. C., Fan, W., Wang, D., Three dimensionally ordered mesoporous carbon as a stable, high-performance Li–O2 battery cathode. *Angewandte Chemie* 2015, *127* (14), 4373–4377.

42. Liu, Y., Meng, X., Wang, Z., Qiu, J., A Li2S-based all-solid-state battery with high energy and superior safety. *Science Advances* 2022, *8* (1), eabl8390.
43. Chakrapani, V., Rusli, F., Filler, M. A., Kohl, P. A., Silicon nanowire anode: Improved battery life with capacity-limited cycling. *Journal of Power Sources* 2012, *205*, 433–438.
44. Yao, Y., Liu, N., McDowell, M. T., Pasta, M., Cui, Y., Improving the cycling stability of silicon nanowire anodes with conducting polymer coatings. *Energy & Environmental Science* 2012, *5* (7), 7927–7930.
45. Xu, W., Vegunta, S. S. S., Flake, J. C., Surface-modified silicon nanowire anodes for lithium-ion batteries. *Journal of Power Sources* 2011, *196* (20), 8583–8589.
46. Kwabi, D. G., Ji, Y., Aziz, M. J., Electrolyte lifetime in aqueous organic redox flow batteries: a critical review. *Chemical Reviews* 2020, *120* (14), 6467–6489.
47. Wang, D., Bie, X., Fu, Q., Dixon, D., Bramnik, N., Hu, Y.-S., Fauth, F., Wei, Y., Ehrenberg, H., Chen, G., Sodium vanadium titanium phosphate electrode for symmetric sodium-ion batteries with high power and long lifespan. *Nature Communications* 2017, *8* (1), 1–7.
48. Kundu, D., Adams, B. D., Duffort, V., Vajargah, S. H., Nazar, L. F., A high-capacity and long-life aqueous rechargeable zinc battery using a metal oxide intercalation cathode. *Nature Energy* 2016, *1* (10), 1–8.
49. Swanson, H. E., *Standard X-ray diffraction powder patterns*. US Department of Commerce, National Bureau of Standards: 1953; Vol. 1.
50. Garnaes, J., Kragh, F., Mo/rch, K. A., Thölén, A., Transmission electron microscopy of scanning tunneling tips. *Journal of Vacuum Science & Technology A: Vacuum, Surfaces, and Films* 1990, *8* (1), 441–444.
51. Zhou, W., Apkarian, R., Wang, Z. L., Joy, D., Fundamentals of scanning electron microscopy (SEM). In *Scanning microscopy for nanotechnology*, Springer: 2006; pp. 1–40.
52. Binnig, G., Rohrer, H., Scanning tunneling microscopy. *Surface Science* 1983, *126* (1-3), 236–244.
53. Lv, B., Qian, T., Ding, H., Angle-resolved photoemission spectroscopy and its application to topological materials. *Nature Reviews Physics* 2019, *1* (10), 609–626.
54. Hipps, K., Scanning tunneling spectroscopy (STS). In *Handbook of applied solid state spectroscopy*, Springer: 2006; pp. 305–350.
55. Sun, G., Kürti, J., Rajczy, P., Kertesz, M., Hafner, J., Kresse, G., Performance of the Vienna ab initio simulation package (VASP) in chemical applications. *Journal of Molecular Structure: THEOCHEM* 2003, *624* (1–3), 37–45.
56. Hafner, J., Kresse, G., The vienna ab-initio simulation program VASP: An efficient and versatile tool for studying the structural, dynamic, and electronic properties of materials. In *Properties of Complex Inorganic Solids*, Springer: 1997; pp. 69–82.
57. Ernzerhof, M., Scuseria, G. E., Assessment of the Perdew–Burke–Ernzerhof exchange-correlation functional. *The Journal of Chemical Physics* 1999, *110* (11), 5029–5036.
58. Blöchl, P. E., Projector augmented-wave method. *Physical Review B* 1994, *50* (24), 17953.
59. Harl, J., Kresse, G., Cohesive energy curves for noble gas solids calculated by adiabatic connection fluctuation-dissipation theory. *Physical Review B* 2008, *77* (4), 045136.
60. Hine, N. D., Robinson, M., Haynes, P. D., Skylaris, C.-K., Payne, M. C., Mostofi, A. A., Accurate ionic forces and geometry optimization in linear-scaling density-functional theory with local orbitals. *Physical Review B* 2011, *83* (19), 195102.
61. Hornyak, G., Sawitowski, T., Schmid, G., TEM, STM and AFM as tools to study clusters and colloids. *Micron* 1998, *29* (2-3), 183–190.

62. Golberg, D., Costa, P. M., Mitome, M., Bando, Y., Nanotubes in a gradient electric field as revealed by STM TEM technique. *Nano Research* 2008, *1* (2), 166–175.
63. Eithiraj, R., Jaiganesh, G., Kalpana, G., Rajagopalan, M., First-principles study of electronic structure and ground-state properties of alkali-metal sulfides–Li_2S, Na_2S, K_2S and Rb_2S. *Physica Status Solidi (b)* 2007, *244* (4), 1337–1346.
64. Yi, Z., Su, F., Huo, L., Cui, G., Zhang, C., Han, P., Dong, N., Chen, C., New insights into Li2S2/Li2S adsorption on the graphene bearing single vacancy: A DFT study. *Applied Surface Science* 2020, *503*, 144446.
65. Yang, B., Dirac cone metric and the origin of the spin connections in monolayer graphene. *Physical Review B* 2015, *91* (24), 241403.
66. Das, P., Wickramaratne, D., Debnath, B., Yin, G., Lake, R. K., Charged impurity scattering in two-dimensional materials with ring-shaped valence bands: GaS, GaSe, InS, and InSe. *Physical Review B* 2019, *99* (8), 085409.
67. Rybkovskiy, D. V., Osadchy, A. V., Obraztsova, E. D., Transition from parabolic to ring-shaped valence band maximum in few-layer GaS, GaSe, and InSe. *Physical Review B* 2014, *90* (23), 235302.
68. Zhou, M., Zhang, D., Yu, S., Huang, Z., Chen, Y., Yang, W., Chang, K., Spin-charge conversion in InSe bilayers. *Physical Review B* 2019, *99* (15), 155402.
69. Gan, X., Lei, D., Wong, K.-Y., Two-dimensional layered nanomaterials for visible-light-driven photocatalytic water splitting. *Materials Today Energy* 2018, *10*, 352–367.
70. Aierken, Y. *First-principles studies of novel two-dimensional materials and their physical properties*. University of Antwerp, 2017.
71. Feenstra, R. M., Scanning tunneling spectroscopy. *Surface Science* 1994, *299*, 965–979.
72. Chen, C. J., Theory of scanning tunneling spectroscopy. *Journal of Vacuum Science & Technology A: Vacuum, Surfaces, and Films* 1988, *6* (2), 319–322.
73. Olk, C. H., Heremans, J. P., Scanning tunneling spectroscopy of carbon nanotubes. *Journal of Materials Research* 1994, *9* (2), 259–262.
74. Deniz, O., Sánchez-Sánchez, C., Dumslaff, T., Feng, X., Narita, A., Müllen, K., Kharche, N., Meunier, V., Fasel, R., Ruffieux, P., Revealing the electronic structure of silicon intercalated armchair graphene nanoribbons by scanning tunneling spectroscopy. *Nano Letters* 2017, *17* (4), 2197–2203.
75. Brar, V. W., Zhang, Y., Yayon, Y., Ohta, T., McChesney, J. L., Bostwick, A., Rotenberg, E., Horn, K., Crommie, M. F., Scanning tunneling spectroscopy of inhomogeneous electronic structure in monolayer and bilayer graphene on SiC. *Applied Physics Letters* 2007, *91* (12), 122102.
76. Klusek, Z., Waqar, Z., Denisov, E., Kompaniets, T., Makarenko, I., Titkov, A., Bhatti, A., Observations of local electron states on the edges of the circular pits on hydrogen-etched graphite surface by scanning tunneling spectroscopy. *Applied Surface Science* 2000, *161* (3-4), 508–514.

7 Electronic and Magnetic Properties of $LiMnO_2$ Compound

Le Vo Phuong Thuan
Department of Physics, National Cheng Kung University, Tainan City, Taiwan

Ming-Fa Lin
Department of Physics and Hierarchical Green-Energy Material (Hi-GEM) Research Center, National Cheng Kung University, Tainan City, Taiwan

CONTENTS

7.1 INTRODUCTION

The trend of using Lithium in modern life is increasing rapidly. The main reason comes from three essential characteristics including Li batteries can operate in a wide temperature range, self-discharge ability during storage is deficient (about 5% per year), an increased rate of the high energy output due to the specific energy contribution of the Lithium atoms and the high work potential in these systems. In this framework, many metal oxide materials have been used for cathode such as V_20_5 [1–3], Li_xNiO_2 [4–7], Li_xCoO_2 [8–12], Li_xCrO_x [13,14], … among these materials, compounds synthesized from Mn atom are implemented and applied most in previous studies [15–20]. These compounds have a lot of advantages such as the following reasons: (1) manganese is inexpensive relative to other elements. (eg Ni, V, O, …); (2) bulk synthesis is significantly cheaper than synthesis at the higher temperatures required for other popular cathode materials (eg $LiNiO_2$, $LiMn_2O_4$, …); (3) the reported reversible capacity for the first few cycles is high, and capacity loss on the first cycle is usually small; (4) most of the capacity is between 3.0 and 3.8 V, which is low enough that electrolyte decomposition and cell

DOI: 10.1201/9781003367215-7

hardware corrosion problems should be minimal at high temperatures, while still being high enough to achieve competitive energy densities.

Besides, many previous reports using the X-ray and neutron diffraction a-$NaMnO_2$ method to create a $LiMnO_2$ structure with an almost ideal layered arrangement of lithium and manganese ions have shown that $LiMnO_2$ structure is about 3 to 9% manganese ions in the lithium layers [21–24]. The strong asymmetry in the $LiMnO_2$ structure compared with $LiCoO_2$ leads to high spin polarization. The experimental results clearly show that the layered $LiMnO_2$ structure transforms into a spinel-like material during rotation; however, detailed structural information has not been reported on spinel products' nature and structural features microstructure [25]. We use Convergent-beam electron diffraction methods to check for discrete phases in multiphase samples. This method can provide valuable information on interlaced and disordered structures, stack error disorder, and subtle differences in crystallographic symmetry. With the spatial resolution available with focused beam electron diffraction, it is possible to identify individual crystals with monoclinic layered $LiMnO_2$ structures and distinguish them from lithiated-spinel Li_2 $[Mn_2]$ O_4 crystal that has tetragonal symmetry.

In this chapter, we have used the theoretical investigations by using the first-principles calculations and delicate analysis to get a full understanding of the essential properties of the $LiMnO_2$ compound with monoclinic and orthorhombic phases. Our calculated results are successfully achieved, such as the optimized geometric structure with position-dependent chemical bonding, the atom-dominated energy spectrum at various energy ranges, the spatial charge density distribution due to different orbitals, and the atom- and orbital-projected density of states. In addition, to further understand the magnetic properties of the compound, we analyze the spin-degenerate or spin-split energy bands, the spin-decomposed van Hove singularities, and the atom- & orbital-induced magnetic moments.

7.2 COMPUTATIONAL DETAILS

In this chapter, we calculate the electronic and magnetic properties of $LiMnO_2$ compounds, which are based on density functional theory (DFT) [26–31] by using the Vienna Ab-initio Simulation Package (VASP) [32–35] software. To fully explore the spin-dependent magnetic configurations, GGA+U is utilized to deal with the single- and many-particle intrinsic interactions [36,37]. This simulation method principally covers the frequent electron-ion scatterings due to the atom-dependent crystal potentials. The projector-augmented wave (PAW) method [38–41] was used to characterize the valence electron and ionic interactions. The cutoff energy for planar wave base expansion was set to 520 eV. To optimize the structure, the central sampling technique integrated the Brillouin region [42] with a $20 \times 20 \times 20$ unique k-point grid. The ground state convergence condition is 10^{-8} eV between two successive simulation steps. The atoms can fully expand during the geometry optimization until the Hellmann–Feynman force [43–45] acting on each particle is less than 0.01 eV.

7.3 ELECTRONIC PROPERTIES

To evaluate the electronic properties, we optimized the geometric structure of $LiMnO_2$ by using the VASP software package [32,33]. $LiMnO_2$ has three stable phases, including monoclinic, orthorhombic, and tetragonal. However, in this work, we only focus on semiconductor compound, which belongs to monoclinic and orthorhombic phases without the tetragonal phase. Figure 7.1 presents the optimal geometric structure of $LiMnO_2$ in monoclinic and orthorhombic phases in the x, y, and z directions, respectively. Figures 7.1(a–c) depict the monoclinic $LiMnO_2$ configuration. Similarly, Figures 7.1(d–f) show the orthorhombic structure. For the $LiMnO_2$ monoclinic configuration, a unit cell has four atoms comprising one Li atom, one Mn atom, and two O atoms. Besides, the lattice constants of this structure are a = 3.1736 (Å), b = 5.1436 (Å), and c = 5.1574 (Å) for the x, y, and z directions, respectively. The tilted angles are α = 92.657°, β = 85.666°, and γ = 96.072°. The corresponding space group symmetries are Herman Mauguin (Cm [8]), Hall (C $\bar{2}$y), and Point Group 1. The orthorhombic

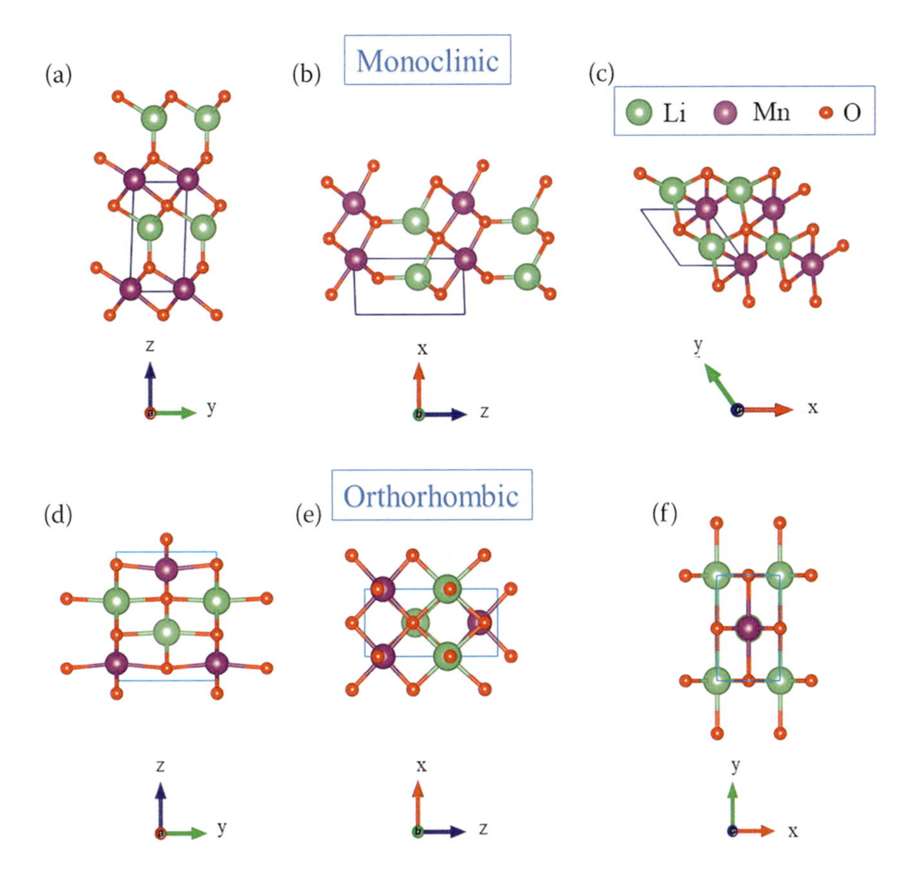

FIGURE 7.1 The optimal geometric structures of $LiMnO_2$ compound with: (a), (b), and (c) monoclinic phase and (d), (e), and (f) orthorhombic phase in the x, y, and z directions.

$LiMnO_2$ configuration has eight atoms in a unit cell, consisting of two Li atoms, two Mn atoms, and four O atoms. The calculated lattice constants are a = 2.8595 (Å), b = 4.6364 (Å), and c = 5.8178 (Å), and tilted angles of $\alpha = \beta = \gamma = 90°$, respectively. The symmetric group space belongs to Herman Mauguin (Pmmn [59]), Hall (P 2 2ab 1¯ab), and Point Group (mmm). Both structures show the existence of a non-uniform and highly asymmetric chemical environment due to the high modulation of chemical bond lengths. The results imply that the geometric structure of $LiMnO_2$ in both phases only exists in chemical bonds of Li-O atoms and Mn-O ones. As we know the unique chemical bonds in a unit cell determine the essential properties of materials. The structure of $LiMnO_2$ includes two types of chemical bonds, respectively, those of Li-O and Mn-O. We calculated their bond lengths in Table 7.1. The orthorhombic configuration has Li-O and Mn-O bond lengths in the range of 2.07–2.32 Å and 1.92–2.32 Å, respectively. Similarly, their bond lengths in the monoclinic structure are 1.88–2.08 Å and 1.94–2.23 Å, respectively. The change in Li-O and Mn-O bond lengths shows that the $LiMnO_2$ structure in both phases has non-uniform bonding strengths. Moreover, the diverse chemical bonds in the unit cell can be explained by the multi-orbital hybridization theory. The chemical bond lengths will directly determine the change of the spatial charge density shape after crystal formation (discussed later in Figures 7.2 and 7.3) and the strengths of the chemical bonds. Therefore, their changes show that non-uniform multi-orbital hybridization exists in the $LiMnO_2$ structure.

TABLE 7.1
The Space-Group Symmetries and Optimal Geometric Parameters of $LiMnO_2$ with Monoclinic and Orthorhombic Phases

		Monoclinic	Orthorhombic
Herman Mauguin		Cm [8]	Pmmn [59]
Hall		C $\bar{2}$y	P 2 2ab $\bar{1}$ab
Point Group		1	mmm
Crystal System		Monoclinic	Orthorhombic
Unit Cell (atom)		4	8
Bond Length (Å)	**Li – O**	1.88–2.08	2.07–2.32
	Mn – O	1.94–2.23	1.92–2.32
Lattice Parameters	**a**	3.1736 (Å)	2.8595 (Å)
	b	5.1436 (Å)	4.6364 (Å)
	c	5.1574 (Å)	5.8178 (Å)
	α	92.657°	90°
	β	85.666°	90°
	γ	96.072°	90°
Bandgap – E_g (eV)		1.1579	1.0114
Magnetic moment (μB)		4	8
Ground states energy (eV)		–26	–52

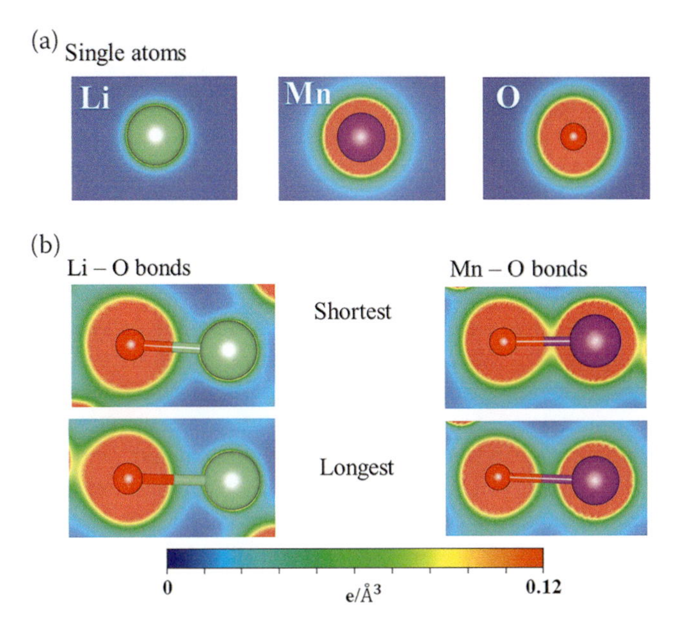

FIGURE 7.2 The charge density of the monoclinic $LiMnO_2$ structure including (a) shows the isolated atoms and (b) under the longest and shortest Li–O and Mn–O bonds, respectively.

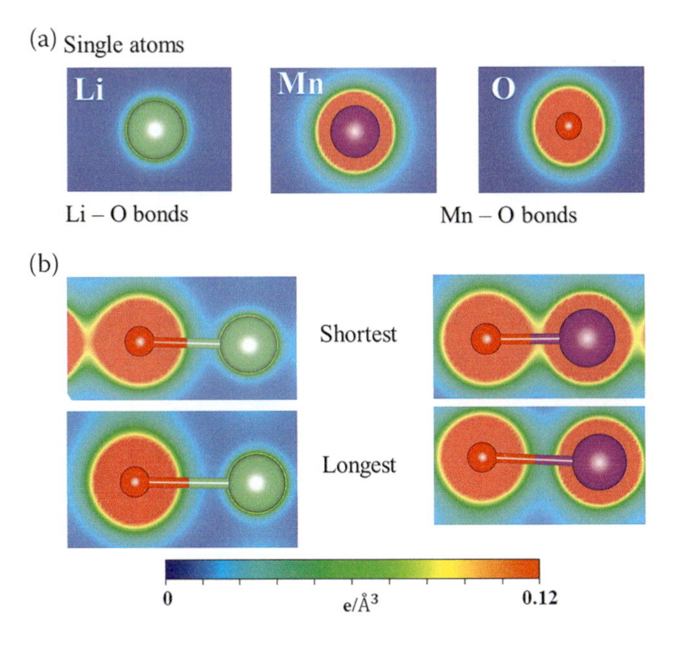

FIGURE 7.3 The charge density of the orthorhombic $LiMnO_2$ structure including (a) shows the isolated atoms and (b) under the longest and shortest Li–O and Mn–O bonds, respectively.

The previous theoretical predictions showed that the optimal geometrical structures were determined experimentally. As we know, X-ray diffraction is the preferred method for measuring 3D symmetric structures. Both monoclinic and orthorhombic phases of the $LiMnO_2$ configuration have been scrutinized by X-ray diffraction [21,22,46–48]. For this structure, we can use this method to thoroughly explore the various binding lengths of Li-O and Mn-O, which help determine the complexity of orbital hybridizations.

The $LiMnO_2$ compounds have many unusual electronic states. Figures 7.4 and 7.5 describe the band structure of $LiMnO_2$ in the monoclinic and orthorhombic phases. Their key features were obtained along with the highly symmetric points in the first Brillouin region. We know the band structure is susceptible to changes in the wave vector. Therefore, it is highly anisotropic. In addition, the contribution from the active orbitals of atoms (Li, Mn, and O) in a specific unit cell leads to the formation of many sub-bands in both the valence and conduction bands. Their high asymmetry around the Fermi level under the complicated intrinsic interactions. The individual atoms dominate at different energy bands. The spin-decoupled and spin-attenuated ferromagnetic electronic states exist in the $LiMnO_2$ structure. Besides, the high anisotropy and non-monotonic energy dispersion are closely related to the non-uniform chemical environment, leading to the formation of many band-edge states and the semiconductor behavior present in the bandgap of this structure.

Specifically, in the monoclinic system (shown in Figure 7.4), an asymmetry between the occupied electron energy spectrum and the unoccupied hole energy

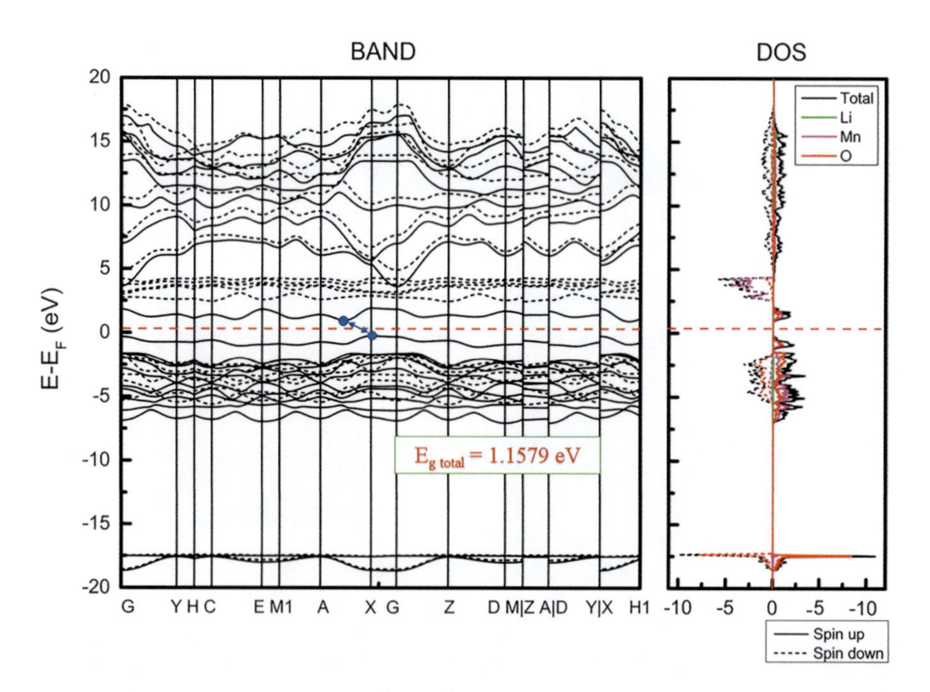

FIGURE 7.4 The band structure (BAND) and density of states (DOS) of the monoclinic $LiMnO_2$ structure.

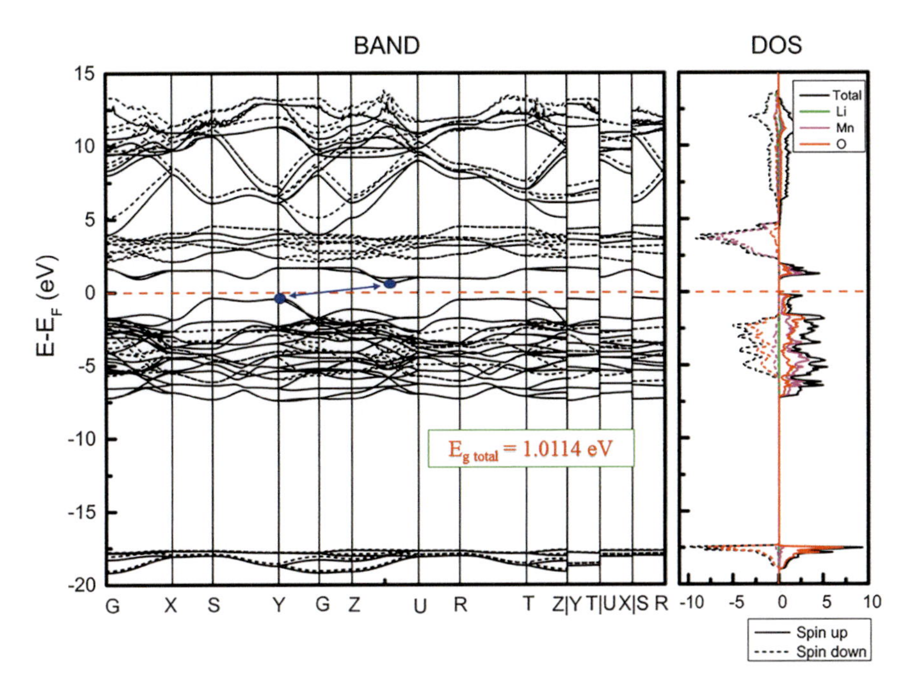

FIGURE 7.5 The band structure (BAND) and density of states (DOS) of the orthorhombic $LiMnO_2$ structure.

spectrum about the Fermi level ($E_F = 0$), mainly because of the difference in multi-orbital hybridization in the non-uniform chemical bonds. The highest valence band state and the lowest conduction band state decide an energy gap of 1.1579 (Å). Moreover, they do not lie at the same wave vector, resulting in an indirect gap semiconductor. Many sub-bands exist in both the valence and conduction bands of the monoclinic $LiMnO_2$ structure because of the contribution of the active orbitals of different atoms in a unit cell. In general, there exists specific energy dispersion. Different sub-bands exhibit other behaviors, including non-crossing, crossing, and anti-crossing. They confirmed that the latter phenomenon occurs when two neighboring sub-bands are compared to components in the anti-crossing region. Besides, a separation between two distinct electronic spin states at the Fermi level, suggesting the existence of spin degradation of the conduction and valence bands around the Fermi level. A ferromagnetic band structure with ground state energy of −26 eV exits corresponds to the expression of the degenerate-spin and spin-spit electronic states. Besides, the spin-up configuration dominated the density of occupied states, meaning there are more spin-up valence bands. As a result, this is consistent with magnetic moments shown in Table 7.1. In the band structure for spin-down (shown in Figure 7.6), the lowest conduction band states and the highest valence band states have an elevation compared to the spin-up band structure (Figure 7.7), leading to a significantly enlarged bandgap (approximately 4.1678 eV).

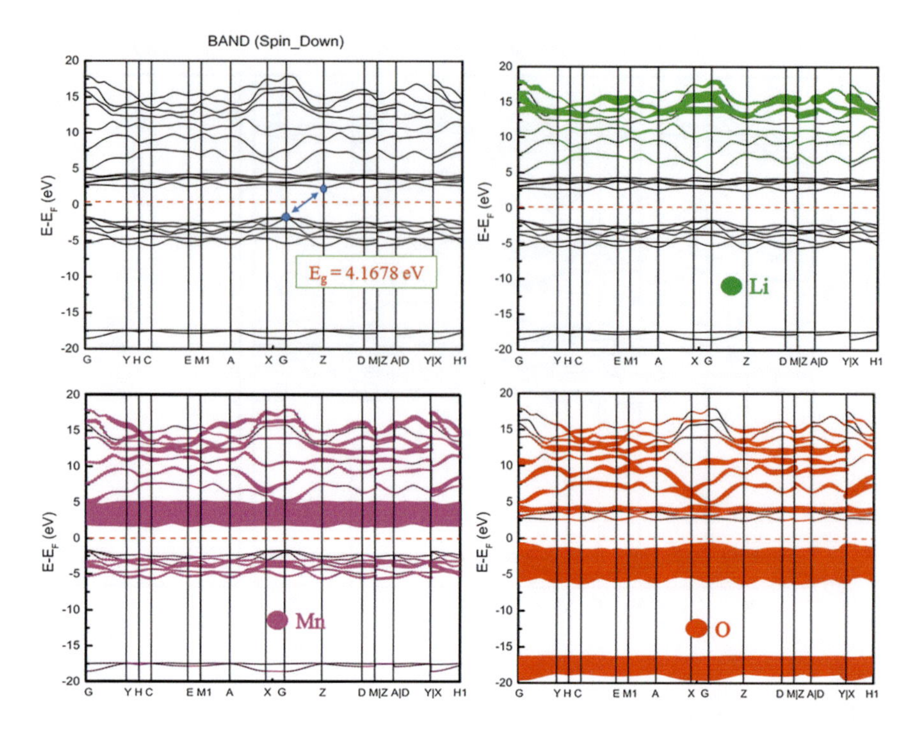

FIGURE 7.6 The electronic energy spectrum with atom dominances of the monoclinic $LiMnO_2$ structure in spin-down states.

To better understand the critical role of chemical bonds in the electron-rich state, we calculated the atom dominance band structure of both spin-up and spin-down states of the monoclinic $LiMnO_2$ system presented in Figures 7.6 and 7.7. We convene that the contribution of Li, Mn, and O atoms are represented by green, pink and red balls, respectively. Li atoms mainly contribute to the conduction band in the band structure and there is no difference between spin-up and spin-down configurations. This result shows that the strength of the chemical bonds of the Li-O bonds presents a small range because their length fluctuates only slightly. Moreover, the Li atom only contributes with s-orbital. In contrast, the Mn and O atoms contribute to both the conduction and valence bands; however, both of these atoms mainly contribute to the valence band (in the −8 – 0 eV energy band). Besides, there are also notable differences when comparing the contributions of atoms in the band structure with spin-up and spin-down. Figure 7.7 shows that in the spin-up band structure, the contribution of the Mn atom is mainly concentrated in the valence band around the Fermi level. Besides, the O atom contributes mainly to the valence band from −20 to 0 eV. In contrast to the spin-down configuration (Figure 7.6), the Mn atom contributes the most to the conduction band near the Fermi level (2 to 5 eV).

Similarly, in the band structure of the orthorhombic $LiMnO_2$ (shown in Figure 7.5), we find that the occupied electron energy spectrum and the unoccupied holes energy spectrum are also highly asymmetric. The highest valence band states

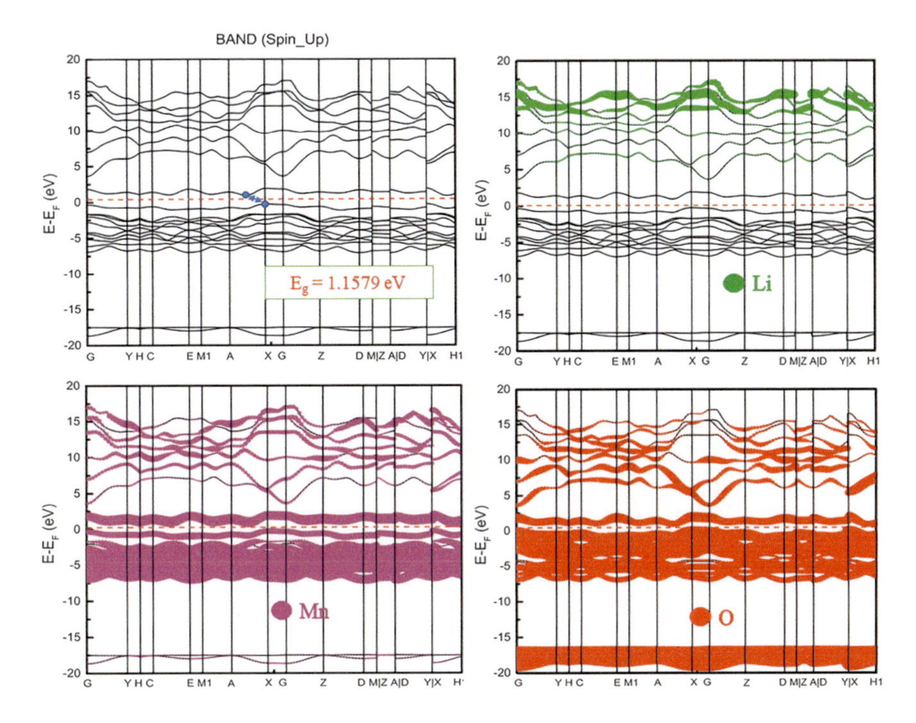

FIGURE 7.7 The electronic energy spectrum with atom dominances of the monoclinic $LiMnO_2$ structure in spin-up states.

and the lowest conduction band states determined the value of the bandgap be 1.0114 eV. Similar to the $LiMnO_2$ structure with the monoclinic phase, the orthorhombic $LiMnO_2$ system exhibits semiconductor properties. In this structure, many sub-bands exist in both conduction and valence bands, which exhibit energy dispersion characteristics. Besides, a lot of sub-band crossing, non-crossing and anti-crossing exist in the energy range from –20 eV to 15 eV. Moreover, their band structure has distinct spin separation. The spin-dependent energy band structure of the $LiMnO_2$ compound is exhibited, especially the remarkable spin splitting in the vicinity of the Fermi energy. Interestingly, the energy band gaps for spin-up states (Figure 7.8) and spin-down states (Figure 7.9) are rather different, corresponding to the indirect gap of 1.0114 eV and 3.8268 eV for the former and the latter, respectively. The spin splitting with a complicated energy band structure reflects the ferromagnetic (FM) configuration of the orthorhombic $LiMnO_2$ compound and then can be responsible for the unusual optical properties.

The atom dominances in the orthorhombic $LiMnO_2$ structure are clearly shown in Figures 7.8 and 7.9, corresponding to the spin-up and spin-down states. We find that the contribution of atoms to the band structure is not significantly different between the monoclinic and orthorhombic phases. The Li atom contributes the most to the conduction band, while the Mn and O atoms contribute the most to the valence band states. The unoccupied electronic states are dominated mainly by the O and Mn atoms. However, the Mn atom has a more significant

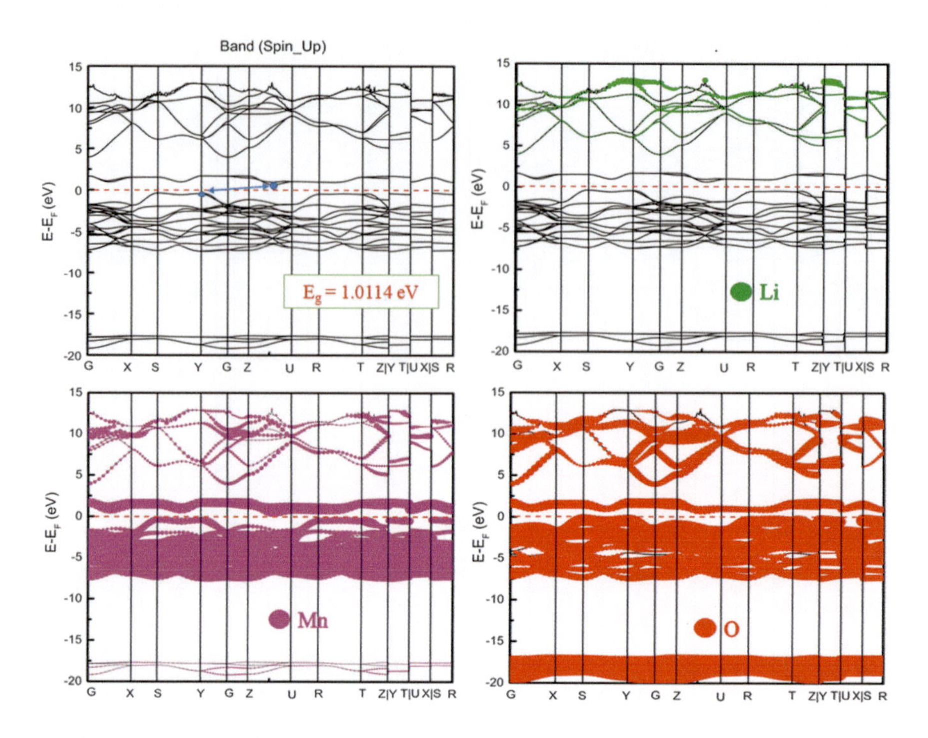

FIGURE 7.8 The electronic energy spectrum with atom dominances of the orthorhombic $LiMnO_2$ structure in spin-up states.

influence than the 0 atom, which means that the Mn atoms contribute more to the chemical bonding. The opposite happens for Li atoms. In short, atom-dominated electronic structures could provide partial information about critical multi-orbital hybridizations.

Exploring the charge density distribution in real space is useful for better understanding the complex hybridization in chemical bonds. According to VASP simulation results, the charge densities of $LiMnO_2$ compounds are presented in Figures 7.2 and 7.3 with monoclinic and orthorhombic phases, respectively. Figures 7.2a and 7.3a depict the charge densities of atoms in an isolated state. We can easily see the strong charge exchange between the atoms before and after forming the $LiMnO_2$ structure, especially in the vicinity. The calculated results are thoroughly analyzed to extract the significant orbital hybridizations through the delicate comparisons between the real and isolated cases. When in isolation, the atoms have a spherical charge density distribution. The red regions are s-orbitals and below, the blue and yellow regions being p_x, p_y, and p_z - orbitals. After forming the $LiMnO_2$ structure (Figures 7.2b and 7.3b), there exists a lot of multi-orbital mixings, namely, 2s - [2s, $2p_x$, $2p_y$, $2p_z$] and [4s, $3d_{x^2-y^2}$, $3d_{z^2}$, $3d_{xy}$, $3d_{yz}$, $3d_{xz}$] – [2s, $2p_x$, $2p_y$, $2p_z$] in Li-O and Mn-O bonds, respectively. The above results are inferred from the change in the shape of the different color regions since the spatial charge density distributions are isotropic and anisotropic forms. Moreover, they mainly cover regions near the Li, Mn, and O atoms. Note that it is complicated to

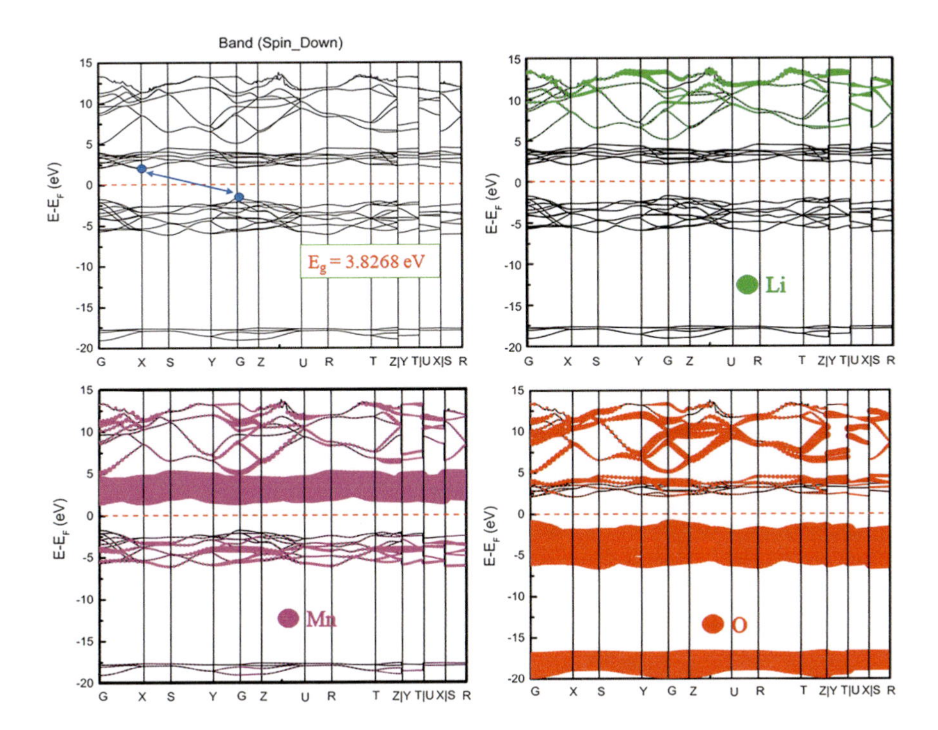

FIGURE 7.9 The electronic energy spectrum with atom dominances of the orthorhombic $LiMnO_2$ structure in spin-down states.

distinguish the p and d orbitals of the O and Mn atoms, suggesting that they can contribute together. The X-ray measurements can verify the charge distributions in a unit cell [46].

The number of energy states characterizes the density of states (DOS) in an infinitely small energy band, which provides all the necessary information about the multi-orbital recombination of chemical bonds in the $LiMnO_2$ structure (shown in Figures 7.10 and 7.11 for monoclinic and orthorhombic phases, respectively). Based on the calculation criteria [49], DOS corresponds to the inverse integral of the group velocity over a constant energy space. It creates the unusual van Hove singularities under the vanishing or singular derivative first of state energy versus wave vector. We see that their form, quantity, energy, and intensity change depending on the states and sizes of the different band edges. In particular, such unique states belong to the critical conditions in the energy-vector wave space, including local maxima or minima, saddle point, constant energy ring, and partially planar energy dispersion. As a result, diverse structures of van Hove singularities exist.

The atomic and orbital state densities of the $LiMnO_2$ structure are shown in Figures 7.10–7.13 for monoclinic and orthorhombic phases, respectively. The dominance of different atoms is divided into many kinds of energy ranges. The orbital- and spin-projected DOS can further identify the significant orbital hybridizations. First, we see that the density of states in a unit cell disappears in

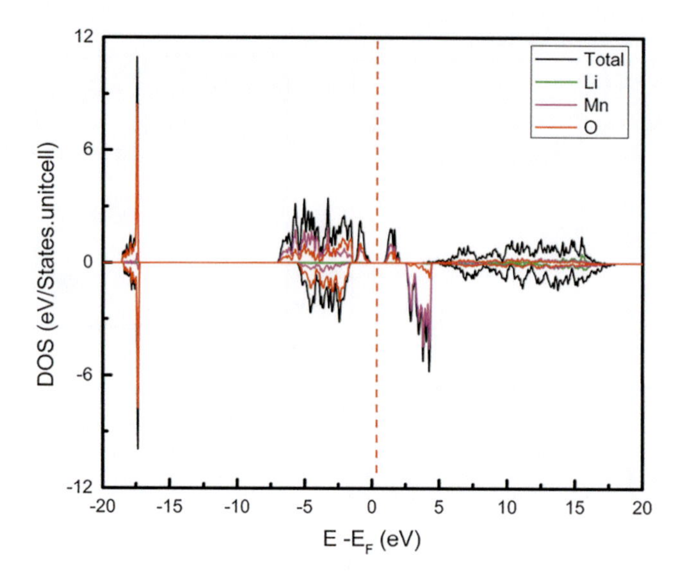

FIGURE 7.10 The total density of state (DOS) of the monoclinic $LiMnO_2$ structure.

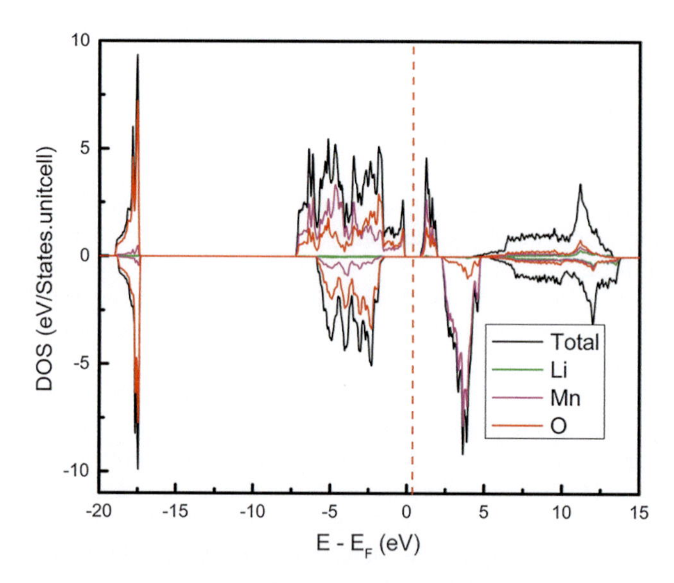

FIGURE 7.11 The total density of state (DOS) of the orthorhombic $LiMnO_2$ structure.

the bandgap of both spin-up and spin-down states. In addition, the energy spectra of the valence and conduction states have asymmetry around the Fermi level. Many asymmetric peaks exist, and it is clear that the valence band states have larger values than the conduction band, which is entirely consistent with the band structure. Such van Hove singularities originate from the band-edge states of energy sub-bands, with the local minimum, maximum, saddle, and almost

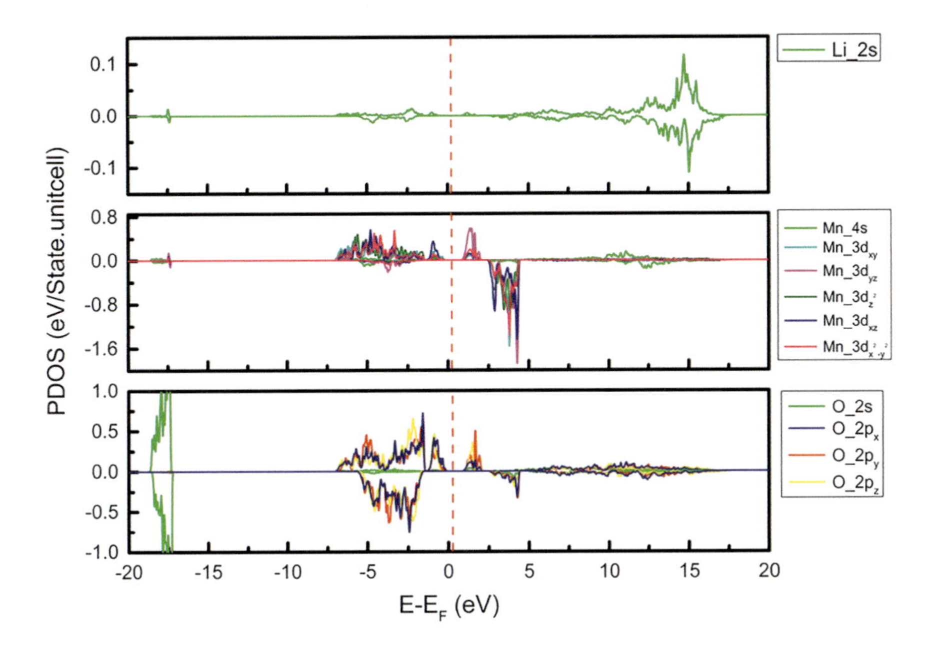

FIGURE 7.12 The project density of states (PDOS) of the monoclinic $LiMnO_2$ structure.

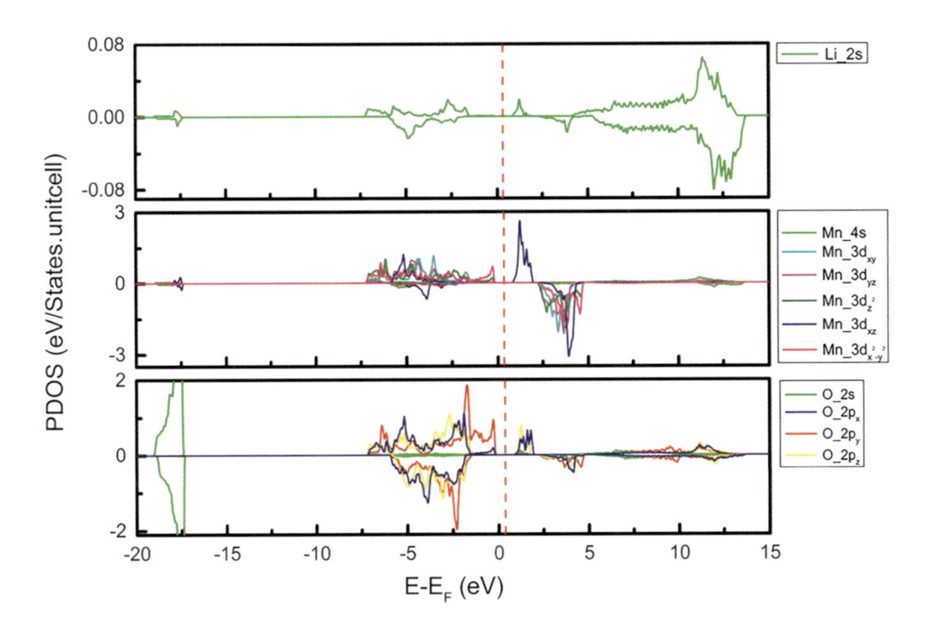

FIGURE 7.13 The project density of states (PDOS) of the orthorhombic $LiMnO_2$ structure.

dispersionless points in the energy wave-vector spaces. They might appear in between the high-symmetry points. Besides, the contribution of the Mn atom in the structure is more dominant than that of the O and Li atoms because the Mn atom contributes more orbitals. Figure 7.12 shows that, for the ns orbital, it is clear that the Li atom dominates, but this contribution is not significant for the entire energy range in different cases. However, the 2s orbitals of the O atom reveal that they have an essential gift to the more profound and more significant energy regions in the $E < -16$ eV energy range through the blue plots. The complex contribution of the p and d orbitals of the Mn and O atoms in the energy range $E > -8$ eV indicates multi-orbital hybridization in the Mn-O bond. In summary, the characteristics mentioned above can define energy-dependent orbital hybridization.

7.4 MAGNETIC PROPERTIES

To predict magnetic properties, we calculated the spin density of the $LiMnO_2$ structure. The spin configuration is carefully examined from properties such as the spin-degenerate or spin-split energy bands (Figures 7.6 and 7.7 for monoclinic phase, Figures 7.8 and 7.9 for orthorhombic phase), the spin-decomposed van Hove singularities (Figures 7.10–7.13 respectively), and the atom- & orbital-induced magnetic moments (Figures 7.14 and 7.15). The property of the spin-degenerate or spin-split energy bands clearly describes that spin-related magnetic properties are directly reflected in the entire energy band. In any case, only the lithium atoms give no evidence of a spin arrangement, as determined from the DOS spin-up & spin-down cancellation and the complete disappearance of the magnetic moment (Figures 7.4 and 7.5). Among the active orbitals of Mn and O atoms, five 3d orbitals contribute the most to the magnetic moment, indicating their dominance over

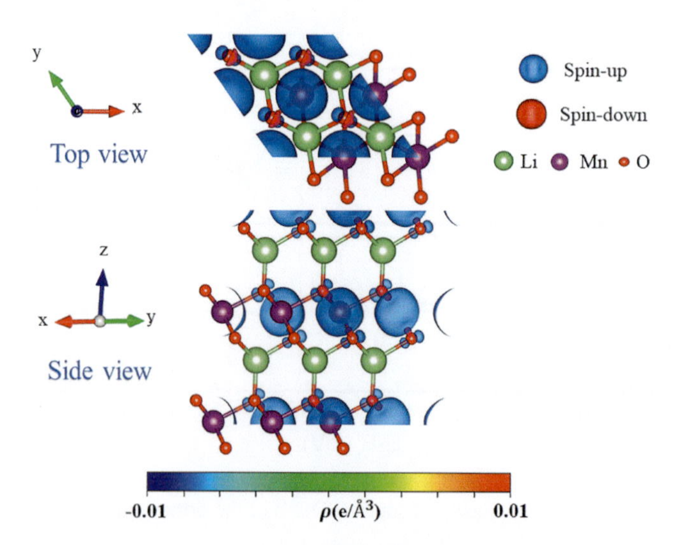

FIGURE 7.14 The spin density of the monoclinic $LiMnO_2$ structure.

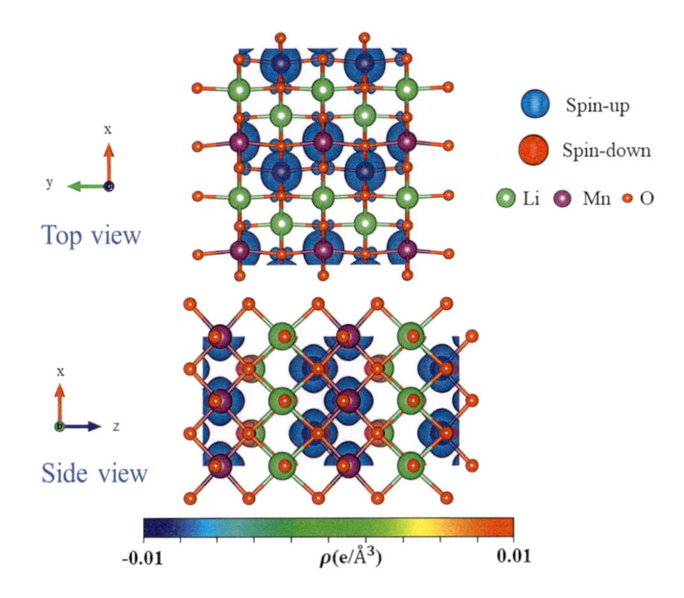

FIGURE 7.15 The spin density of the orthorhombic $LiMnO_2$ structure.

ferromagnetic configurations. The s- and p-orbitals reveal a weak strength in the spin-induced magnetic field properties. Such characteristic magnetic moments in Table 7.1 are entirely consistent with the spin arrangement in space. Spin density displays the spin-up and spin-down distributions [in blue represents spin-up and spin-down shown in red] around the atoms in the structure shown in Figures 7.14 and 7.15. In both phases, the spin-up value completely dominates the spin-up value. Besides, spin-up values are mainly concentrated around Mn atoms, a small part is focused on O atoms, and we can hardly find spin ta deals on Li atoms. This entirely coincides with the charge distribution and energy values in the band structure mentioned earlier.

The diverse phenomena produced by atoms, orbitals, and spins deserve closer investigation. We know that when the value of the magnetic moment is positive, the material exhibits ferromagnetic behavior. The charge density (shown in Figures 7.10 and 7.11) is spin-polarized. Furthermore, the virtual orbitals of the atoms cover different energy regions for the energy states around the Fermi level, in which the orbitals of the Mn and O atoms dominate. Note that the van Hove singularities of the spin-down are much more dominant in the upper Fermi level ($3\ eV < E < 5eV$). The transport behavior of metals or semiconductors is related to spin-down states, while the magnetic configuration mainly arises from occupied spin-up states. This unusual phenomenon is thought to produce unique magnetic properties.

7.5 CONCLUSION

In summary, the properties of $LiMnO_2$ have been studied using VASP calculations. The calculation results confirm that the $LiMnO_2$ compound has fascinating electronic and magnetic properties. Our main result is the $LiMnO_2$ configuration in both monoclinic and orthorhombic phases has multi-orbital hybridization in chemical bonds. The intralayer π, σ, and sp^3 bondings are thoroughly identified from a highly anisotropic chemical environment, the atom-dominated energy spectra within the specific energy ranges, the orbital-related charge density deformations, and the merged van Hove singularities across the Fermi level. The Li-, Mn-, and O bonds, respectively, have active multi-orbital hybridization of 2s- [2s, $2p_x$, $2p_y$, $2p_z$], [4s, $3d_{x^2-y^2}$, $3d_{z^2}$, $3d_{xy}$, $3d_{yz}$, $3d_{xz}$] – [2s, $2p_x$, $2p_y$, $2p_z$]. The different orbitals and atoms dominate at distinct regions; however, both configurations have indirect bandgaps and exhibit semiconductor behavior. Both monoclinic and orthorhombic $LiMnO_2$ compounds are predicted to possess a 1.15 eV direct bandgap and a 1.01 eV direct one. The low-lying hole and electron energy spectra are, respectively, dominated by Li, Mn, and O atoms, since the former and the latter have more occupied and occupied orbitals, as seen in a periodical table. Very interestingly, the various van Hove singularities are capable of providing rather useful messages such as energies, corresponding prominent structures, indicating the atom- and orbital-dependent ionization ones, and more orbital contributions become close to the Fermi level through the interlayer orbital mixings. In addition, spin polarization in the band structure and state density graph; positive magnetic moment values indicate that the $LiMnO_2$ structure exhibits ferromagnetic behavior.

REFERENCES

1. Y. Sato, T. Nomura, H. Tanaka, and K. Kobayakawa, "Charge-discharge characteristics of electrolytically prepared V_2O_5 as a cathode active material of lithium secondary battery," *Journal of the Electrochemical Society*, vol. 138, no. 9, p. L37, 1991.
2. S. C. Nam *et al.*, "The effects of Cu-doping in V_2O_5 thin film cathode for micro-battery," *Korean Journal of Chemical Engineering*, vol. 18, no. 5, pp. 673–678, 2001.
3. F. Coustier, S. Passerini, and W. Smyrl, "A 400 mAh/g aerogel-like V_2O_5 cathode for rechargeable lithium batteries," *Journal of the Electrochemical Society*, vol. 145, no. 5, p. L73, 1998.
4. M. Broussely *et al.*, "Li_xNiO_2, a promising cathode for rechargeable lithium batteries," *Journal of Power Sources*, vol. 54, no. 1, pp. 109–114, 1995.
5. J. Dahn, E. Fuller, M. Obrovac, and U. Von Sacken, "Thermal stability of Li_xCoO_2, Li_xNiO_2 and λ-MnO_2 and consequences for the safety of Li-ion cells," *Solid State Ionics*, vol. 69, no. 3-4, pp. 265–270, 1994.
6. T. Miyashita, H. Noguchi, K. Yamato, and M. Yoshio, "Preparation of layered $LiNiO_2$ by alcoholate method and its behavior as a cathode in lithium secondary battery," *Journal of the Ceramic Society of Japan*, vol. 102, no. 1183, pp. 258–261, 1994.
7. H. Li, Z. Lu, H. Huang, X. Huang, and L. Chen, "Electrochemical impedance spectroscopic study of the rate-determining step of Li ion intercalation and deintercalation in Li_xNiO_2 cathodes," *Ionics*, vol. 2, no. 3, pp. 259–265, 1996.
8. Y. Baba, S. Okada, and J.-i. Yamaki, "Thermal stability of Li_xCoO_2 cathode for lithium ion battery," *Solid State Ionics*, vol. 148, no. 3-4, pp. 311–316, 2002.
9. J.-i. Yamaki, Y. Baba, N. Katayama, H. Takatsuji, M. Egashira, and S. Okada, "Thermal stability of electrolytes with Li_xCoO_2 cathode or lithiated carbon anode," *Journal of Power Sources*, vol. 119, pp. 789–793, 2003.
10. A. Veluchamy *et al.*, "Thermal analysis of Li_xCoO_2 cathode material of lithium ion battery," *Journal of Power Sources*, vol. 189, no. 1, pp. 855–858, 2009.

11. Y. Matsuda, N. Kuwata, T. Okawa, A. Dorai, O. Kamishima, and J. Kawamura, "In situ Raman spectroscopy of Li_xCoO_2 cathode in $Li/Li_3PO_4/LiCoO_2$ all-solid-state thin-film lithium battery," *Solid State Ionics*, vol. 335, pp. 7–14, 2019.
12. V. Malavé, J. Berger, H. Zhu, and R. J. Kee, "A computational model of the mechanical behavior within reconstructed Li_xCoO_2 Li-ion battery cathode particles," *Electrochimica Acta*, vol. 130, pp. 707–717, 2014.
13. Y. N. Ko, S. H. Choi, Y. C. Kang, and S. B. Park, "Characteristics of Li_2TiO_3 –$LiCrO_2$ composite cathode powders prepared by ultrasonic spray pyrolysis," *Journal of Power Sources*, vol. 244, pp. 336–343, 2013.
14. Y. Lyu *et al.*, "Probing reversible multielectron transfer and structure evolution of $Li_{1.2}Cr_{0.4}Mn_{0.4}O_2$ cathode material for Li-ion batteries in a voltage range of 1.0–4.8 V," *Chemistry of Materials*, vol. 27, no. 15, pp. 5238–5252, 2015.
15. S.-h. Wu and M.-t. Yu, "Preparation and characterization of o-$LiMnO_2$ cathode materials," *Journal of Power Sources*, vol. 165, no. 2, pp. 660–665, 2007.
16. F. Kong *et al.*, "*Ab initio* study of doping effects on $LiMnO_2$ and Li_2MnO_3 cathode materials for Li-ion batteries," *Journal of Materials Chemistry A*, vol. 3, no. 16, pp. 8489–8500, 2015.
17. X. Zhu *et al.*, "$LiMnO_2$ cathode stabilized by interfacial orbital ordering for sustainable lithium-ion batteries," *Nature Sustainability*, vol. 4, no. 5, pp. 392–401, 2021.
18. S. Chen *et al.*, "Facile hydrothermal synthesis and electrochemical properties of orthorhombic $LiMnO_2$ cathode materials for rechargeable lithium batteries," *RSC Advances*, vol. 4, no. 26, pp. 13693–13703, 2014.
19. H. Ji, G. Yang, X. Miao, and A. Hong, "Efficient microwave hydrothermal synthesis of nanocrystalline orthorhombic $LiMnO_2$ cathodes for lithium batteries," *Electrochimica Acta*, vol. 55, no. 9, pp. 3392–3397, 2010.
20. C. Li *et al.*, "Cathode materials modified by surface coating for lithium ion batteries," *Electrochimica Acta*, vol. 51, no. 19, pp. 3872–3883, 2006.
21. I. Kötschau and J. Dahn, "In situ X-ray study of $LiMnO_2$," *Journal of the Electrochemical Society*, vol. 145, no. 8, p. 2672, 1998.
22. X. Tu and K. Shu, "X-ray diffraction study on phase transition of orthorhombic $LiMnO_2$ in electrochemical conversions," *Journal of Solid State Electrochemistry*, vol. 12, no. 3, pp. 245–249, 2008.
23. A. R. Armstrong and P. G. Bruce, "Synthesis of layered $LiMnO_2$ as an electrode for rechargeable lithium batteries," *Nature*, vol. 381, no. 6582, pp. 499–500, 1996.
24. Y. Shao-Horn *et al.*, "Structural characterization of layered $LiMnO_2$ electrodes by electron diffraction and lattice imaging," *Journal of the Electrochemical Society*, vol. 146, no. 7, p. 2404, 1999.
25. G. Vitins and K. West, "Lithium intercalation into layered $LiMnO_2$," *Journal of the Electrochemical Society*, vol. 144, no. 8, p. 2587, 1997.
26. R. G. Parr, "Density functional theory," *Annual Review of Physical Chemistry*, vol. 34, no. 1, pp. 631–656, 1983.
27. M. Orio, D. A. Pantazis, and F. Neese, "Density functional theory," *Photosynthesis Research*, vol. 102, no. 2, pp. 443–453, 2009.
28. E. K. Gross and R. M. Dreizler, *Density functional theory*. Springer Science & Business Media, 2013.
29. E. Engel and R. M. Dreizler, "Density functional theory," *Theoretical and Mathematical Physics*, pp. 351–399, 2011.
30. W. Kohn, "Density functional theory," *Introductory Quantum Mechanics with MATLAB: For Atoms, Molecules, Clusters, and Nanocrystals*, 2019.
31. N. Argaman and G. Makov, "Density functional theory: An introduction," *American Journal of Physics*, vol. 68, no. 1, pp. 69–79, 2000.

32. G. Sun, J. Kürti, P. Rajczy, M. Kertesz, J. Hafner, and G. Kresse, "Performance of the Vienna ab initio simulation package (VASP) in chemical applications," *Journal of Molecular Structure: THEOCHEM*, vol. 624, no. 1–3, pp. 37–45, 2003.
33. J. Hafner, "Ab-initio simulations of materials using VASP: Density-functional theory and beyond," *Journal of Computational Chemistry*, vol. 29, no. 13, pp. 2044–2078, 2008.
34. R. J. Bartlett, V. F. Lotrich, and I. V. Schweigert, "Ab initio density functional theory: The best of both worlds?" *The Journal of Chemical Physics*, vol. 123, no. 6, p. 062205, 2005.
35. R. J. Bartlett, I. Grabowski, S. Hirata, and S. Ivanov, "The exchange-correlation potential in ab initio density functional theory," *The Journal of Chemical Physics*, vol. 122, no. 3, p. 034104, 2005.
36. V. Jafarova and G. Orudzhev, "Structural and electronic properties of ZnO: A first-principles density-functional theory study within LDA (GGA) and LDA (GGA)+ U methods," *Solid State Communications*, vol. 325, p. 114166, 2021.
37. S. Bo and Z. Ping, "First-principles local density approximation (LDA)+ U and generalized gradient approximation (GGA)+ U studies of plutonium oxides," *Chinese Physics B*, vol. 17, no. 4, p. 1364, 2008.
38. G. Kresse and D. Joubert, "From ultrasoft pseudopotentials to the projector augmented-wave method," *Physical Review B*, vol. 59, no. 3, p. 1758, 1999.
39. M. Gajdoš, K. Hummer, G. Kresse, J. Furthmüller, and F. Bechstedt, "Linear optical properties in the projector-augmented wave methodology," *Physical Review B*, vol. 73, no. 4, p. 045112, 2006.
40. J. J. Mortensen, L. B. Hansen, and K. W. Jacobsen, "Real-space grid implementation of the projector augmented wave method," *Physical Review B*, vol. 71, no. 3, p. 035109, 2005.
41. N. Holzwarth, A. Tackett, and G. Matthews, "A projector augmented wave (PAW) code for electronic structure calculations, Part I: Atompaw for generating atom-centered functions," *Computer Physics Communications*, vol. 135, no. 3, pp. 329–347, 2001.
42. A. Griffin, "Brillouin light scattering from crystals in the hydrodynamic region," *Reviews of Modern Physics*, vol. 40, no. 1, p. 167, 1968.
43. P. Politzer and J. S. Murray, "The Hellmann–Feynman theorem: A perspective," *Journal of Molecular Modeling*, vol. 24, no. 9, pp. 1–7, 2018.
44. M. Di Ventra and S. T. Pantelides, "Hellmann–Feynman theorem and the definition of forces in quantum time-dependent and transport problems," *Physical Review B*, vol. 61, no. 23, p. 16207, 2000.
45. V. Bakken, T. Helgaker, W. Klopper, and K. Ruud, "The calculation of molecular geometrical properties in the Hellmann—Feynman approximation," *Molecular Physics*, vol. 96, no. 4, pp. 653–671, 1999.
46. Y. Takahashi, J. Akimoto, Y. Gotoh, K. Dokko, M. Nishizawa, and I. Uchida, "Structure and electron density analysis of lithium manganese oxides by single-crystal X-ray diffraction," *Journal of the Physical Society of Japan*, vol. 72, no. 6, pp. 1483–1490, 2003.
47. Y.-I. Jang, W. D. Moorehead, and Y.-M. Chiang, "Synthesis of the monoclinic and orthorhombic phases of $LiMnO_2$ in oxidizing atmosphere," *Solid State Ionics*, vol. 149, no. 3-4, pp. 201–207, 2002.
48. X. Li, Z. Su, and Y. Wang, "Electrochemical properties of monoclinic and orthorhombic $LiMnO_2$ synthesized by a one-step hydrothermal method," *Journal of Alloys and Compounds*, vol. 735, pp. 2182–2189, 2018.
49. F. Wang and D. P. Landau, "Efficient, multiple-range random walk algorithm to calculate the density of states," *Physical Review Letters*, vol. 86, no. 10, p. 2050, 2001.

8 Surface Property of High-Voltage Cathode $LiNiPO_4$ in Lithium-Ion Batteries

A First-Principles Study

Chien-Ke Huang and Wen-Dung Hsu
Department of Materials Science and Engineering and Hierarchical Green-Energy Materials (Hi-GEM) Research Center, National Cheng Kung University, Tainan City, Taiwan

CONTENTS

8.1 LITHIUM-ION BATTERY

In the initial development of lithium-ion batteries, the safety and cycle life of lithium-ion batteries were doubtful due to the unstable anode and cathode materials, so they could not effectively be used in commercial products. In 1980, Professor Goodenough's team proposed to use $LiCoO_2$ as the positive electrode material and graphite as the negative electrode material to form the lithium-ion secondary battery, which has greatly improved the safety and stability and also has the advantages of good electrochemical reaction characteristics and high energy density at the same

DOI: 10.1201/9781003367215-8

time. From then lithium-ion batteries gradually become the dominant battery in commercial products [1]. For example, in the 1990s, Sony launched the first lithium-ion battery product for cell phones and cameras with the composition of cobalt cathode($LiCoO_2$) and graphite anode [2].

8.2 GENERAL CATHODE MATERIALS AND THE STABILITY AT HIGH VOLTAGE

Cathode materials play an important role in Li-ion batteries because they decide the cell potential and capacity. Besides, they contribute more than 30% cost of the lithium battery system [3]. Currently, cathode materials can be classified into three types according to their structure: layer, spinel, and olivine type [4].

The most representative layer type cathode is $LiCoO_2$, which is used in early commercial applications. It has an operating voltage of 4.2V and a capacity of 140mAhg^{-1}, which is 50% of the $LiCoO_2$ theoretical capacity [5]. Under the operation up to 4.5V potential, however, it shows the rapid decay of capacity during cycling due to phase transformation [6]. Recently a ternary system layer type cathode material $LiNi_xMn_yCo_{1-x-y}O_2$ (NMC) is being developed, by combining three transition metal elements, nickel, manganese, and cobalt, of different chemical proportions to mitigate the cyclic issue and achieve the balance between battery capacity, cycle stability, and thermal stability [7]. It also reduces the cost by lowering the cobalt content. However, NMC materials suffer from surface phase transformation and particle cracking during high-potential charging and discharging [8,9].

Mohanty D. et al. explore the structure change during the charging and discharging process in Mn-rich NMC cathode material by neutron-diffraction analysis [10] they found that the transition metal ions move to the lithium layer during the highly charged lithium-deficient process, which is presumed to induce the oxygen vacancy formed by the loss of oxygen atoms during the charging process. Further analysis by HR-TEM (high-resolution transmission electron microscopy) reveals that the surface region forms a spinel-like and layer-like two-phase structure due to the movement of cations, and this phase change not only causes a decrease in structural stability but also increases the interfacial transfer impedance of lithium ions during the cycling, resulting in capacity loss. Therefore, when testing the NMC cathode with graphite negative electrode for the charging and discharging process from 3.0V to 4.7V, an irreversible capacity loss of 60mAhg^{-1} can be observed in the first cycle.

Yan et al. discuss the effect of particle cracking in NMC cathode [11] they investigated the changes of the particle structure, morphology, and cycling efficiency of NMC333 ($LiNi_{0.3}Co_{0.3}Mn_{0.3}O_2$) material with different charging potentials of 4.2V, 4.5V, and 4.7V, by the SEM and STEM-HAADF (high-angle annular dark-field scanning transmission electron microscopy). They found that the NMC333 particles do not show intergranular cracking when charging and discharging below 4.5V. The particles start cracking from inside when charging and discharging above 4.5V and the degree of rupture increases as the charging voltage increase. They pointed out that this phenomenon is due to a highly

Lithium-deficient state when charging at high potential, which exacerbates the instability of the structure and increases the strain inside the particle as well, resulting in the formation of cracks. The particle destroyed from the inside causes severe capacity fading. By comparing the cycle efficiency of these three charging potentials, it can be found that 4.2V and 4.5V can effectively maintain the capacity after 100 cycles, while 4.7V charging and discharging can have a higher capacity in the first cycle, but after 100 cycles its capacity has already lost 25%.

$LiMn_2O_4$ is the first developed spinel-type cathode material, which has good electrochemical properties due to the three-dimensional diffusion of lithium ions [4,12]. When charged to 4.3V, however, it produces a lithium-deficient phase of Mn_3O_4 that accelerates the dissolution of manganese ions and causes the instability of the structure and the decrease in capacity [13,14]. The $LiNi_{0.5}Mn_{1.5}O_4$(LNMO) cathode is a new commercialized spinel cathode in industrial applications, by exchanging the manganese with nickel to stabilize its structure and enhance the operating voltage during the charging and discharging process [13]. Another advantage for LNMO is its relatively high operating potential and capacity, which can reach 4.7V and 140 $mAhg^{-1}$, so it has a higher energy density. According to STEM HAADF analysis, the transition metal ions would enter the lithium tetrahedral site leaving the empty octahedral site during the initial charging and discharging cycle [15]. This induces the formation of Mn_3O_4 structure right at the surface and a rock salt structure at the subsurface. The phase transformation eventually leads to severe capacity degradation while cycling.

The earliest and most famous olivine structured cathode material is $LiFePO_4$ (LFP), already commercialized nowadays, which has a theoretical capacity of 170$mAhg^{-1}$ and also excellent structure stability during the cycling process. With these features, in the LFP full cell, experiments show that it can maintain 97%–98% of the initial capacity after 1000 cycles [16]. The other advantage of LFP is no toxic compound is contained, which is friendly to the environment compared to the cobalt-content material. The relatively low operating potential of 3.45V is the drawback, which limits the usage of LFP cathode in industrial applications due to low energy density. In order to solve this problem, based on LFP, different transition metal ions have been used to replace iron ions to enhance the oxidation-reduction potential in the electrochemical reaction process to achieve a higher working potential, thus resulting in the development of $LiMnPO_4$ (LMP) (4.2V), $LiCoPO_4$ (LCP) (4.8V) and $LiNiPO_4$ (LNP) (5.2V) these kinds of high voltage cathode, which have a similar capacity to LFP but with much higher operation potential and energy density [17].

Although LCP and LNP cathode materials have the advantages of high potential, high capacity, and structural stability during cycling, they also have some problems that need to be overcome, such as easy-to-form impurity phase during synthesis [18]. Low electronic conductivity and ionic conductivity of the material are other important issues. From the observation of the cyclic voltammetry test between 3.2V–5.6V of LNP, it shows that the oxidation and reduction peak don't appear obvious, which indicates that LNP cathode material has low performance on electrochemical reaction and lithium-ion mobility. Therefore, although it has a theoretical capacity of 170 $mAhg^{-1}$, only 80 $mAhg^{-1}$ is utilizable during cycling. It also cannot achieve good performance during charging and discharging at high C-rate conditions [19,20]. The main reason for these phenomena can attribute to the structural composition of the olivine material itself.

LCP and LNP have the same olivine structure as LFP, with the orthorhombic structure and Pnma space group, lithium ions, and transition metal ions occupy the 4a and 4c site, respectively, and the tetrahedral of PO_4 and octahedral of MO_6 are connected each other to form a three-dimensional periodic structure. This three-dimensional structure makes them relatively stable during the charging and discharging process compared to the layer-type and spinel-type cathode, but this structure also makes lithium-ion transport difficult. Theoretical calculations comparing the migration energy barrier of lithium ions in different directions in LMP(M is transition metal ion) bulk material. The results show that the olivine structure allows lithium ions to conduct only in the [010] direction, unlike the cathodes of layer and spinel structures which have two-dimensional and three-dimensional ion transfer directions, so it is not sufficient to conduct the ions compared with the cathode of other structures [21]. The molecular dynamics simulation under 800k running with 4 nanoseconds also shows that lithium ions only path along the [010] direction of the LMP cell [22].

Moreover, during the synthesis process, the transition metal ions that should be located at the 4c site tend to occupy the position of lithium ions at the 4a site, forming an anti-site defect of cations inside the cell [23,24], which will cause the diffusion mechanism changes [25]. Specifically, the original pure exchange between lithium ions and lithium vacancies that make lithium-ion diffusion is now shared with the exchange between transition metal ions and lithium vacancies at some specific position. The migration energy barrier required for the exchange of transition metal ions with lithium vacancies, based on the calculation, is higher than the energy required for the exchange of lithium with lithium vacancies by about 0.15–0.7 eV [21], especially for nickel ions, which has the highest energy requirement. These transition ions play as rolling pebbles in the creek that slow down the water stream, therefore, reducing the ionic conductivity of Li-ions in the material.

At the same time, due to the high potential operating range, conventional liquid electrolytes and salts are prone to decompose and form hydrofluoric acid during cycling, which tends to react with the cathode surface, resulting in the dissolution of the cathode transition metal ion, which in turn leads to the instability of the structure and cause the serious decrease in cycle efficiency. Therefore, the development of electrolyte systems that can operate at high potential windows is also a major focus of future lithium-ion battery development [17]. Discovering a new electrolyte is another field of lithium-ion battery research that is out of the scope of this chapter. We do not discuss it here.

Through the comparison of the potential and capacity of different cathode materials measured by experiment and the consideration of the current development trend [17], LNP has higher potential and capacity than other kinds of cathode materials, it has good safety properties during cycling as well, and does not contain metallic cobalt, which is in line with the future development trend of the ideal lithium-ion battery. Based on the above discussion, we would like to select the LNP cathode for future investigation.

8.3 CATION DOPING AND SURFACE COATING EFFECT ON THE HIGH VOLTAGE CATHODE

In order to effectively solve the problem of poor electrical and ionic conductivity, doping with cations and modifying surface layers are common techniques applied to synthesizing olivine cathode materials [26,27]. The dopant should be able to increase the transport efficiency of lithium ions which usually has different charge and size with cations to increase the lattice size and create sufficient defects. This concept, however, has been criticized due to the literature reporting that not all the dopants improve the properties of LMP system based on simulation results [21], but only several experimental works observed better cycling performance when the proper dopant was selected [28]. Take LNP as an example, Örnek et al. utilized cobalt to replace the nickel in the LNP cathode by the synthesis of $LiNi_{1-x}Co_xPO_4/C$ cathode material. The results show that with doping cobalt, the lattice size increases due to the larger ionic radius of cobalt, and the conductivity increases significantly from 10^{-9} Scm^{-1} to 10^{-4} Scm^{-1}. Therefore, the capacity of $LiNi_{1-x}Co_xPO_4/C$ in the initial cycle increases to 140 $mAhg^{-1}$, while the un-doped LNP/C has only 90 $mAhg^{-1}$. Besides, the EIS measurement also shows a decrease in interfacial resistance in the doped materials [19]. Although the literature reported that cobalt doping improves the performance of LNP material, however, all the doped samples have been modified with carbon material as well. Their data also show significant improvement of the conductivity (10^{-4} Scm^{-1}) for the un-doped LNP modified with carbon material. Therefore, no obvious evidence could confirm whether the improvement of conductivity is mainly attributed to the doping of cobalt or the modified carbon layer on the surface. Other experiment works reported by Kim et al. demonstrate the effect of doping on LNP cathode. Their data shows the electronic conductivity can be increased from 10^{-9} Scm^{-1} to 10^{-8} Scm^{-1} as manganese ions be doped into LNP forming $LiNi_{1-x}Mn_xPO_4$ and without surface carbon layer modification. Their discharging test also shows improvement in the first discharge capacity, which increased from 76 $mAhg^{-1}$ to 94 $mAhg^{-1}$. This capacity, however, is far lower than the theoretical capacity of LNP which is about 170 $mAhg^{-1}$, and the capacity retention is only about 50% after 100 cycles [29]. Based on the above two papers, it can be found that although doping with cations can help to improve the conductivity, the improvement is about an order of magnitude. Surface carbon materials modification is the key contributor to conductivity improvement. Another issue for cation doping is that when dopes other ions with LNP cathode, the dopant will average the redox potential. For example, nickel has the highest redox potential, so doping will result in a decrease in the operating potential, which leads to a lower energy density. This is not the phenomenon that people want to see in the high-voltage cathode [30].

In contrast to the cation doping method, surface layer modification is an alternative method to enhance the electrical conductivity of LNP, which would also help the lithium-ion conductivity at the interface. Besides those, the particle size of cathode materials is also a key factor in the performance. Generally, a small particle would show better conductivity and cycle life. Örnek et al. synthesized the LNP and LNP/C sample by using the microwave heating method. They found that the particle size of LNP could reduce from 100–400nm to 20–100nm after the addition of carbon to form the core-shell structure as observed by the SEM. The HR-TEM analysis indicates that the carbon layer on the surface of LNP/C is amorphous. The cyclic voltammetry shows that the oxidation and reduction peak of LNP/C is more than 2 times higher than that of LNP, and the resistance of the interface observed from the EIS analysis is also significantly reduced, indicating that the carbon layer modification can effectively improve the kinetic performance of lithium ions. The cyclic performance of LNP/C can reach a capacity of 150 $mAhg^{-1}$ which is close to the theoretical capacity of LNP 170 $mAhg^{-1}$ with 92% capacity retention after 100 cycles [20]. From the above discussion, it can be found that the modification of LNP cathode material by the amorphous carbon layer can be a more effective way to solve the low electrochemical performance issue.

8.4 CATHODE SURFACE AND INTERFACE PROPERTIES

In order to investigate the effect of surface modification on LNP, it is necessary to have a good understanding of its surface properties. The simulation methods are usually used to explore the microscopic phenomena that are difficult to observe in experiments. The surface properties and structure of LNP are not well investigated at the present stage, however, there is literature that discusses the surface properties of LFP which has the same crystalline structure as LNP.

G. Ceder et al. and M. Saiful Islam et al. [31,32] utilized the theoretical simulation method to establish the surface model of LFP. They calculated the surface energy of the low and high index surfaces, and then build up the Wulff shape of LFP to determine its particle's surface morphology. Their results indicate that most of the LFP surfaces are TASKA III type surfaces, meaning that the polarity in the same plane of LFP is not zero and the polarity cannot be eliminated by repeated layers. Therefore, surface reconstruction should occur at the LFP surface to cancel the surface dipole. LFP surface reconstruction is mainly achieved by taking 50% of the lithium-ion from the bottom layer and placing it on the top layer so that the polarities of the top and bottom are offset with the center of the surface model as the symmetric axis [33]. Besides, not only do surface lithium ions need to reconstruct, but the oxygen atoms need to reconstruct as well to form complete PO_4 bonding to reduce the instability of the surface due to the surface dangling bonds.

After calculating the surface energy and establishing the Wulff shape of LFP, G. Ceder et al. and M. Saiful Islam et al. [31,32] found that (010) (100) (011) (201) and (101) are thermodynamically stable surfaces of LFP in the equilibrium condition. Also, by calculating the lithium intercalated redox potential of different surfaces and considering the Wulff shape, the (010) surface turns out to be the main surface for lithium-ion intercalating. Also, the exposure of (010) surface is important for olivine-type cathode since the facile pathway for lithium-ion transport is along [010] direction. Based on their work, literature reported the methods to synthesize the (010) prefer-facet olivine type cathode materials and the investigation of the LFP interface focusing on the (010) surface.

W. T. Geng et al. investigated the structure of the carbon layer adsorbed on the (010) surface of LFP by DFT calculations [34]. First, he used a single layer of graphene to investigate the adsorption direction of the graphene on the surface, he found that the graphene had twice as large binding energy as it is standing on the surface than being flat over the surface. He also found that the bond length between carbon-oxygen is 17% shorter than that between carbon-iron indicating surface oxygen site is the more stable adsorption site. The long carbon-iron bond distance is presumed to be caused by the steric barrier generated by the surface structure. These results indicate that the carbon layer adsorbed on surface oxygen sites might be the more stable structure. The works done for LFP discussed above provide a reference for the research of other olivine-type high-voltage cathode materials.

8.5 CALCULATION METHOD

To study the effect of carbon layer modification on the LNP surface, one must first investigate the LNP surface property and structure. Here, the first-principles calculation method was chosen. The surface model of LNP referred to the surface model of LFP in the previous chapter and then built and calculated the surface energy of the five low-index surfaces (010), (100), (011), (201), and (101). The surface energies were then used to establish the Wulff shape of LNP by the Wulffmaker software [35]. The electronic properties of the surface models were compared with those of the bulk models to understand the surface effect.

First-principles calculation is one of the most widely used simulation methods in computational chemistry and solid-state physics. Its potential literally does not require experimental data while developing. The method can be used for the prediction of the material properties which can reduce the time for the try-and-error process in developing new materials. The main issue of the First-principles calculations method is the cost of computation time and the size of the calculated model. Nowadays, with computing power growing dramatically, the First-principles calculation becomes an indispensable tool in the materials research field.

The density-functional theory is an important method in the first-principles calculation, it makes the solving of the multi-electron Schrodinger equation feasible by transforming the corresponding equation to the Kohn-Sham equation. From that, one could obtain the electronic structure in the ground state, then various physical properties could be derived from it.

In this study, all the calculations were performed with a commercially available Vienna ab initio simulation package (VASP), based on a plane wave basis set [36]. For the interactions between ionic cores and valence electrons, the projector augmented wave (PAW) potential was utilized with Hubbard U correction (DFT+U) [37]. The exchange-correlation function used in this study is generalized gradient approximation (GGA) with parameterization of Perdew-Burke-Ernzerhof (PBE) [38]. The DFT-D3 semi-empirical correction is considered for the effect of interatomic van der Waals forces [39]. The plane wave cut-off energy is expanded until a kinetic energy convergence of 520eV is attained. Surface relaxation and structural optimization of the slab models were carried out on a $3 \times 3 \times 1$, $3 \times 3 \times 1$, $2 \times 3 \times 1$, $1 \times 3 \times 1$, $2 \times 2 \times 1$, k-point mesh for (010), (100), (201), (101), (011) surface tested by the convergence test. The criterion for geometry optimization was defined as the action of the Hellman-Feynman forces on atoms that are less than 0.01 eV/Å and the external pressure less than 0.8 kB.

8.6 LNP SURFACE PROPERTIES

8.6.1 LNP Bulk and Surface Model

A $2 \times 1 \times 1$ supercell of LNP bulk model with an orthorhombic structure and the Pnma space group, as reported in the experiment [40,41] was built and optimized by considering the magnetic moment along the b-axis. The optimized bulk model is shown in Figure 8.1. The result of lattice constants when compared with the experimental data has an error within 1%, as shown in Table 8.1 [20].

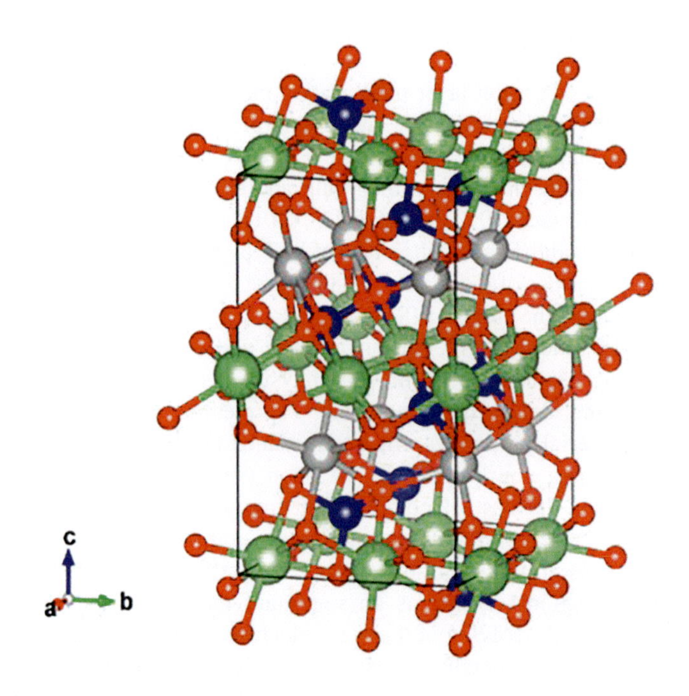

FIGURE 8.1 The optimized LNP bulk model.

TABLE 8.1
Lattice Constant Obtained in This Study and the Experiment Results [20]

	Experiment	Simulation	Error
a	10.05	10.06	0.050%
b	5.85	5.86	0.102%
c	4.68	4.69	0.107%

The surface (slab) models of (010), (100), (011), (201), and (101) were established by cutting the bulk model into different facets. The surface reconstruction was applied according to the aforementioned literature. The slab model is then optimized again to release the artificial stress. The thickness of slab models was tested by comparing the density of states (DOS) of center layers with the bulk model. The vacuum layer of 15 *Å* was chosen after the convergency test. The reconstructed and optimized surface models are shown in Figure 8.2.

8.6.2 Property Analysis

The surface energy was calculated by the optimized energy after slab relaxation and the surface area of the relaxed slab by Equation (8.1).

$$\frac{E_{unrelax} - n * E_{bulk}}{2A_1} + \frac{E_{relax} - E_{unrelax}}{2A_2} \tag{8.1}$$

Where

E_{bulk}: energy of bulk per formula
n: number of formulas in the surface model
A1: surface area before optimization
A2: surface area after optimization
Erelax: energy after optimization
Eunrelax: energy before optimization

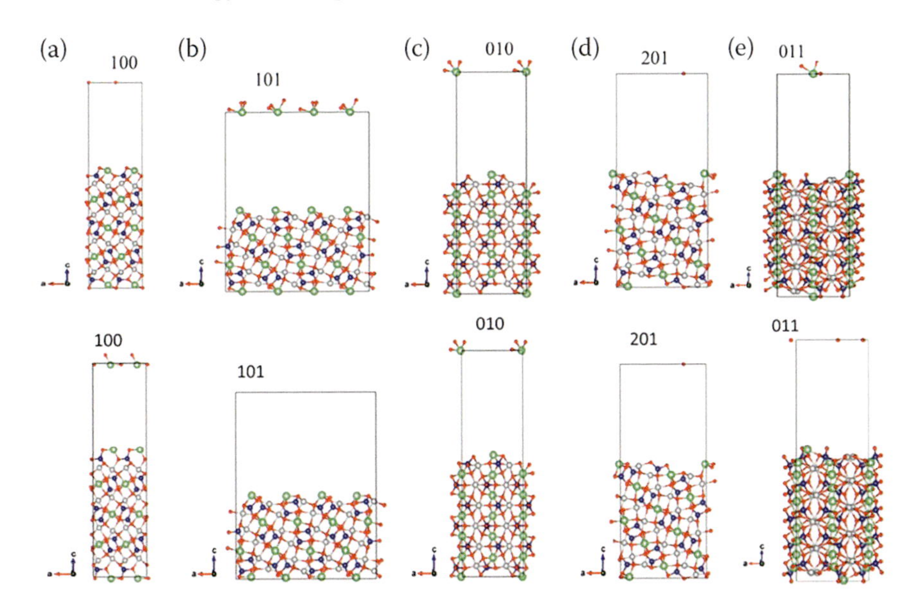

FIGURE 8.2 The reconstructed and optimized LNP surface (slab) model. (a) (100) slab model (b) (101) slab model (c) (010) slab model (d) (201) slab model and (e) (011) slab model. The up figure is initial reconstructed surface models, the down figure are the reconstructed models after optimization.

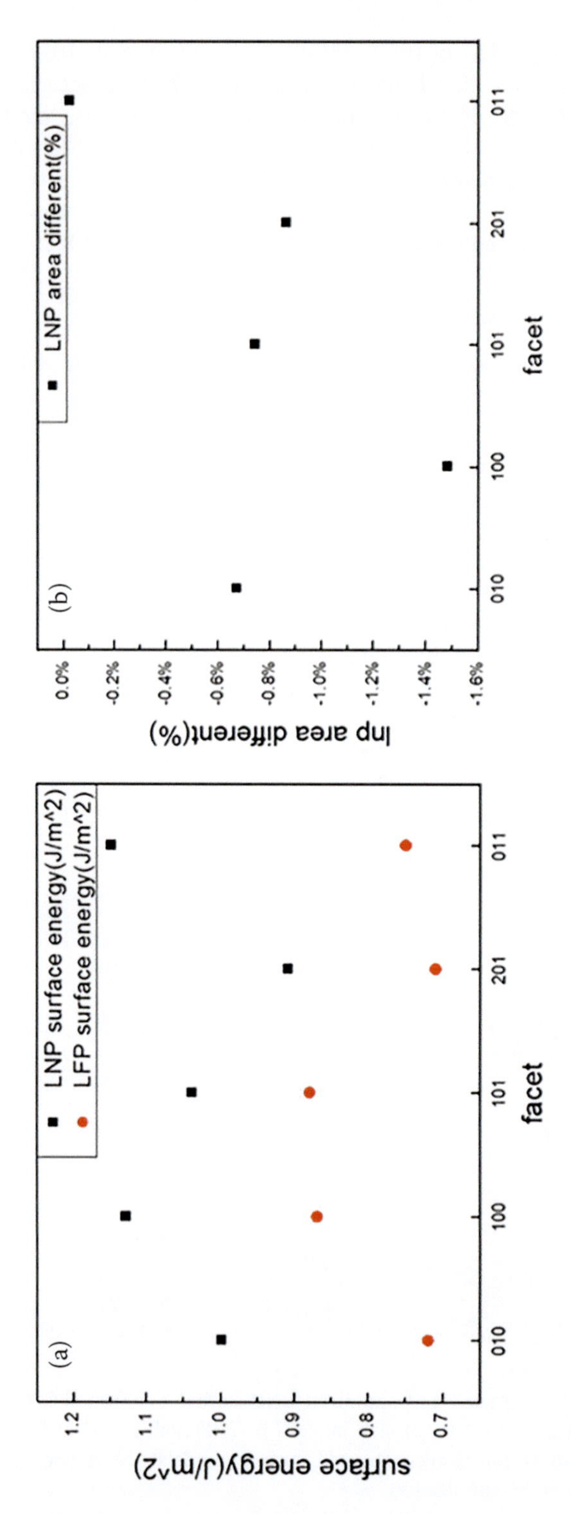

FIGURE 8.3 (a) Surface energy of LNP slab models and (b) the percentage of surface area difference before and after relaxation.

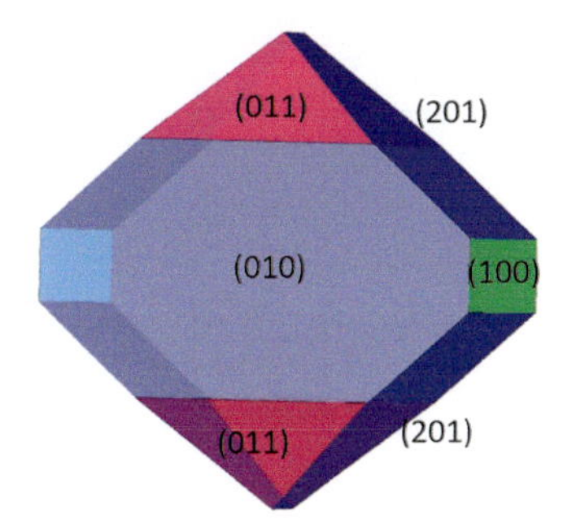

FIGURE 8.4 The Wulff shape of LNP.

Since the LNP slab models were built based on the results of LFP reported in the literature, the surface energy was compared with that of the LFP as shown in Figure 8.3. From the figure, one can observe that the surface energy trends of LNP and LFP are similar, specifically, (201) and (010) surfaces have the lowest and second lowest surface energy, respectively. There is also no significant change in the surface area of LNP slab model before and after surface optimization.

The Wulff shape can be constructed with the information of bulk lattice constants, space group, and surface energies of various surfaces by Wulffmaker software [35]. The result is shown in Figure 8.4. From the LNP Wulff shape, it can be found that the (010) and (201) surfaces are the two dominant facets indicating that they are the most exposed surfaces in the thermodynamic equilibrium condition. This result is consistent with the plate-like LNP particles observed in the experiment [24].

The electronic properties of different surfaces can be obtained by conducting density of state analysis of each surface by using the Gaussian smearing method and the width of the smearing is 0.1eV, as shown in Figure 8.5, where it can be observed that the energy gap of the LNP surface decreases from about 4eV to 2.57eV, 2.75eV, 2.03eV, 2.43eV, 1.97eV for (010),(100),(101),(201),(011) surface compared to the bulk and the fermi level of the (010) surface shift near to the conduction band, indicating that the (010) surface should have better conductivity during charging and discharging process.

8.7 CONCLUSION

This study used the first-principles calculation method to study the $LiNiPO_4$ surface properties. The LNP Wulff shapes that build from the five low-index surfaces (010), (100), (011), (201), and (101) show that the LNP surface is mainly exposed with (010) and (201) at equilibrium condition. The (010) surface has a smaller bandgap than other surfaces indicating it might have better electronic conductivity properties.

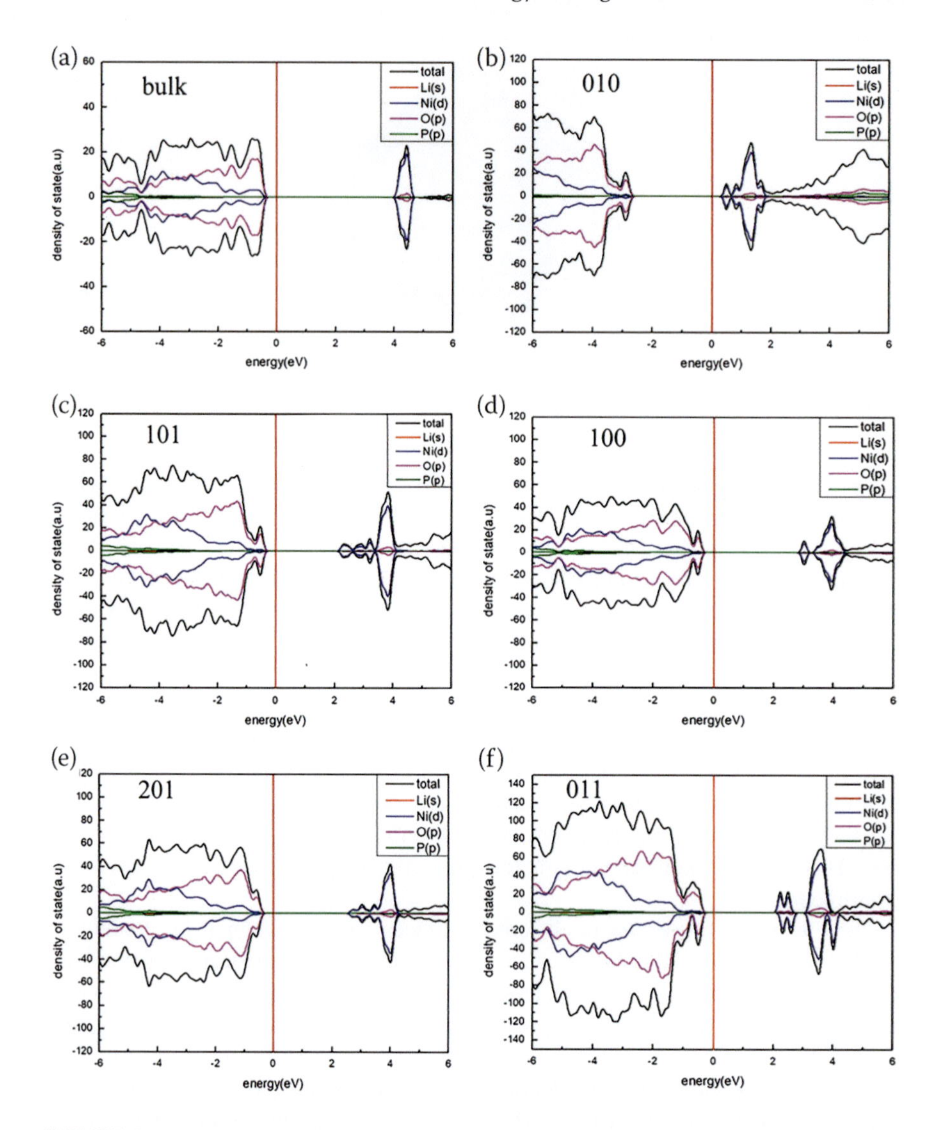

FIGURE 8.5 The density of state (DOS) of the bulk and slab models of LNP. (a) Bulk PDOS (b) 010 PDOS (c) 101 PDOS (d) 100 PDOS (e) 201 PDOS (f) 011 PDOS. The Fermi level is shift to zero.

ACKNOWLEDGMENTS

This work was financially supported by the Hierarchical Green-Energy Materials (Hi-GEM) Research Center.

REFERENCES

1. Mizushima, K., et al., Li_xCoO_2 (0<x<-1): a new cathode material for batteries of high energy density. *Materials Research Bulletin*, 1980. **15**(6): p. 783–789.

2. Goodenough, J.B. and K.-S. Park, The Li-ion rechargeable battery: a perspective. *Journal of the American Chemical Society*, 2013. **135**(4): p. 1167–1176.
3. Chung, D., E. Elgqvist, and S. Santhanagopalan, Automotive lithium-ion cell manufacturing: Regional cost structures and supply chain considerations. 2016, National Renewable Energy Lab.(NREL), Golden, CO (United States).
4. Meng, Y.S. and M.E. Arroyo-de Dompablo, Recent advances in first principles computational research of cathode materials for lithium-ion batteries. *Accounts of Chemical Research*, 2013. **46**(5): p. 1171–1180.
5. Nitta, N., et al., Li-ion battery materials: present and future. *Materials Today*, 2015. **18**(5): p. 252–264.
6. Yazami, R., et al., Mechanism of electrochemical performance decay in $LiCoO_2$ aged at high voltage. *Electrochimica Acta*, 2004. **50**(2-3): p. 385–390.
7. Rozier, P. and J.M. Tarascon, Review—Li-rich layered oxide cathodes for next-generation Li-Ion batteries: chances and challenges. *Journal of the Electrochemical Society*, 2015. **162**(14): p. A2490–A2499.
8. Ryu, H.-H., et al., Capacity fading of Ni-Rich $Li[Ni_xCo_yMn_{1-x-y}]O_2$ ($0.6 \leq x \leq 0.95$) cathodes for high-energy-density lithium-ion batteries: bulk or surface degradation? *Chemistry of Materials*, 2018. **30**(3): p. 1155–1163.
9. Xia, Y., et al., Designing principle for Ni-rich cathode materials with high energy density for practical applications. *Nano Energy*, 2018. **49**: p. 434–452.
10. Mohanty, D., et al., Resolving the degradation pathways in high-voltage oxides for high-energy-density lithium-ion batteries; Alternation in chemistry, composition and crystal structures. *Nano Energy*, 2017. **36**: p. 76–84.
11. Yan, P., et al., Intragranular cracking as a critical barrier for high-voltage usage of layer-structured cathode for lithium-ion batteries. *Nature Communications*, 2017. **8**: p. 14101.
12. Chakraborty, A., et al., Layered cathode materials for lithium-ion batteries: review of computational studies on $LiNi_{1-x-y}Co_xMn_yO_2$ and $LiNi_{1-x-y}Co_xAl_yO_2$. *Chemistry of Materials*, 2020. **32**(3): p. 915–952.
13. Julien, C.M. and A. Mauger, Review of 5-V electrodes for Li-ion batteries: status and trends. *Ionics*, 2013. **19**(7): p. 951–988.
14. Shin, Y. and A. Manthiram, Factors influencing the capacity fade of spinel lithium manganese oxides. *Journal of the Electrochemical Society*, 2004. **151**(2): p. A204.
15. Lin, M., et al., Insight into the atomic structure of high-voltage spinel $LiNi_{0.5}Mn_{1.5}O_4$ cathode material in the first cycle. *Chemistry of Materials*, 2015. **27**(1): p. 292–303.
16. Wang, B., et al., Desired crystal oriented $LiFePO_4$ nanoplatelets in situ anchored on a graphene cross-linked conductive network for fast lithium storage. *Nanoscale*, 2015. **7**(19): p. 8819–8828.
17. Wu, F.X., J. Maier, and Y. Yu, Guidelines and trends for next-generation rechargeable lithium and lithium-ion batteries. *Chemical Society Reviews*, 2020. **49**(5): p. 1569–1614.
18. Rommel, S.M., et al., Challenges in the synthesis of high voltage electrode materials for lithium-ion batteries: a review on $LiNiPO_4$. *Monatshefte für Chemie – Chemical Monthly*, 2014. **145**(3): p. 385–404.
19. Örnek, A., E. Bulut, and M. Can, Influence of gradual cobalt substitution on lithium nickel phosphate nano-scale composites for high voltage applications. *Materials Characterization*, 2015. **106**: p. 152–162.
20. Örnek, A. and M.Z. Kazancioglu, A novel and effective strategy for producing core-shell $LiNiPO_4$/C cathode material for excellent electrochemical stability

using a long-time and low-level microwave approach. *Scripta Materialia*, 2016. **122**: p. 45–49.

21. Fisher, C.A.J., V.M. Hart Prieto, and M.S. Islam, Lithium battery materials $LiMPO_4$ (M = Mn, Fe, Co, and Ni): insights into defect association, transport mechanisms, and doping behavior. *Chemistry of Materials*, 2008. **20**(18): p. 5907–5915.
22. Kutteh, R. and M. Avdeev, Initial assessment of an empirical potential as a portable tool for rapid investigation of Li+ diffusion in Li+-battery cathode materials. *The Journal of Physical Chemistry C*, 2014. **118**(21): p. 11203–11214.
23. Ramana, C.V., et al., Structural characteristics of lithium nickel phosphate studied using analytical electron microscopy and Raman spectroscopy. *Chemistry of Materials*, 2006. **18**(16): p. 3788–3794.
24. Kempaiah Devaraju, M., et al., Synthesis, characterization and observation of antisite defects in $LiNiPO_4$ nanomaterials. *Scientific Reports*, 2015. **5**(1): p. 11041.
25. Shi, J., Effect of antisite defects on Li Ion diffusion in $LiNiPO_4$: a first principles study. *International Journal of Electrochemical Science*, 2016: p. 9067–9073.
26. Tolganbek, N., et al., Current state of high voltage olivine structured $LiMPO_4$ cathode materials for energy storage applications: a review. *Journal of Alloys and Compounds*, 2021. **882**: 160774.
27. Ornek, A., The synthesis of novel $LiNiPO_4$ core and Co_3O_4/CoO shell materials by combining them with hard-template and solvothermal routes. *Journal of Colloid and Interface Science*, 2017. **504**: p. 468–478.
28. Karthickprabhu, S., et al., Electrochemical and cycling performance of neodymium (Nd^{3+}) doped $LiNiPO_4$ cathode materials for high voltage lithium-ion batteries. *Materials Letters*, 2019. **237**: p. 224–227.
29. Karthikprabhu, S., et al., Electrochemical performances of $LiNi_{1-x}Mn_xPO_4$ ($x = 0.05$–0.2) olivine cathode materials for high voltage rechargeable lithium ion batteries. *Applied Surface Science*, 2018. **449**: p. 435–444.
30. Snydacker, D.H. and C. Wolverton, Transition-metal mixing and redox potentials in $Li_x(M_{1-y}M'_y)PO_4$ (M, M′ = Mn, Fe, Ni) olivine materials from first-principles calculations. *The Journal of Physical Chemistry C*, 2016. **120**(11): p. 5932–5939.
31. Wang, L., et al., First-principles study of surface properties of $LiFePO_4$: surface energy, structure, wulff shape, and surface redox potential. *Physical Review B*, 2007. **76**(16): p. 165435.
32. Fisher, C.A.J. and M.S. Islam, Surface structures and crystal morphologies of $LiFePO_4$: relevance to electrochemical behaviour. *Journal of Materials Chemistry*, 2008. **18**(11): p. 1209–1215.
33. Tasker, P.W., The stability of ionic crystal surfaces. *Journal of Physics C: Solid State Physics*, 1979. **12**(22): p. 4977–4984.
34. Geng, W.T., et al., Formation of perpendicular graphene nanosheets on $LiFePO_4$: a first-principles characterization. *The Journal of Physical Chemistry C*, 2012. **116**(33): p. 17650–17656.
35. Zucker, R.V., et al., New software tools for the calculation and display of isolated and attached interfacial-energy minimizing particle shapes. *Journal of Materials Science*, 2012. **47**(24): p. 8290–8302.
36. Kittel, C. and P. McEuen, Kittel's introduction to solid state physics. 2018: John Wiley & Sons.
37. Dudarev, S.L., et al., Electron-energy-loss spectra and the structural stability of nickel oxide: An LSDA+U study. *Physical Review B*, 1998. **57**(3): p. 1505–1509.
38. Perdew, J.P., K. Burke, and M. Ernzerhof, Generalized gradient approximation made simple. *Physical Review Letters*, 1996. **77**(18): p. 3865–3868.

39. Grimme, S., et al., A consistent and accurate ab initio parametrization of density functional dispersion correction (DFT-D) for the 94 elements H-Pu. *The Journal of Chemical Physics*, 2010. **132**(15): p. 154104.
40. Shang, S.L., et al., Lattice dynamics, thermodynamics, and bonding strength of lithium-ion battery materials $LiMPO_4$ (M = Mn, Fe, Co, and Ni): a comparative first-principles study. *Journal of Materials Chemistry*, 2012. **22**(3): p. 1142–1149.
41. Rui, X., et al., Olivine-type nanosheets for lithium ion battery cathodes. *ACS Nano*, 2013. **7**(6): p. 5637–5646.

9 Introductory to Machine Learning Method and Its Applications in Li-Ion Batteries

Zhe-Yun Kee
Department of Materials Science and Engineering, National Cheng Kung University, Tainan City, Taiwan

Ngoc Thanh Thuy Tran
Hierarchical Green-Energy Materials (Hi-GEM) Research Center, National Cheng Kung University, Tainan City, Taiwan

CONTENTS

DOI: 10.1201/9781003367215-9

9.1 INTRODUCTION

The readily available large databases through the ubiquity of the internet and tremendous improvement in computing capability from the past decades such as the generalization of dedicated graphics cards and the Compute Unified Device Architecture (CUDA) parallel computing platform have provided a great foundation for the popularization of ML as it speeds up data processing of machine learning (ML) models. Furthermore, the most important factor that helps in democratizing ML is the rise of Python following the open-source libraries of AI applications that enhance the learning curve of ML. Not to be confused with the high-throughput calculation in which the machine was used merely as a tool to manipulate data following a predefined set of instructions, instead, ML is a method that enables the machine itself to extract specific sets of rules from large amounts of data focusing on the desired output that could be utilized for future application.

Broadly speaking, ML algorithms can be categorized into three types depending on their purpose, namely, supervised learning, unsupervised learning, and reinforced learning. Firstly, the supervised learning algorithm consists of a target/output variable (corresponding/dependent variable) which is to be predicted from a given set of predictors/inputs (manipulated/independent variable), with these so-called labeled data, we generate a function that maps the inputs to the desired output. As more data is used to train and calibrate the function, it eventually achieves the desired level of accuracy on the training data. The more diverse the data points, the more accurate the model will be in predicting the properties of unknown materials. We could then use this resulting function on novel material to realize properties prediction.

On the other hand, unsupervised learning focuses on clustering the data into groups and classes with the given properties. This method is favorable when the grouping of the data is more important than the exact properties of the data, such as customers' segmentation by age for different ads promotion.

The third type is reinforced learning, in which the ML model is trained to make decisions based on a set of rewarding mechanisms. The model is placed in an environment where it trains itself continuously via trial and error, and no specific restriction is imposed on the model, except a rewarding mechanism that keeps track of the score of the model throughout the training. A score is granted when the behavior of the model leads to the desired result or the behavior itself is desired and deducted from the undesired one. The ultimate target of the training is to train the model to achieve the highest score possible.

Within the supervised learning regime, further classification of the model is possible depending on the type of output result. When the output variable is a continuous value of some properties, the model is known as a regressor. The process of regression (hence the name) is carried out in the model to map the correlation between the inputs and the continuous output variables and ultimately find the best fitting function which can predict the output accurately with the provided information, e.g., weather prediction, house price prediction, etc.

While the output result is a discrete value that represents some classes or groups, it is known as a classifier. The classification process is done within the model to find a function that can divide the dataset into classes based on the given parameters. The goal of the classifier is to find a decisive boundary that can divide the dataset into different classes as confidently/distinctively as possible, for example, the identification of spam email, cancer cell, etc.)

In this chapter, a general introduction to machine learning (ML) and its application to the field of batteries will be given. In order to provide a better understanding of the supervised learning method which is the main application in the field of lithium-ion batteries (LIBs), further details will be presented.

9.2 METHODOLOGY

9.2.1 Target Identification

The first step in establishing an ML model is to identify the goals and the prediction targets. This is arguably the most important part as the identified target must be potentially predictable from information that is cost-effective in terms of time and resources, as the ideal of ML is to speed up productivity, it defeats the purpose if the time spent on collecting data for the ML model is on par with acquiring the target property directly with first-principles method such as density functional theory (DFT) or even experiments (1).

9.2.2 Data Collection

Following the choice of prediction target, is the collection of training data, either experimentally or more commonly, computationally, as they are usually less time-consuming and cost-effective. Data can be self-generated by conducting experiments and performing high-throughput computation via various software, or from open-source databases that are available throughout the internet. The final ML results are directly affected by the volume and reliability of the data, thus datasets that are sufficient in quantity and high quality are very critical. The quality of the data is determined by the coverage of chemical space and also the diversity associated with the data. The training data should be able to represent the future data and have similar properties distribution, as the distribution bias on the predictors may have adverse effects on the prediction result if the future predicting sets have a significantly different biases on properties distribution or chemical space. It should be noted that the data uncertainty arises from the experimental errors or computational errors in the training datasets that are expected to present in the ML result, as ML can only produce results that are as accurate as the inputs to the best extent. Experimental Crystal/molecular structure databases have been steadily growing over the past decades, such as the Pauling File Database (2), Inorganic Crystal Structure Database (ICSD) (3), Pearson Crystal Data, Cambridge Structural Database (4), Crystal Open Database (5), CRYSTMET (6), ZINC database (7), PubChem (8), etc. Computational databases, particularly based on DFT, have been emerging in recent years due to the Materials Genome Initiative, e.g., Materials Project (9), AFLOWLIB (10), the computational materials repository (11), open quantum materials data (OQMD) (12), AiiDA (13), JAVIS- DFT (14), CatApp (15), etc.

Although ab initio calculation has only been on the rise for the past few decades, owing to its nature, computed data has unmatched quantity and reliability/consistency compared to experimental data. Many existing experimental data repositories are too small and inconsistent (different experiment conditions and measurement techniques) for high-quality ML models.

9.2.3 Data Processing

Before getting into the actual training session of the ML model, preprocessing on the dataset is required to ensure the quality of the training result. Many problems may still present in the as-collected data or any other readily available datasets, such as abnormal values and duplicate data points. Therefore, data processing is necessary for reducing the amount of futile calculation and obtaining an efficient ML model. Data cleaning can be defined as data operation consisting of the following steps: abnormal data processing, data discretization, data normalization, and data sampling (16).

Abnormal data processing is imperative for the accuracy of the ML model, as the model learns from the data that it is provided with. This step includes the elimination of duplicate data as a result of data mishandling, and abnormal data (outliers) that exhibit extreme values that lie outside the expected range. Besides, one may need to resolve inconsistent data values if the training data is concatenated or merged from multiple datasets.

In addition, data discretization can significantly reduce the number of possible values of continuous features and speed up the overall learning process since a large number of possible feature values is inefficient for inductive ML algorithms. In a nutshell, data discretization is a method of converting continuous feature values into a finite set of intervals with minimum data loss. Generally, data discretization is preceded by an algorithm that groups the data into intervals based on different sets of rules, such as supervised/unsupervised discretization which depends on whether the class labels of data are considered. Discretization algorithms are also divided into top-down and bottom-up methods based on how the process took place. Top-down methods start from a single interval and recursively split into smaller intervals, while bottom-up methods start from a set of single-value intervals and iteratively merge neighboring intervals (17).

Furthermore, data normalization can be used to adjust the magnitudes of data to a suitable level, which is crucial for many ML algorithms, for example, the deep learning family, as the values increase exponentially propagating down the network.

Finally, in the preprocessing step is data sampling. For a large enough dataset, data sampling ensures researchers can complete the ML training with fewer data, thus reducing cost and greater speed, without compromising the predicting accuracy. Three sampling methods are commonly used in ML, namely, random sampling, systematic sampling, and stratified sampling. Random sampling is rather straightforward, in which the samples are drawn from the dataset with a uniform probability. For systematic sampling, the samples are drawn from the dataset with a predefined pattern. In the case of stratified sampling, samples are drawn within a predefined stratum, which are subsets divided from the whole dataset using different rules. Although data sampling is generally beneficial to ML, poor sampling criteria may introduce sampling errors into data samples. Common errors include selection bias and sampling error, which emanate from the random nature that is inherent to any sampling method due to not working with the full datasets.

Data sampling is favorable in the field of pharmaceuticals and medicine as the dataset available are mostly very large and with great diversity, which is not the case in material sciences. Currently, the major problem of ML in the field of material sciences is the lack of a dataset that is good in quality as well as quantity. Performing data sampling on a relatively small dataset may lead to large bias or introduce large sampling errors as the resulting data sample may fail to represent the whole population. Thus, it is advised to omit this step on a small dataset unless it is necessary and supported by rigorous statistical analysis performed with meticulous care.

At this point, the data is prepared to be transformed into useful features or descriptors that are processable by the computer.

9.2.4 Feature Engineering

Feature engineering is the process of extracting the most appropriate numerical values that provide important information relating to the goal of the ML model, and at the same time distinguishes between different materials from a given data. These numerical values have been typically referred to as descriptors, features of fingerprints in the literature. In this chapter, the term 'descriptor' will be used to imply the representation vector of the materials while the term 'feature' will be used to imply the characteristic property of the materials. Descriptor refers to syntax or values that describe a mechanism or a property of a material. For example, the rows and columns in the periodic table can be considered as a set of descriptors, as the elements are arranged into the rows and columns following a set of rules based on their physical and chemical properties.

The appropriate descriptors integrate essential information that constitutes the materials, which allows the ML model to effectively learn and identify the hidden relationships among each independent variable and ultimately associate them with the output. Descriptors have a major impact on the model performance since the quality of the descriptors tremendously affects the final result, thus it is also the step that requires the most expertise. A suitable descriptor should fulfill the following criteria described by Luca M. Ghiringhelli (18) and the others (19–21) for optimal performance to be expected:

1. Compact: The descriptor should contain sufficient information about the system/material while keeping the dimension (feature count) as low as possible.
2. Uniqueness: The descriptor should characterize each material along with its relevant property or mechanism uniquely. (i.e., one-to-one)
3. Continuous: Materials that have similar structure/properties should have similar descriptor values and vice versa. (Materials that resemble one another should not have significantly different descriptor values)
4. Computationally cheap: The determination of the descriptor should not involve calculation as intensive as evaluating the property to be predicted.

Additional requirements are imposed to atomistic/bottom-up descriptors, i.e.:

1. Invariant with respect to:
   a. Spatial translation of coordinate system
   b. Rotation of coordinate system
   c. Permutation of atomic indices

In general, the descriptor sets should not be excessive in size compared to the data sets to avoid overfitting, this is particularly important in materials science as the available data sets tend to be small. Furthermore, descriptors that are highly correlated to one another should be avoided for efficient training as they only provide slightly more information at a much higher computational cost. A different strategy is adopted when working with datasets of different scales. When working with a small dataset, which is often the case in material science, Daniel C. Elton believes that characterization of the data is more important than the ML model itself, as a set of highly efficient descriptors that are hand selected with prior knowledge can ensure the accuracy of prediction (22). While dealing with large datasets, ML algorithms are able to extract complex and underlying features from the traits of the data. However, due to the large amount of data to be processed, descriptors with superior computational efficiency or experimental performance are suggested. Besides, abstract descriptors are preferred to ensure accuracy in the case of large datasets as abstract descriptors usually contain highly compact information in contrast to fundamental descriptors that contain explicit or straightforward information.

The descriptors can be roughly categorized into three types based on how they are assembled, i.e., chemical/composition-based, structure/distance-based, and bottom-up approach. The composition-based descriptor. Composition-based descriptors consist of any descriptors that are the known properties of composing elements and their derivatives, such as electronegativity, atomic number, etc. However, the drawback of this kind of descriptor is ostensible, as they are unable to extract any information related to position and distance. Thus, composition-based descriptors should be used under structural constraint, i.e., when the method is applied to materials of certain crystal/molecular structures of interest. As such, a descriptor set that provides full structural information is generally preferred in most cases. A functional crystal/molecular structural descriptor should be able to identify the symmetry present in the structure, i.e., invariant to translation, rotation, and permutation of identical atoms. Descriptors such as Coulomb matrix, Ewald sum matrix, sine matrix, etc. fall into this category. For molecules, representations that encode partial structure information (connectivity between molecular fragments) exist, such as Simplified molecular-input line-entry systems (SMILES) and Extended-connectivity fingerprint (ECFP). These descriptors can only distinguish isomers but not conformers as a portion of the three-dimension information is lost through encoding.

In recent years, the bottom-up approach has become popular in machine-learning-driven material discovery, as most of the information from the material is retained in the descriptor constructed from the local environment and iteratively merges into crystal level. These descriptors together with ML models constitute the Machine learning potential (MLP). Such descriptors include Many-body tensor representation (MBTR), Smooth overlap of atomic positions (SOAP), Atom-centered symmetry functions (ACSF), classical force-field inspired descriptors (CFID), etc. The approximation of locality does impose a limit on the accuracy of resulting MLP since any interactions outside the cutoff radius are neglected, such as long-range electrostatic interactions. However, these descriptors benefit from the locality of properties, i.e., each property can be considered as the summation of local contribution, enabling the potential to be readily applied to systems with any number of atoms. More specific descriptors may also be extracted from computational results to perform machine learning on investigated materials, e.g., the d-band center descriptor for metals.

9.2.5 ML Algorithms and Models

After having the dataset and the determination of appropriate descriptors, it is required to select a suitable ML algorithm based on the problem type. As the purpose of this chapter is to provide a brief introduction and promote the utilization of ML in LIBs, the following discussion will focus mainly on supervised algorithms as they are by far the most commonly used algorithm type in material-based research.

9.2.5.1 Regression Algorithm

Regression analysis algorithm as illustrated in Figure 9.1 can evaluate the magnitude in which the target variable is affected by the input variables by analyzing a large volume of data and a depending model used (linear or nonlinear), regression equations are fitted, and properties could be predicted for new materials discovery, such algorithm includes Gaussian process regression (GRP) (23), ordinary squares regression (OLSR) (24), partial least-square regression (PLSR) (25), etc. Regression model.

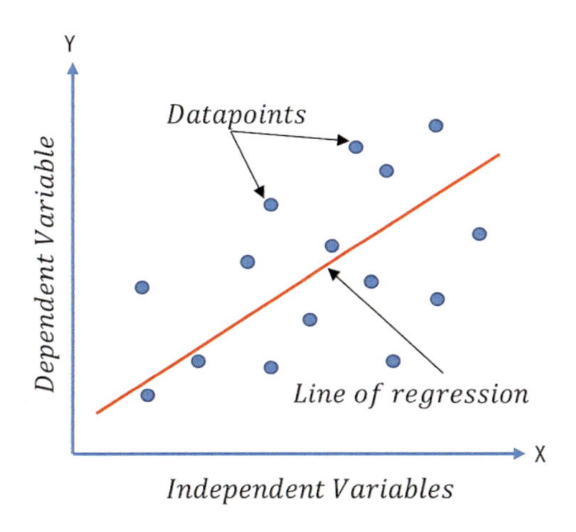

FIGURE 9.1 Schematic representation of regression model.

9.2.5.2 Support Vector Machine (SVM)

Support vector machine (26) is available for both regression and classification, either in the form of linear or nonlinear. Each data point is viewed as a P-dimensional vector in the support vector machine, the support vector machine will try to find a set of hyperplanes of P-1 dimension (the omitted dimension represents the class/group of the data point) to separate the data into predefined classes (Figure 9.2). For the case with two classes, many hyperplanes might be able to classify the data, but the one that represents the largest separation between the two classes (the distance to nearest data point on each class is maximized), so called maximum-margin hyperplane is chosen as the functional margin, since the larger the margin, the lower the generalization error is achieved on the classifier. SVM is created originally as a linear classifier, but it turns out that the datasets involved are often not linearly separable in the original space. Thus, modification (27) is suggested by mapping the original input space into a higher-dimensional space by replacing the dot product with a nonlinear kernel function, allowing the algorithm to fit the maximum-margin hyperplane in the transformed feature space.

9.2.5.3 K Nearest Neighbor (k-NN) Algorithm

k-NN algorithm is employed when the intrinsic data relationships are similar within a class or group. The “k” in the name indicates the number of neighbors to be considered, and “neighbors”’ are data points with known label (class or property value), e.g., an unlabeled data is classified according to a “majority voting” of its nearest neighbor (Euclidean distance is generally used to determine the “nearest” neighbor), with the resulting class being assigned to the most common among its k nearest neighbors (Figure 9.3). In the case of regression, the resulting output is the property value that is the average, or weighted average (the reciprocal of Euclidean

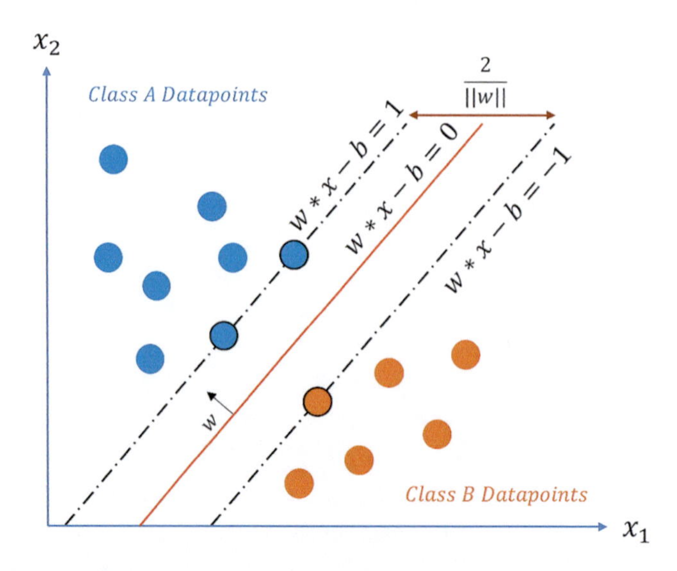

FIGURE 9.2 Schematic representation of support vector machine.

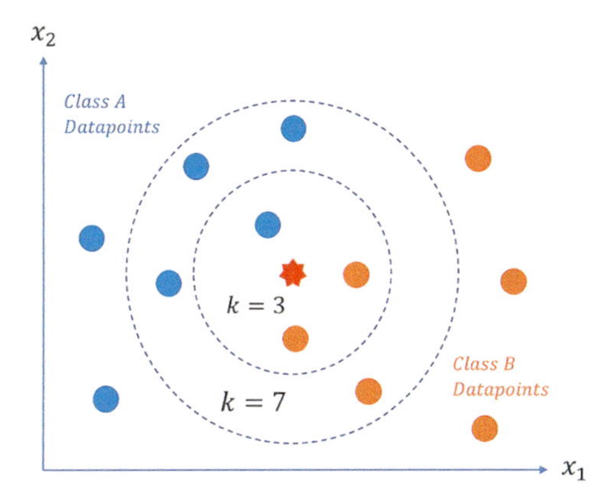

FIGURE 9.3 Schematic representation of k-NN model.

distance is commonly used so that nearer neighbors contribute more than the distant ones) from its k nearest neighbors. The advantage of the k-NN algorithm is that no explicit training step is required since no parameter is trained throughout the whole process. However, there are a number of caveats to applying the k-NN algorithm to the ML model. As the nature of k-NN algorithms considers only local information around the testing point, it is very sensitive to the local structure of the data, hence normalizing should be performed if vastly different scales are presented in the data. Besides, skewness (predominant of specific class) in the overall class distribution also causes the predicted result to be biased, which up/downscaling of data should be performed before applying the algorithm.

9.2.5.4 Tree-Based Algorithm

The tree-based algorithm is a type of rule-based model that classifies or performs regression on the dataset by splitting the source dataset into multiple subsets according to a series of splitting criteria based on input descriptors (Figure 9.4).

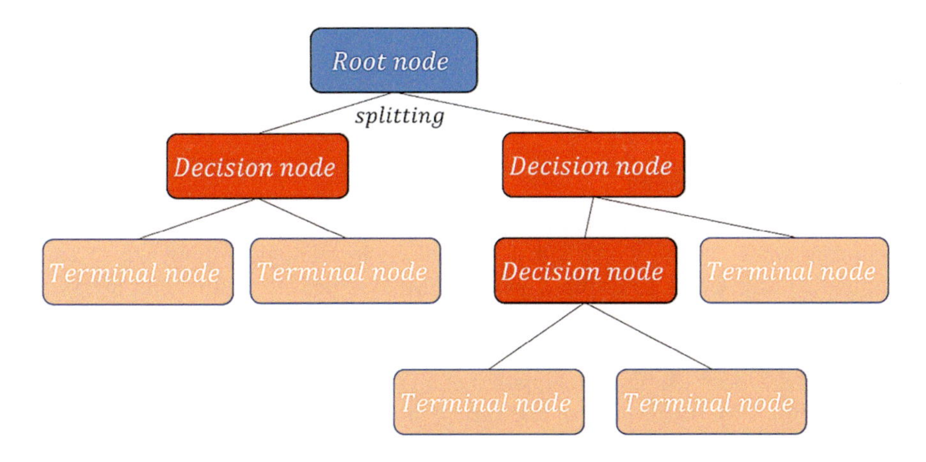

FIGURE 9.4 Schematic representation of decision tree.

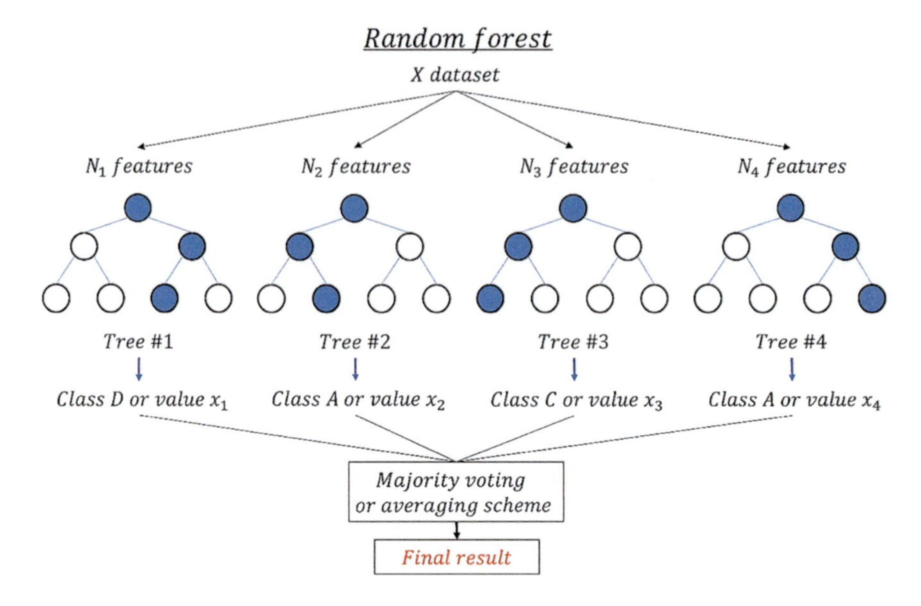

FIGURE 9.5 Schematic representation of random forest.

The fundamental form of the tree-based algorithm is binary decision tree, in which by repeating the above splitting process, a tree with two branches on each node (where decisions are made) is grown. However, the performance of primitive decision trees is limited by overfitting and generalizing weakness. Therefore, variation of such algorithms is created to overcome these shortcomings, such as Random Forest (RF) (28), Extremely randomized Trees (Extra trees) (29), lifting tree algorithm, etc. The random forest algorithm, composed of multiple decision trees, performs the splitting process individually, and the final decision is made based on the majority voting of all trees, hence remarkably improving the generalizing capability (Figure 9.5). While the Extra trees algorithm is very similar to the random forest algorithm, some modifications to the sampling method (whole original sample vs. bootstraps replica in the random forest) and cut point selection in node splitting (random split vs. optimal split) are made to reduce bias and variance.

9.2.5.5 Deep Learning

Deep learning originated from multilayer perceptron (MLP), also known as artificial neural network (ANN) which aims to simulate the biological brain where each computing unit works as artificial neurons. MLP is defined as multiple layers of neurons that progressively extract the underlying features with higher levels of abstraction from the inputs as the data pass down the hierarchy. In its conceptually simple model, deep learning is composed of an input layer, hidden layer, and output layer (Figure 9.6). As the number and variety of layers increase (hence the "deep" in the nomenclature), the features obtained by the algorithm will be more abstract as the information is condensed and combined as it passes through each layer.

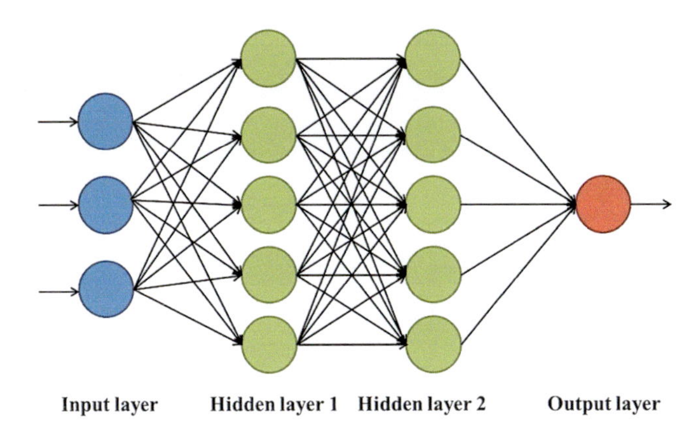

FIGURE 9.6 Artificial neural network model.

One major advantage of deep learning is that it is capable of learning the feature representation from the data and hence, no explicit feature engineering is required on the input data.

Since the proposal of deep learning, many important algorithms have been developed, for example, convolution neural networks (CNN), recurrent neural network (RNN), restricted Boltzmann machine (RBM), automatic encoder, etc. Convolution neural network consists of an additional feature learning layer on top of the traditional MLP, which comprises convolution filters and pooling layers that extract features from images and convert the information into computer-readable data. CNN is commonly used in image recognition, computer vision, etc. Recently, the modified version of CNN, graph-CNN could be used with graph representation of molecules or crystals, as the graph provides atoms and bonds information of the structure (30). On the other hand, RNN considers the temporal sequence of input and the information from the previous time step will contribute to the current and future time steps. RNN works well on sequential data such as simplified molecular-input line-entry system (SMILES) string-like molecule representation (31).

9.2.5.6 Cross-Validation (CV)

Since the main goal of ML is material prediction, the viability and reliability of the model should be tested before any practical applications. In order to verify this, the dataset is first split into 2 parts, i.e., training and test sets (Figure 9.7). The training set is resampled several times(iteration), with part of the dataset left out of the training and is used as a validation set on each iteration, which provides guidance on model selection, features selection, hyperparameter tuning, etc. The performance across each split is averaged to provide an assessment of the model reliability, and to ensure the constructed ML model is reasonably general. Finally, the test set is used to provide an unbiased evaluation of the finalized model. However, CV could also lead to overconfidence in the predictive power of the model, as material discovery often involves the search for materials that possess extraordinary properties which lie beyond the scope of

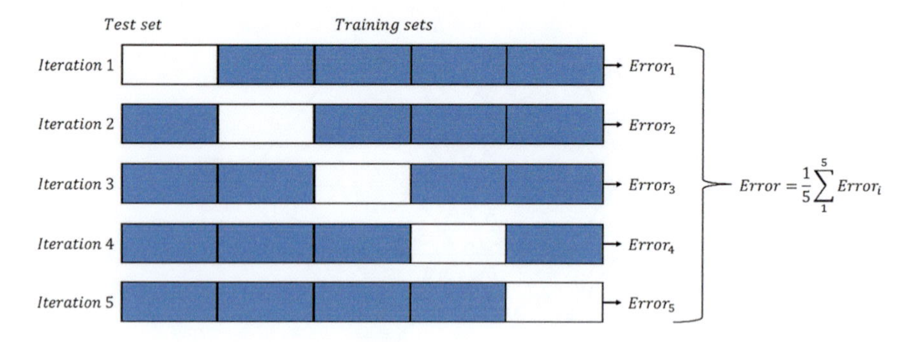

FIGURE 9.7 General working schemes of Cross-validation.

available training data, similar to an "outlier". Thus, the reliability of CV is established on the fact that the training sets and validation sets are capable of representing the entire data space in interest.

9.2.5.7 Leave-p-Out (LPO) Cross-Validation

P number of data points is left out from the training on each iteration, and is cycled throughout the whole dataset, with a resulting calculation of C_n^P for the CV (32). The high computational cost with a large dataset is impractical. A special form of LPO with the number p = 1, namely the Leave-one-out (LOO) (33), significantly reduces the number of subsets for CV from C_n^p to n. Although the computation load is greatly reduced, underestimation of predicting error and overfitting may occur.

9.2.5.8 K-Fold Cross-Validation

Currently the most commonly used CV method for generalization error estimation. The training set is split equally into k non-overlapping folds, one-fold is selected as the validation set and the remaining as training sets on each iteration, with only k calculation required for the CV (34). Caution should be taken on choosing the k number as large biases present on relatively small k numbers.

9.2.5.9 Bootstrap Cross-Validation

Multiple subsets are created from the original training set using the bootstrap method, in which each subset is resampled with replacement, leading to a dataset that contains multiple instances of the same cases while completely omitting other original cases (35). A different subset is used to train the model on each iteration. Bootstrap CV has less variability and biases under small amounts of sample, but computational cost increases sharply in large datasets.

9.3 APPLICATIONS

Before entering into the applications of ML on LIBs, a brief introduction to LIB is given as follows for the general idea of how it works and its basic structure.

9.3.1 Introduction to LIBs

LIB is a type of rechargeable battery that provides storage to disruptive energy sources and is a preeminent candidate to power our electrified transportation and portable devices. LIB comprises four different parts, namely, the cathode, anode, separator, and electrolyte system. The cathode/anode pair being the fundamental part of LIBs, provide a redox reaction pair for the conversion of energy and thus enables the storing of energy. The separator separates the cathode and anode while allowing the movement of Li ions through the cell; additional functions such as a failsafe to prevent thermal runaway are also commonly incorporated nowadays. Electrolyte systems could be categorized into liquid (major part of commercialized battery), all-solid-state, and quasi-solid-state electrolyte. The liquid electrolyte system is mainly composed of electrolytes, lithium salt, and functional additives. Electrolytes provide a stable migration route for the Lithium ions, and functional additives aim to enhance the overall performance of the battery via various mechanisms, such as promoting the formation of stable SEI or acting as a sacrificial additive to prevent the degradation of electrolytes. The lithium salt produces excessive Lithium ions in the battery system to facilitate the exchange of Lithium ions between the cathode/anode pair, i.e., the excessive ions could directly incorporate into either the anode or cathode, when possible, without actually waiting for the ions to diffuse across the whole pathway.

LIBs store energy by converting electric energy to chemical energy through the redox reaction of active species (transition metal ions) in the cathode material and the migration of Lithium ions. The energy is converted via different pathways according to different converting mechanisms, i.e., intercalation, conversion, and alloying. The most commercially successful intercalation type battery is given as an example to elucidate the process of energy conversion. During charging, a reverse bias is applied across the cathode and anode, forcing the oxidation of transition metal (TM) ions within the cathode, accompanied by the flow of electrons to anode via external circuit. At the same time, the Lithium ions migrate toward the anode material and intercalate into the anode. While during discharging, an external load is connected to the battery, the inherent potential difference between the cathode and anode causes a spontaneous.

9.3.2 ML on LIBs

The system of crystal structure has a major effect on the physical and chemical properties of Li-ion silicate cathodes. Combining DFT calculations from Materials Project with ML, Shandiz et al. (36) studied the correlation of the crystal system with various calculated properties and attempted to predict the crystal structure based on the. A wide variety of classification ML algorithms are trained on 339 silicate-based cathodes with Li–Si– (Mn, Fe, Co)–O compositions composed of three major crystal systems (monoclinic, orthorhombic, and triclinic). The DFT calculated properties that are used in the feature descriptor include volume unit cell, number of sites, formation energy, energy above hull and band gap, with crystal structure as the response vector. The correlation between each property and the distribution of features for each crystal structure are plotted and no evident correlation is directly observed, which suggests that the complexity of correlation is beyond the scope of conventional methods. Ensemble methods including random forests and extremely randomized trees provided the highest accuracy of prediction among other classification methods in the Monte Carlo cross-validation tests. Based on the feature importance evaluation in extremely randomized trees, the volume of crystal and number of sites contribute the most to determine the type of crystal system in the dataset.

Recently, all-solid-state LIBs (ASSLIBs) have received resurgence of interest as a potentially safer alternative to traditional liquid-based LIBs. Instead of liquid-based electrolyte, ASSLIBs uses the superionic conducting solid electrolyte. In order to allow comparable performance to liquid-based LIBs, very high ionic conductivity and electrochemical stability towards electrodes should be achieved. Fujimura et al. (37) used Support Vector Regression (SVR) algorithm with a Gaussian kernel to predict the ionic conductivity at 373 K (σ_{373}) of the LISICON-type superionic conductors with formula γ-Li8–cAaBbO4 (A = Zn, Mg, Al, Ga, P, As; B = Si, Ge). Calculated diffusivity from first-principles molecular dynamics (FPMD) simulations at 1600K (D_{1600}), the order-disorder phase transition temperature (T_c), (estimated by the temperature at which the DFT energies of the ordered and disordered phases equalized) and the averaged volume of disordered structure (V_{dis}) are used as the training dataset. The resulting model predicts that systems with high D_{1600} and low T_c tend to have high conductivity at 373K.

Besides the conductivity, the performance of ASSLIBs can also be improved by integrating the interfacial coating materials as it could overcome the poor electrochemical stability towards the electrodes. but computationally identifying materials with sufficiently high lithium-ion conductivity can be challenging. Methods such as first-principles molecular dynamics possibly are immensely expensive when used on materials that have lower ionic conductivity but are still potentially useful as interfacial coatings. Thus, Wang et al. (38) demonstrated a way to address this problem using machine-learned interatomic potential models in the form of moment tensor potentials (MTP). To prevent the potentials from significantly deviating from DFT calculations, molecular dynamics simulations were coupled with on-the-fly ML. This approach uses FPMD trajectory as the initial training set with DFT-calculated forces weighted 10 times over the energies. After the initial training, MTP molecular dynamics is performed. As the structure evolves dynamically, the retraining of MTP is activated by D-optimality criterion, which tags "unlabeled" configuration with an extrapolation grade. When the selection threshold (typically 2–11 for efficiency-vs.-accuracy performance) for extrapolation grade is exceeded, the MTP molecular dynamic is terminated for retraining. Any configurations that are tagged with an extrapolation grade of over 1.5 are selected from the MTP trajectory to be calculated via the DFT method. The results from DFT calculations are then added to the training set and MTP was retrained. This method increases the efficiency of the calculations by 7 orders of magnitude compared to purely first-principles molecular dynamics, significantly reducing the uncertainty in calculated migration energies, and improving agreement with experimentally determined activation energies. Wang et al. identifies two particularly promising materials ($Li_3Sc_2(PO_4)_3$ and $Li_3B_7O_{12}$) for use as coatings in batteries.

Nevertheless, the microstructure of a composite electrode determines how individual battery particles are charged and discharged in a LIB. It is a challenge to experimentally visualize and understand the electrochemical consequences of de(attachment) of a battery particle with the conductive matrix. Hereby, Jiang et al. (39) tackle this issue with a unique combination of multiscale experimental approaches, ML-assisted statistical analysis, and experiment-informed mathematical modeling. Mask regional convolution neural network (Mask R-CNN) is used in tandem with phase contrast tomography for the machine-based identification and segmentation of cracked particles. Slices of tomography data at different heights are annotated manually to be used as training data. Their results suggest that the degree of particle detachment is positively correlated with the charging rate and that smaller particles exhibit a higher degree of uncertainty in their detachment from the carbon/binder matrix. The balanced diffusion kinetics for both Li-ions and electrons are also emphasized for optimal battery performance.

9.4 CONCLUSION

As well as providing an introduction to the ML method, this chapter highlights how it can be applied to materials science, particularly rechargeable batteries. All in all, ML has recently gained a reputation as one of the most important tools in materials science. This collection of statistical methods is useful for both basic and applied research. However, a major challenge facing ML on batteries is the lack of databases. The high-throughput experiments along with large-scale simulations and calculations are expected to be conducted in the near future, generating an enormous amount of data for ML methods in materials science.

REFERENCES

1. Chen C, Zuo Y, Ye W, Li X, Deng Z, Ong SP. A critical review of machine learning of energy materials. *Advanced Energy Materials.* 2020;10(8):1–36.
2. Villars P, Onodera N, Iwata S. The linus pauling file (LPF) and its application to materials design. *Journal of Alloys and Compounds [Internet].* 1998;279(1):1–7. Available from: https://www.sciencedirect.com/science/article/pii/S0925838898006057
3. Bergerhoff G, Hundt R, Sievers R, Brown ID. The inorganic crystal structure data base. *Journal of Chemical Information and Computer Sciences [Internet].* 1983 May 1;23(2):66–69. Available from: 10.1021/ci00038a003
4. Allen FH. The Cambridge structural database: a quarter of a million crystal structures and rising. *Acta Crystallographica Section B [Internet].* 2002 Jun;58(3 Part 1):380–388. Available from: 10.1107/S0108768102003890
5. Graulis S, Chateigner D, Downs RT, Yokochi AFT, Quirós M, Lutterotti L, et al. Crystallography open database – an open-access collection of crystal structures. *Journal of Applied Crystallography.* 2009;42(4):726–729.
6. White P, Rodgers J, Page Y. CRYSTMET: a database of the structures and powder patterns of metals and intermetallics. *Acta Crystallographica Section B, Structural Science.* 2002 Jul 1;58:343–348.
7. Irwin JJ, Shoichet BK. Zinc-A Free Database of Commercially Available Compounds for Virtual Screening. 2005; Available from: http://chembank.med.harvard.edu

8. Kim S, Thiessen PA, Bolton EE, Chen J, Fu G, Gindulyte A, et al. PubChem substance and compound databases. *Nucleic Acids Research*. 2016;44(D1):D1202–D1213.
9. Jain A, Ong SP, Hautier G, Chen W, Richards WD, Dacek S, et al. Commentary: the materials project: a materials genome approach to accelerating materials innovation. Vol. 1, *APL Materials*. American Institute of Physics Inc.; 2013.
10. Taylor RH, Rose F, Toher C, Levy O, Yang K, Buongiorno Nardelli M, et al. A RESTful API for exchanging materials data in the AFLOWLIB.org consortium. *Computational Materials Science [Internet]*. 2014;93:178–192. Available from: https://www.sciencedirect.com/science/article/pii/S0927025614003322
11. Landis DD, Hummelshøj JS, Nestorov S, Greeley J, Dułak M, Bligaard T, et al. The computational materials repository. *Computing in Science & Engineering*. 2012; 14(6):51–57.
12. Kirklin S, Saal JE, Meredig B, Thompson A, Doak JW, Aykol M, et al. The open quantum materials database (OQMD): assessing the accuracy of DFT formation energies. *npj Computational Materials [Internet]*. 2015;1(1):15010. Available from: 10.1038/npjcompumats.2015.10
13. Pizzi G, Cepellotti A, Sabatini R, Marzari N, Kozinsky B. AiiDA: Automated Interactive Infrastructure and Database for Computational Science [Internet]. 2015. Available from: http://www.elsevier.com/open-access/userlicense/1.0/
14. Choudhary K, Zhang Q, Reid ACE, Chowdhury S, van Nguyen N, Trautt Z, et al. Computational screening of high-performance optoelectronic materials using OptB88vdW and TB-mBJ formalisms. *Scientific Data*. 2018 May 8;5.
15. Hummelshøj JS, Abild-Pedersen F, Studt F, Bligaard T, Nørskov JK. CatApp: a web application for surface chemistry and heterogeneous catalysis. *Angewandte Chemie – International Edition*. 2012 Jan 2;51(1):272–274.
16. Cai J, Chu X, Xu K, Li H, Wei J. Machine learning-driven new material discovery. *Nanoscale Advances*. 2020;2(8):3115–3130.
17. Kotsiantis SB, Kanellopoulos D. Data preprocessing for supervised leaning. *International Journal of Computer Science*. 2006;1(2):1–7.
18. Ghiringhelli LM, Vybiral J, Levchenko SV, Draxl C, Scheffler M. Big data of materials science: critical role of the descriptor. *Physical Review Letters*. 2015; 114(10):1–5.
19. Faber F, Lindmaa A, Von Lilienfeld OA, Armiento R. Crystal structure representations for machine learning models of formation energies. *International Journal of Quantum Chemistry*. 2015;115(16):1094–1101.
20. Huo H, Rupp M. Unified Representation of Molecules and Crystals for Machine Learning. 2017;(i).
21. Himanen L, Jäger MOJ, Morooka EV, Federici Canova F, Ranawat YS, Gao DZ, et al. DScribe: library of descriptors for machine learning in materials science. *Computer Physics Communications*. 2020;247:106949.
22. Elton DC, Boukouvalas Z, Butrico MS, Fuge MD, Chung PW. Applying machine learning techniques to predict the properties of energetic materials. *Scientific Reports*. 2018;8(1):1–12.
23. Eberhard J, Geissbuhler V. Konservative und operative therapie bei harninkontinenz, deszensus und urogenital-beschwerden. *Journal fur Urologie und Urogynakologie*. 2000;7:32–46.
24. 200: Quantitative economic methods and data. *Journal of Economic Literature*. 1985 Feb 16;23(2):849–857.
25. Esposito Vinzi V, Russolillo G. Partial least squares algorithms and methods. *Wiley Interdisciplinary Reviews: Computational Statistics*. 2013;5(1):1–19.
26. Hearst MA, Dumais ST, Osuna E, Platt J, Scholkopf B. Support vector machines. *IEEE Intelligent Systems and their Applications*. 1998;13(4):18–28.

27. Boser BE, Guyon IM, Vapnik VN. A training algorithm for optimal margin classifiers. In: Proceedings of the fifth annual workshop on Computational learning theory – COLT '92. ACM Press; 1992.
28. Liaw A, Wiener M. Classification and regression by randomForest. *R News*. 2002;2(3):18–22.
29. Geurts P, Ernst D, Wehenkel L. Extremely randomized trees. *Machine Learning*. 2006;63(1):3–42.
30. Chen C, Ye W, Zuo Y, Zheng C, Ong SP. Graph networks as a universal machine learning framework for molecules and crystals. *Chemistry of Materials*. 2019 May; 31(9):3564–3572.
31. Gómez-Bombarelli R, Wei JN, Duvenaud D, Hernández-Lobato JM, Sánchez-Lengeling B, Sheberla D, et al. Automatic chemical design using a data-driven continuous representation of molecules. *ACS Central Science*. 2018;4(2):268–276.
32. Celisse A, Robin S. Nonparametric density estimation by exact leave-p-out cross-validation. *Computational Statistics & Data Analysis*. 2008 Jan;52(5):2350–2368.
33. Kearns M, Ron D. Algorithmic stability and sanity-check bounds for leave-one-out cross-validation. *Neural Computation*. 1999;11(6):1427–1453.
34. Rodriguez JD, Perez A, Lozano JA. Sensitivity analysis of k-fold cross validation in prediction error estimation. *IEEE Transactions on Pattern Analysis and Machine Intelligence*. 2010;32(3):569–575.
35. Fu WJ, Carroll RJ, Wang S. Estimating misclassification error with small samples via bootstrap cross-validation. *Bioinformatics*. 2005 May 1;21(9):1979–1986.
36. Attarian Shandiz M, Gauvin R. Application of machine learning methods for the prediction of crystal system of cathode materials in lithium-ion batteries. *Computational Materials Science [Internet]*. 2016;117:270–278. Available from: 10.1016/j.commatsci.2016.02.021
37. Fujimura K, Seko A, Koyama Y, Kuwabara A, Kishida I, Shitara K, et al. Accelerated materials design of lithium superionic conductors based on first-principles calculations and machine learning algorithms. *Advanced Energy Materials* 2013;3:980–985. 10.1002/aenm.201300060
38. Wang C, Aoyagi K, Wisesa P, Mueller T. Lithium ion conduction in cathode coating materials from on-the-fly machine learning. *Chemistry of Materials*. 2020;32(9): 3741–3752.
39. Jiang Z, Li J, Yang Y, Mu L, Wei C, Yu X, et al. Machine-learning-revealed statistics of the particle-carbon/binder detachment in lithium-ion battery cathodes. *Nature Communications [Internet]*. 2020;11(1). Available from: 10.1038/s41467-020-16233-5

10 SnO_x (x = 0,1,2) and Mo Doped SnO_2 Nanocomposite as Possible Anode Materials in Lithium-Ion Battery

Sanjaya Brahma
Department of Materials Science and Engineering and Hierarchical Green-Energy Materials (Hi-GEM) Research Center, National Cheng Kung University, Tainan City, Taiwan

Jow-Lay Huang
Department of Materials Science and Engineering, Center for Micro/Nano Science and Technology, and Hierarchical Green-Energy Materials (Hi-GEM) Research Center, National Cheng Kung University, Tainan City, Taiwan

CONTENTS

DOI: 10.1201/9781003367215-10

10.1 INTRODUCTION

Lithium-ion batteries (LIBs) have a wide range of applications ranging from traditional/modern electronic devices and electric hybrid vehicles due to their advantages, such as high energy density, long life cycle, and environment-friendly [1–4]. LIBs have advantages such as high energy density, long cycle life, and wide working temperature range. Although the main commercial graphite anode material is widely used as the anode material in commercial LIBs, the low theoretical capacity (372 $mAhg^{-1}$) [5] limits its applications in higher-capacity devices. Therefore, scientists have been dedicated to improving the capacity as well as the stability of materials for LIB applications. Alloying materials, which react with Li^+ by alloying/de-alloying mechanism, have been considered as alternative anode materials for LIBs because of their high specific capacity.

Compared with conventional carbon-based materials, tin dioxide (SnO_2) has been regarded as one of the promising anode materials for the next-generation LIBs due to its high theoretical capacity (782 mAh g^{-1}), low cost, nontoxicity, and abundance [6–8]. However, the poor electrical conductivity and the rapid capacity fading due to the large volume change during charging and discharging process is the major problem for its application [6–8]. There are several strategies to enhance the performance of the anode, such as reducing size, composite, and doping [2]. Many ways of synthesis methods are followed to produce tin nanostructures varying shapes and sizes, such as nanoparticles, nanorods, nanowires, nanotubes, and hollow structures, to reduce the volume expansion. However, the effect of reducing size is not enough.

Recently, the combination of tin nanoparticles and carbon materials, graphene, and graphene oxides has proved to be a promising strategy to enhance the stability of tin-based LIB anode, mainly due to the excellent electrical conductivity and flexible mechanical properties of carbon materials. The combination of tin nanoparticles and carbon-based materials has been proven to be a promising way to enhance the stability and electronic conductivity of tin-based anode [9–17]. There is a synergistic effect that the layer structure of reduced graphene oxide(rGO) can buffer the volume change of tin dioxide resulting in an improvement of the capacity/cyclic stability of the anode materials. Also, tin dioxide nanoparticles on rGO sheets can minimize the restacking of rGO. This improves the capacity and cyclic stability of the anode. However, the majority of these methods to produce MOs/carbon composites are energy and time-consuming, involving complex synthesis procedures at high temperatures followed by annealing for a longer duration at elevated temperatures. Room temperature synthesis of SnO_2 graphene nanocomposite was done before but the composite delivered a capacity of <400 mAh g^{-1} after 50 cycles at a current rate of 100 mA.g^{-1} which is much lower than the theoretical capacity of SnO_2 [18,19]. Some of the composites also achieve good reversible capacity greater than >800 mAh g^{-1} [20–27]. For example,

SnO_2-graphene nanoribbon composite prepared by unzipping multiwalled carbon nanotube through a complicated procedure could retain 825 mAh g^{-1} after 50 cycles [20]. Similarly, SnO_2-nitrogen doped graphene as reported by Wang et al [21], could deliver a significant capacity of 905 mAh g^{-1} (@2 $A.g^{-1}$) after 1000 cycles. Liu et al., [22] described the synthesis of SnO_2-graphene nanocomposite by a microwave method at high temperatures that showed excellent battery performance achieving a capacity of 1359 mAh g^{-1} (@100 mAg^{-1}) after 100 cycles.

In this study, a facile and simple chemical reduction method with low cost and low toxicity is designed to synthesize RGO/SnO_x composite. The relationship between morphology and electrochemical performance of the composites, as well as the role of reducing agents in the chemical procedure have been investigated and discussed. Doping is another facile and effective approach to improving the electrochemical performance of SnO_2-based materials [28–33]. Experimental results showed that Mo doping could reduce the nanocrystal size and enhance the electrical conductivity of SnO_2-based materials, resulting in improved cycling capability and coulombic efficiency. We have also prepared Mo-SnO_2/rGO the composite of through same and compared with the undoped SnO_2/rGO composite, the 5% Mo-SnO_2/rGO composite anode exhibits a higher capacity.

10.2 EXPERIMENTAL

10.2.1 Synthesis of Graphene Oxide

Graphene oxide(GO) was prepared by modified Hummer's method [34]. Preparing graphene oxide can mainly divide into four parts:

a. 3g graphite dissolved into 200 ml H_2SO_4 and 3 g $NaNO_3$ mixing solution and then stirred in an ice bath for 10 minutes. Then, slowly add 15g $KMnO_4$ into the reactive solution and keep stirring in an ice bath for 30 minutes.
b. Remove the ice bath, and keep the reactive solution stirring in 35°C water bath for 2 hours.
c. Increase the reactive temperature of water bath and then slowly add 150 ml de-ionized water into the reactive solution stirred in a 75°C water bath for 30 minutes.
d. Cool down the reactive solution, then add 50 ml H_2O_2 and 50 ml HCl into the reactive solution to reduce the unexpected metal ions.
e. Finally, use the centrifuge to clean the reactive solution with de-ionized water for more than 10 times until the pH is neutral.

10.2.2 Synthesis of RGO/SnO_x by Chemical Treatment

All chemicals employed in this work are of analytical grade. In preparing RGO/SnO_x composites, first, 1.00 g of GO powders were dispersed in 200 ml de-ionized water and then stirred for 30 minutes. Then, add 15 g of $Sn(BF_4)_2$ and keep the solution stirred. Then, different concentrations of $NaBH_4$ (0 mole – 0.06 mole) dissolved in 50 ml deionized water were added into the solution. This mixture was kept at room temperature under stirring for 30 minutes in order to produce an RGO/SnO_x composite.

10.2.3 Synthesis of Mo-SnO_2/rGO Composite

To prepare Mo-SnO_2/rGO composite, first, 1.00 g of GO powders were dispersed in 200 ml deionized water and then stirred for 10 minutes. Then, add 15 g of $Sn(BF_4)_{2(s)}$ and keep stirring. Then, add different weights of $Na_2MoO_{4(s)}$ (0.27g, 0.55g, 1.17g) and keep stirring for 30 minutes. The resultant precipitate was collected by centrifugation, washed with deionized water several times, and then dried at 80°C for 12 hours. Finally, the obtained powder was annealed at 500°C for 2 hours in a tube furnace under Ar atmosphere. For comparison, SnO_2/rGO composite was prepared by the same procedure without the addition of $Na_2MoO_{4(s)}$.

10.2.4 Characterization of RGO/SnO_x and Mo-SnO_2/rGO Composite

X-ray diffraction (XRD, RIGAKU D/MAX2500) was used to determine the crystal structure of all the composites by using Cu Kα radiation at an angular speed of 3° (2θ)/min with 2θ from 5° to 80°. The microstructure was observed by high resolution field emission scanning electron microscopy (HR-FESEM) and ultrahigh-resolution analytical electron spectroscopy (HR-AEM). The variation of the oxygen-containing functional groups of RGO/SnO_x composite were confirmed by Raman spectrometer. The energy level of the carbon and tin of the RGO/SnO_x composite was confirmed by electron Spectroscopy for chemical analysis (ESCA).

10.2.5 Electrochemical Analysis

The charge/discharge performance of RGO/SnO_x and Mo-SnO_2/rGO composites were determined by using a coin-type cell composed of metallic lithium foil electrode, a separator (Celgard, 2300), an electrolyte of 1 mole/dm^3 LiPF6 in EC/DMC(1:1 by wt%) and a composite electrode. The coin-type cells were assembled in a glove box. The coin-type cells were tested for charge/discharge performance in a constant current mode between 0.2 and 3.0V, and cycle life test for 50 cycles.

10.3 RESULTS AND DISCUSSION

10.3.1 RGO-SnOx Nanocomposite

Figure 10.1 shows the XRD pattern of the graphene oxide and RGO/SnO_x composite with the increase of the reductant concentration (0 mole, 0.020 mole, 0.040 mole, and 0.060 mole). XRD pattern of GO shows a strong (001) peak at 11° which indicates a high level of oxidation of graphite. After chemical reduction, this peak disappears and a broad peak at 23–30° is observed after adding the tin precursor to the GO solution (without any reducing agent). This indicates an adequate

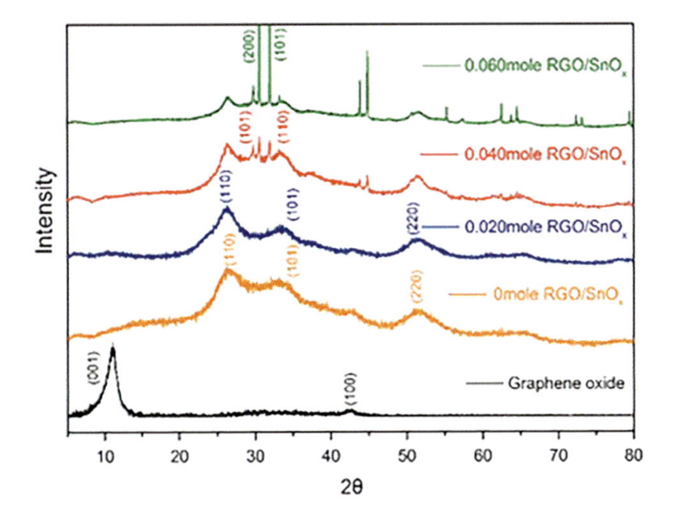

FIGURE 10.1 X-ray diffraction pattern of reduced graphene oxide and SnOx powder nanocomposite prepared at different $NaBH_4$ concentration (graphene oxide; 0 mole $NaBH_4$; 0.02 mole $NaBH_4$; 0.04 mole $NaBH_4$; 0.06 mole $NaBH_4$)

reduction of GO leading to the removal of the oxygen-containing functional groups. The addition of the reducing agent results in the crystallization of SnO_2, SnO, and even metallic Sn. At lower reductant concentration conditions (0.02 mole), only SnO_2 peaks in the composite were observed. When the reductant concentration increased to 0.04 and 0.06 mole, SnO (101) and (110) peaks as well as metallic tin (200) and (101) crystal phases were observed. The results showed that higher reductant concentration facilitated the reduction of tin monoxide and metal tin into the composite.

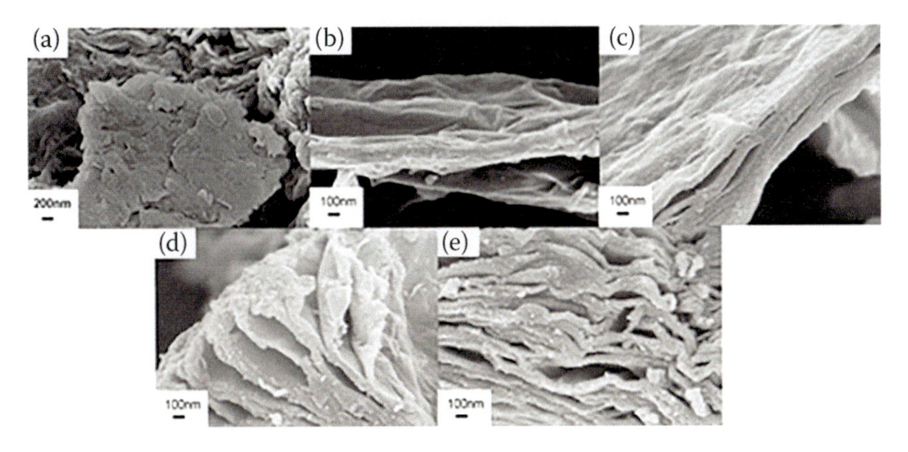

FIGURE 10.2 Scanning electron microscopy images of reduced graphene oxide and SnOx powder nanocomposite prepared at different $NaBH_4$ concentration, (graphene oxide; 0 mole $NaBH_4$; 0.02 mole $NaBH_4$; 0.04 mole $NaBH_4$; 0.06 mole $NaBH_4$).

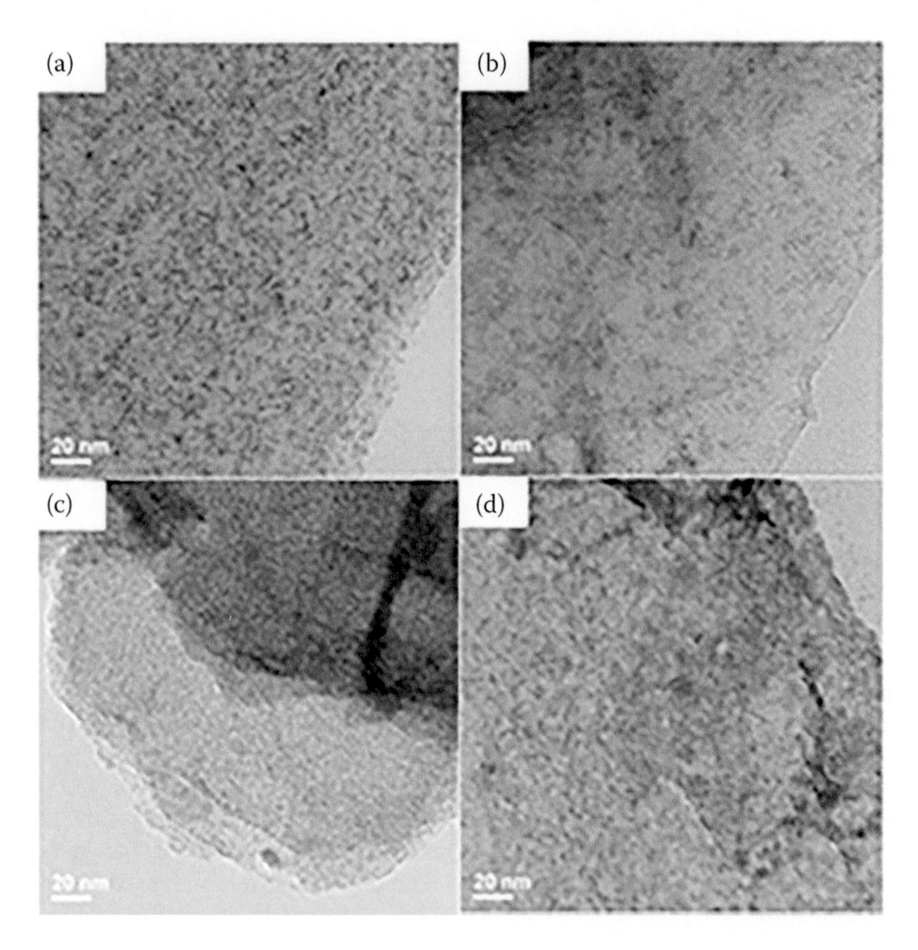

FIGURE 10.3 Transmission electron microscopy images of reduced graphene oxide and SnOx powder nanocomposite prepared at different $NaBH_4$ concentration, (graphene oxide; 0 mole $NaBH_4$; 0.02 mole $NaBH_4$; 0.04 mole $NaBH_4$; 0.06 mole $NaBH_4$).

The corresponding SEM images of GO and RGO/SnO_x is shown in Figure 10.2. Graphene oxide shows layer-like morphology (Figure 10.2a) and some nanoparticles are attached to the GO sheets (Figure 10.2b-e). Though, SEM is difficult to see the growth of tin oxides or metallic tin in RGO/SnO_x composites.

Figure 10.3 describes the TEM analysis of the RGO/SnO_x composite. Figure 10.3a shows a TEM image of the RGO/SnO_x composite without using a reducing agent, and there still forms SnO_2 crystal onto graphene sheets. According to the literature, tin ions from the tin precursor react with oxygen containing functional groups of the GO and form fine SnO_2 nanoparticles. In contrast, after adding reductant, a large number of nanoparticles decorate the RGO sheet. With the increase in reductant concentration, the density, as well as the aggregation of nanoparticles, increases over the RGO surface (Figure 10.3b-d) and it seems that the RGO is completely covered at a higher reductant concentration (Figure 10.3c-d). The result indicates that the reductant

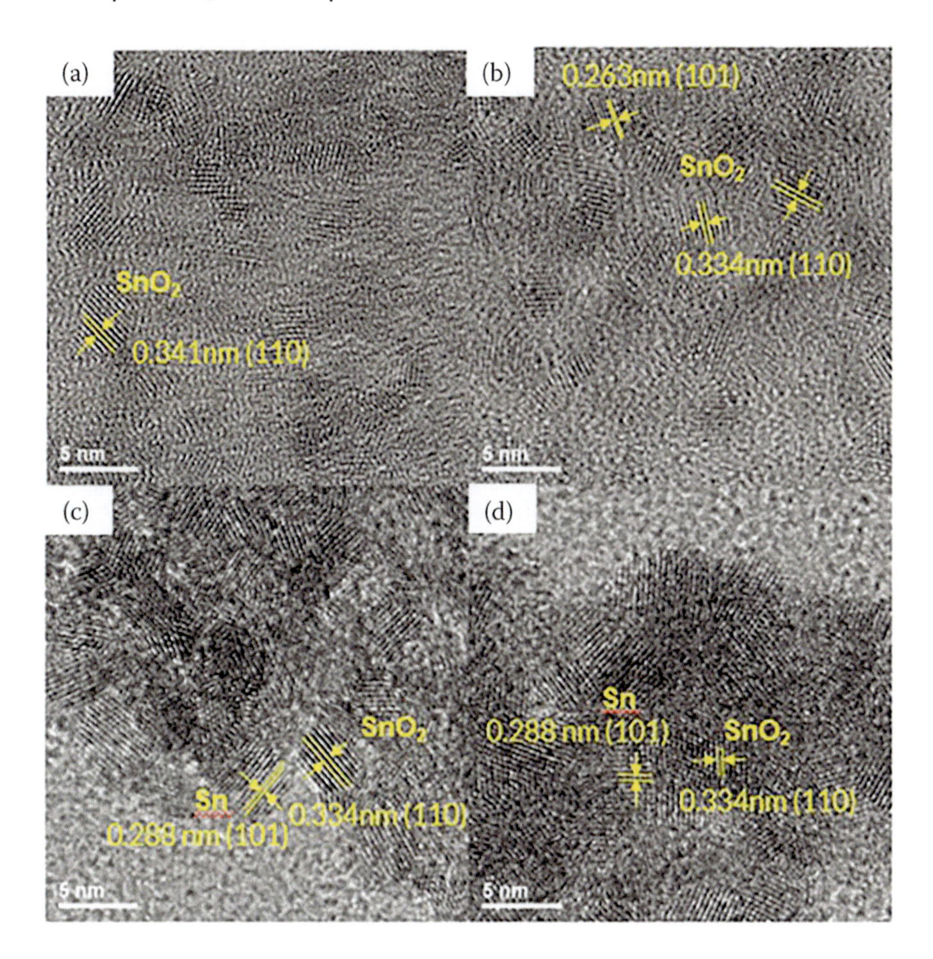

FIGURE 10.4 High-resolution transmission electron microscopy images of reduced graphene oxide and SnOx powder nanocomposite prepared at different $NaBH_4$ concentration, (graphene oxide; 0 mole $NaBH_4$; 0.02 mole $NaBH_4$; 0.04 mole $NaBH_4$; 0.06 mole $NaBH_4$).

concentration significantly affects the density and growth of nanoparticles over the GO sheets. The high-resolution TEM image (Figure 10.4) of tin oxide and metallic tin nanoparticles disperse on RGO sheets. We observed the crystal phase (110) and (101) of tin dioxide in every composite, which corresponds to the results of XRD pattern. Furthermore, metallic tin nanoparticles can be found in higher reductant concentration condition. From HR-TEM analysis, the average size of nanoparticles is observed and about 5 nm in RGO/SnOx composites.

Figure 10.5 shows charge/discharge performance of first cycle of the RGO/SnOx composite. The condition without reductant shows large irreversible capacity due to the residual oxygen functional groups. Then, the discharge capacity of RGO/SnO_x composite with 0.02 mole reductant is 632.5 $mAhg^{-1}$.In comparison, the capacities increases to 937.9 $mAhg^{-1}$ with the addition of 0.04 mole reductant and that decreases to 892.8 $mAhg^{-1}$ with the increase in the concentration 0.06 mole

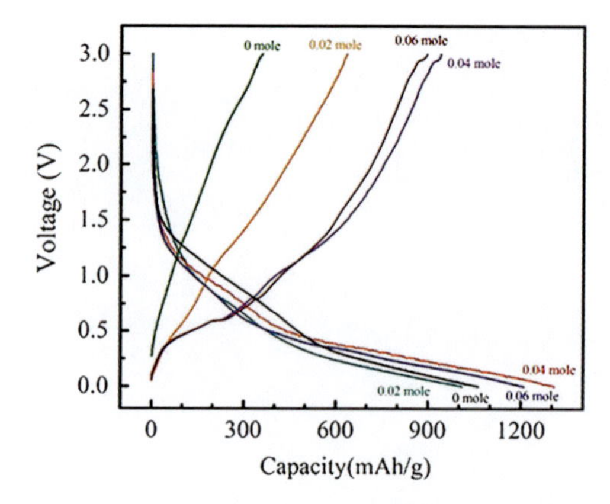

FIGURE 10.5 Charge/Discharge performance of reduced graphene oxide and SnOx powder nanocomposite prepared at different $NaBH_4$ concentration, (graphene oxide; 0 mole $NaBH_4$; 0.02 mole $NaBH_4$; 0.04 mole $NaBH_4$; 0.06 mole $NaBH_4$).

reductant. The irreversible capacity loss of first cycle with 0.02 mole of reductant is 38%. However, after increase in reductant concentration, the irreversible capacity of first cycle is reduced to 29% and 27% of the reductant to 0.04 mole and 0.06 mole respectively. The results show that adding a reductant can improve the Coulombic efficiency for first cycle of materials. Figure 10.6 shows the cycling test results of RGO/SnO_x composites for 50 cycles. The RGO/SnO_x composite's initial up and down in the capacity for the first few cycles may be attributed to the instability in the Li^+ insertion/extraction in RGO/SnO_x composite. The 0 mole composite shows the graphene-like behavior, low capacity but stable cycle life. It is clearly observed

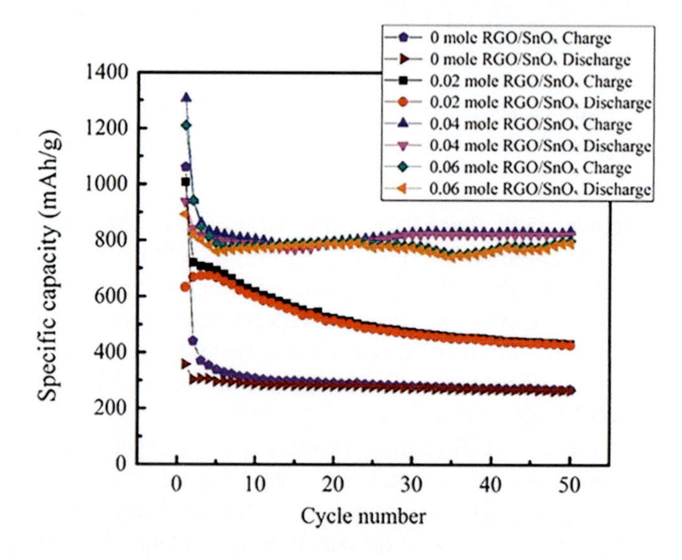

FIGURE 10.6 Cycling performance of reduced graphene oxide and SnOx powder nanocomposite prepared at different $NaBH_4$ concentration, (graphene oxide; 0 mole $NaBH_4$; 0.02 mole $NaBH_4$; 0.04 mole $NaBH_4$; 0.06 mole $NaBH_4$).

that the RGO/SnO_x composite prepared by using 0.04 mole of reductant not only shows a high capacity (824.0 at 50 cycles) among all other composites but also maintains good cycle life stability.

10.3.2 Mo Doped SnO_2-RGO Nanocomposite

Figure 10.7 shows the XRD pattern of the Mo-SnO_2/rGO composites with the increase of the Mo at% (0%, 2.5%, 5%, 10%). Most of the peaks correspond to a SnO_2 rutile structure. We can observe a peak around 38 degree which corresponds to the MoO_2 structure. The intensity of this peak increases with the increase of Mo at%. The addition of the Mo results in the crystallization of SnO_2 and the shift of (110) peak to the higher degree. Because Mo is smaller than Sn, the d-spacing of (110) plane will decrease. From Bragg's law, we can know that when d-spacing decreases, the peak will shift to a higher angle.

Figure 10.8 shows the SEM images of the Mo-SnO_2/rGO composite with the increase of the Mo at% (0%, 2.5%, 5%, 10%). All of the images can see the layer-like morphology of reduced graphene oxide, and some nanoparticles are attached to the rGO sheets. Although SEM shows the rGO morphology very well, it is difficult to see the growth of Mo-SnO_2 nanoparticles in Mo-SnO_2/rGO composite.

Figure 10.9 shows the TEM images of the Mo-SnO_2/rGO composite with the increase of the Mo at% (0%, 2.5%, 5%, 10%). It is obvious that numerous NCs have been successfully anchored onto the rGO sheets with high dispersity, and the rGO sheets are quite large in dimension compared with the nanoparticles. These

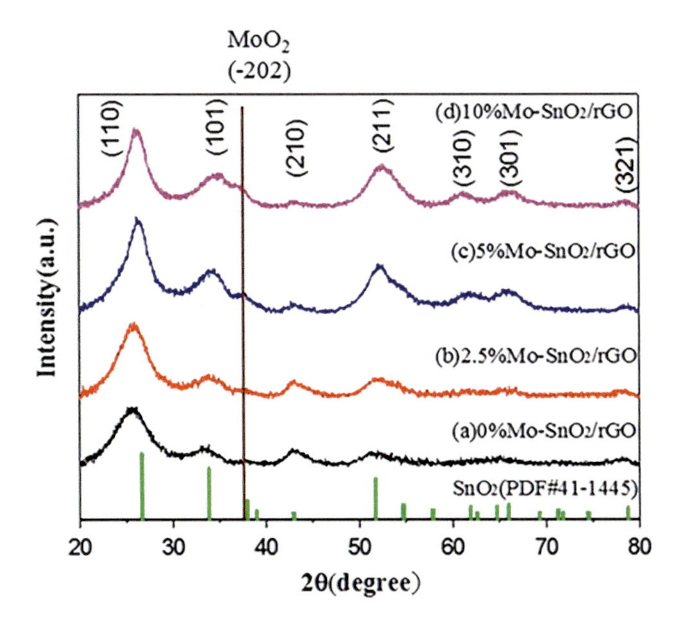

FIGURE 10.7 Powder X-ray diffraction pattern of reduced graphene oxide and Mo doped SnO2 powder nanocomposite prepared at different Mo doping concentration: (a)0%; (b) 2.5%; (c) 5%; (d) 10%.

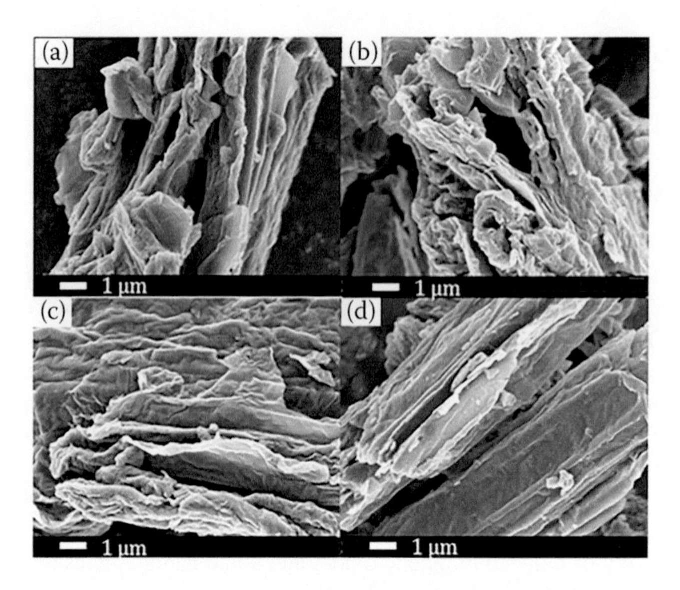

FIGURE 10.8 Scanning electron microscopy images of reduced graphene oxide and Mo doped SnO2 powder nanocomposite prepared at different Mo doping concentrations: (a) 0%; (b) 2.5%; (c) 5%; (d) 10%.

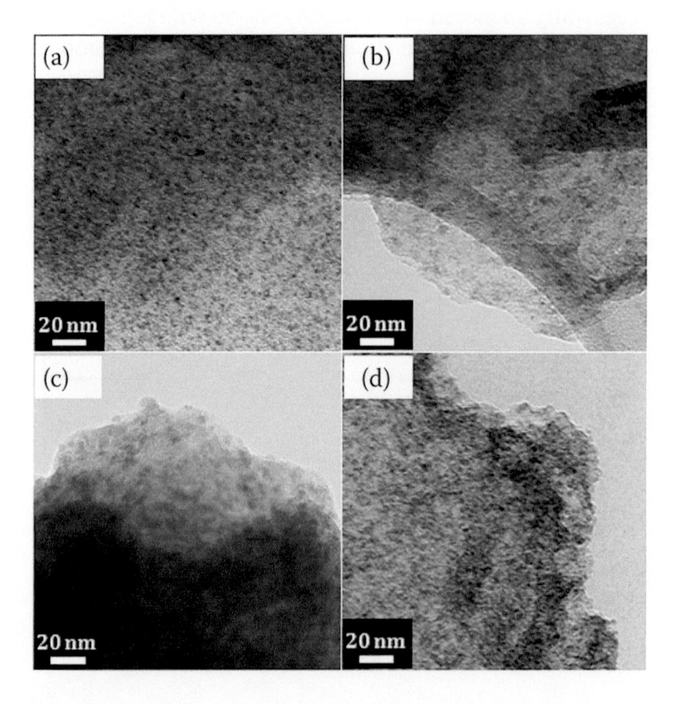

FIGURE 10.9 Transmission electron microscopy images of reduced graphene oxide and Mo-doped SnO2 powder nanocomposite prepared at different Mo doping concentrations: (a) 0%; (b) 2.5%; (c) 5%; (d)10%.

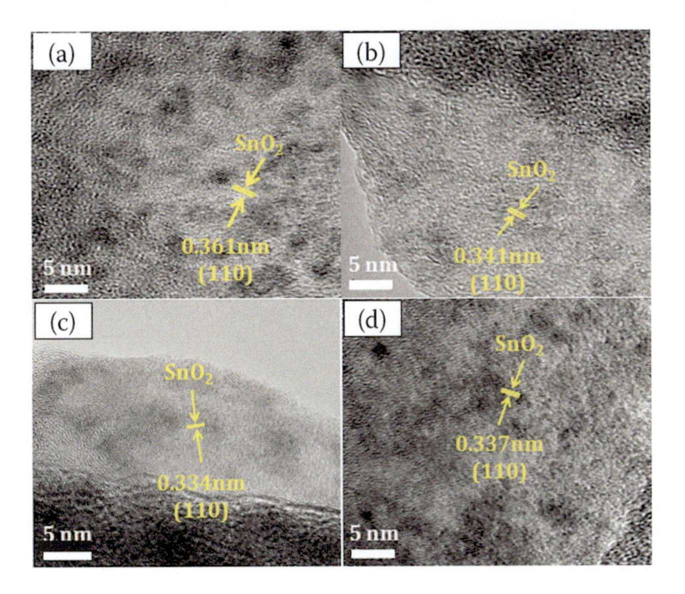

FIGURE 10.10 High-resolution transmission electron microscopy images of reduced graphene oxide and Mo doped SnO2 powder nanocomposite prepared at different Mo doping concentration: (a) 0%; (b) 2.5%; (c) 5%; (d) 10%.

nanoparticles are still firmly attached to the rGO sheets even after the sonication for TEM, indicating the strong interaction between SnO_2 nanoparticles and rGO sheets. Figure 10.10 shows the high-resolution TEM images of the Mo-SnO_2/rGO composite with the increase of the Mo at% (0%, 2.5%, 5%, 10%). We observed the crystal phase (110) of tin dioxide in every composite, which corresponds to the results of XRD pattern, and the trend of d-spacing is the same with the XRD data, too. Furthermore, the average size of nanoparticles in our composite is about 3 nm. Such fine nanoparticles can accommodate the volume change and shorten the diffusion pathway of Lithium ions, which would improve the electrochemical performance of SnO_2-based materials for LIBs.

Figure 10.11 shows Lithiation and delithiation of first cycle with the variation Mo at% (0%, 2.5%, 5%, 10%). All of them are synthesized without adding reductant, so they show large irreversible capacity at the first cycle due to the residual oxygen functional groups. Then, the charge capacity of 0%Mo-SnO_2/rGO is 1057 mAh g^{-1}. In comparison, the capacity of 2.5%Mo-SnO_2/rGO increases to 1128 mAh g^{-1}, and the capacity of 5%Mo-SnO_2/rGO decreases to 1195 mAh g^{-1}, then the capacity of 10%Mo-SnO_2/rGO decreases to 1093 mAh g^{-1}. 5% has the highest capacity at the first three cycles, so the result shows that the appropriate Mo at% has the highest capacity. In the first cycle charging curve, there is a typical plateau at 1.5–0.7V, which can be attributed to the formation of solid electrolyte interphase (SEI) and the transformation from SnO_2 to Sn and Li_2O. Another plateau can be found below 0.7V, which corresponds to the alloying of Li-Sn and the insertion of Li^+ ions into rGO sheets. Subsequently, there are three plateaus in the discharging curve. One plateau below 0.5V is attributed to the dealloying of Li-Sn and the extraction of Li^+ ions from

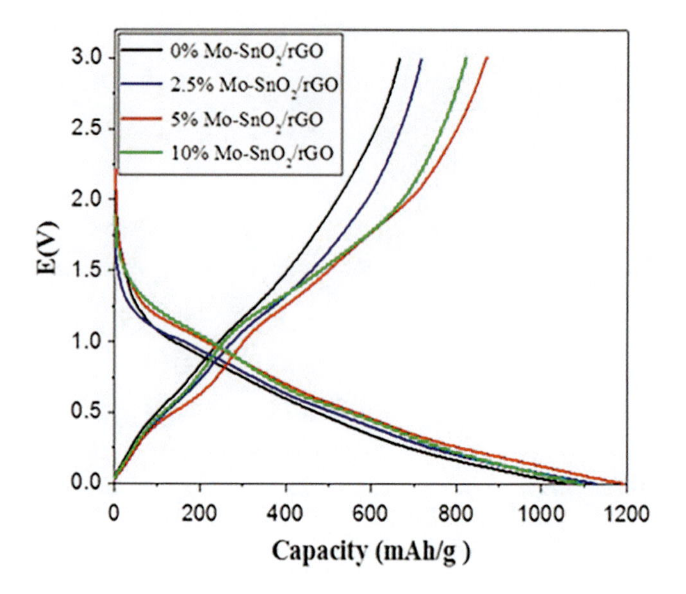

FIGURE 10.11 Lithiation and delithiation of reduced graphene oxide and Mo-doped SnO2 powder nanocomposite prepared at different Mo doping concentration: (a) 0%; (b) 2.5%; (c) 5%; (d) 10%.

rGO sheets, and another plateau at 0.5–1.25V corresponds to the transformation of Sn and Li_2O to SnO. Then, the plateau above 1.25V corresponds to the transformation of Sn and Li_2O to SnO_2 and the transformation of SnO and Li_2O to SnO_2. Figure 10.12 shows the cyclic stability of Mo-SnO_2/rGO composites with variation in

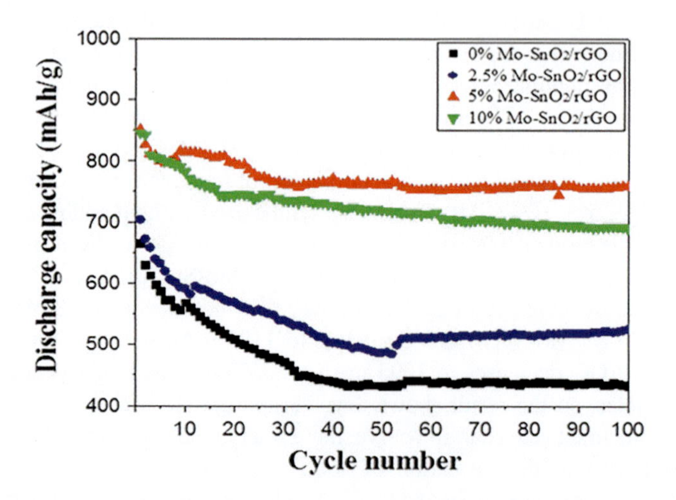

FIGURE 10.12 Cycling performance of reduced graphene oxide and Mo-doped SnO2 powder nanocomposite prepared at different Mo doping concentrations: (a) 0%; (b) 2.5%; (c) 5%; (d) 10%.

Mo at% (0%, 2.5%, 5%, 10%) for 150 cycles. It is clearly observed that 5% not only shows a high capacity (758 mAh g^{-1} at 100 cycles) among all other composites but also maintains good cycle life stability, and the retention after 100 cycles is 89.1%. Besides, it can be seen that Mo can significantly affect the electrochemical performance of SnO_2/rGO composite.

10.4 CONCLUSIONS

A facile and simple chemical reduction process is designed to synthesize RGO/SnO_x and Mo doped SnO_2–RGO nano-composite which can co-reduce GO and tin precursors to form composite simultaneously. The concentration of reductant can control the reduction degree of composite. Moreover, Sn/SnO_x nanoparticles crystallize and aggregate over the RGO sheets with the increase of the reductant concentration. The best electrochemical property of the 0.04 mole RGO/SnO_x composite shows high Li storage capacity and stable cycling performance (824.0 mAh g^{-1} with 87% retention after 50 cycles).

We have also successfully synthesized Mo-SnO_2/rGO composites by the identical method. SnO_2 nanoparticles (3 nm average particle size) have been successfully anchored onto the rGO sheets with high dispersity, which is good for the electrochemical performance of composites. Adding Mo does not change the SnO_2 crystal structure, indicating that the Mo ions are doped in the lattice of SnO_2 rather than simply attaching to the surface. Via electrochemical tests, the results show that 5%Mo-SnO_2/rGO has the best electrochemical performance corresponding to high capacity (1117/483 mAh/g at first cycle), good cycling stability of about 70.3% retention (339.6 mAh/g after 150 cycles), high retention (93.7%) after fast charge/discharge.

ACKNOWLEDGMENT

This work was financially supported by the Hierarchical Green-Energy Materials (Hi-GEM) Research Center, from The Featured Areas Research Center Program within the framework of the Higher Education Sprout Project by the Ministry of Education (MOE) and the Ministry of Science and Technology (MOST 111-2634-F-006-008) in Taiwan.

REFERENCES

1. J. H. B. Scrosati, and Y. K. Sun, "Lithium-ion batteries. A look into the future," *Energy Environ. Sci.*, vol. 4, pp. 3287–3295, 2011.
2. N. Nitta, F. Wu, J. T. Lee, and G. Yushin, "Li-ion battery materials: present and future," *Materials Today*, vol. 18, no. 5, pp. 252–264, 2015.
3. W. S. Taehoon Kim, Dae-Yong Son, Luis K. Ono, and Y. Qi, "Lithium-ion batteries: outlook on present, future, and hybridized technologies," *J. Mater. Chem. A*, vol. 7, pp. 2942–2964, 2019.
4. V. Etacheri, R. Marom, R. Elazari, G. Salitra, and D. Aurbach, "Challenges in the development of advanced Li-ion batteries: a review," *Energy Environ. Sci.,* vol. 4, no. 9, 2011.

5. M. Yoshio, H. Wang, and K. Fukuda, "Spherical carbon-coated natural graphite as a lithium-ion battery-anode material," *Angew. Chem. Int. Ed. Engl.*, vol. 42, no. 35, pp. 4203–4206, Sep 15 2003.
6. J. Li, Y. Zhao, N. Wang, and L. Guan, "A high performance carrier for SnO_2 nanoparticles used in lithium ion battery," *Chem. Commun. (Camb)*, vol. 47, no. 18, pp. 5238–5240, May 14 2011.
7. Cen Wang, Yun Zhou, Mingyuan Ge, Xiaobin Xu, Zaoli Zhang, and J. Z. Jiang 2009, "<Large-scale synthesis of SnO_2 nanosheets with high lithium storage.pdf>," *J. Am. Chem. Soc.*, vol. 132, pp. 46–47 2010.
8. M. S. Park, G. X. Wang, Y. M. Kang, D. Wexler, S. X. Dou, and H. K. Liu, "Preparation and electrochemical properties of SnO_2 nanowires for application in lithium-ion batteries," *Angew. Chem. Int. Ed. Engl.*, vol. 46, no. 5, pp. 750–753, 2007.
9. Y. Z. L. L. Sun, "Significant impact of 2D graphene nanosheets on large volume change tin-based anodes in lithium-ion batteries A review," *J. Power Sources*, vol. 274, pp. 869–884, 2015.
10. Z. Li, Y. Wang, H. Sun, W. Wu, M. Liu, J. Zhou, G. Wu, and M. Wu, "Synthesis of nanocomposites with carbon–SnO_2 dual-shells on TiO_2 nanotubes and their application in lithium ion batteries," *J. Mater. Chemi. A*, vol. 3, no. 31, pp. 16057–16063, 2015.
11. Y. Chen, B. Song, R. M. Chen, L. Lu, and J. Xue, "A study of the superior electrochemical performance of 3 nm SnO_2 nanoparticles supported by graphene," *J. Mater. Chem. A*, vol. 2, no. 16, pp. 5688–5695, 2014.
12. C.-C. Hou, S. Brahma, S.-C. Weng, C.-C. Chang, and J.-L. Huang, "Facile, low temperature synthesis of SnO_2/reduced graphene oxide nanocomposite as anode material for lithium-ion batteries," *Appl. Surface Sci.*, vol. 413, pp. 160–168, 2017.
13. M. Zhang, D. Lei, Z. Du, X. Yin, L. Chen, Q. Li, Y. Wang, and T. Wang, "Fast synthesis of SnO_2/graphene composites by reducing graphene oxide with stannous ions," *J. Mater. Chem.*, vol. 21, no. 6, pp. 1673–1676, 2011.
14. J. Yao, X. Shen, B. Wang, H. Liu, and G. Wang, "In situ chemical synthesis of SnO_2–graphene nanocomposite as anode materials for lithium-ion batteries," *Electrochem. Commun.*, vol. 11, no. 10, pp. 1849–1852, 2009.
15. C.-C. Hou, S. Brahma, S.-C. Weng, C.-C. Chang, and J.-L. Huang, "Multi-layer graphene/SnO_2 nanocomposites as negative electrode materials for lithium-ion batteries," *J. Electrochem. Energy Conversion Storage*, vol. 17, no. 3, 2020.
16. J. Liang, W. Wei, D. Zhong, Q. Yang, L. Li, and L. Guo, "One-step in situ synthesis of SnO_2/graphene nanocomposites and its application as an anode material for Li-ion batteries," *ACS Appl. Mater. Interfaces*, vol. 4, no. 1, pp. 454–459, Jan 2012.
17. Y. Deng, C. Fang, and G. Chen, "The developments of SnO_2/graphene nanocomposites as anode materials for high performance lithium ion batteries: A review," *J. Power Sources*, vol. 304, pp. 81–101, 2016.
18. Y. Li, X. Lv, J. Lu, and J. Li, "Preparation of SnO_2-nanocrystal/graphene-nanosheets composites and their lithium storage ability," *J. Phys. Chem. C.*, vol. 114, pp. 21770–21774, 2010.
19. J. Liang, Y. Zhao, L. Guo, and L. Li, "Flexible free-standing graphene/SnO_2 nanocomposites paper for Li-ion battery," *ACS Appl. Mater. Interfaces*, vol. 4, pp. 5742–5748, 2012.
20. J. Lin, Z. Peng, C. Xiang, G. Ruan, Z. Yan, D. Natelson, and J. M. Tour, "Graphene nanoribbon and nanostructured SnO_2 composite anodes for lithium ion batteries," *ACS Nano*, vol. 7, no. 7, pp. 6001–6006, 2013.
21. R. Wang, C. Xu, J. Sun, L. Gao, and H. Yao, "Solvothermal-induced 3D macroscopic SnO_2/nitrogen-doped graphene aerogels for high capacity and long-life lithium storage," *ACS Appl. Mater. Interfaces*, vol. 6, pp. 3427–3436, 2014.

22. L. Liu, M. An, P. Yang, and J. Zhang, "Superior cycle performance and high reversible capacity of SnO_2/graphene composite as an anode material for lithium-ion batteries," *Sci. Rep.*, vol. 5, pp. 9055 (1–10), 2015.
23. Y. G. Zhu, Y. Wang, J. Xie, G.-S. Cao, T.-J. Zhu, X. Zhao, and H. Y. Yang, "Effects of graphene oxide function groups on SnO_2/graphene nanocomposites for lithium storage application," *Electrochim. Acta*, vol. 154, pp. 338–344, 2015.
24. X. Liu, X. Zhong, Z. Yang, F. Pan, L. Gu, and Y. Yu, "Gram-scale synthesis of graphene-mesoporous SnO_2 composite as anode for lithium-ion batteries," *Electrochim. Acta*, vol. 152, pp. 178–186, 2015.
25. L. Chen, X. Ma, M. Wang, C. Chen, and X. Ge, "Hierarchical porous SnO_2/reduced graphene oxide composites for high-performance lithium-ion battery anodes," *Electrochim. Acta*, vol. 215, pp. 42–49, 2016.
26. W. Xu, K. Zhao, C. Niu, L. Zhang, Z. Cai, C. Han, L. He, T. Shen, M. Yan, L. Qu, and L. Mai, "Heterogeneous branched core–shell SnO_2–PANI nanorod arrays with mechanical integrity and three dimentional electron transport for lithium batteries," *Nano Energy*, vol. 8, pp. 196–204, 2014.
27. K. Zhao, L. Zhang, R. Xia, Y. Dong, W. Xu, C. Niu, L. He, M. Yan, L. Qu, and L. Mai, "SnO_2 quantum dots@graphene oxide as a high-rate and long-life anode material for lithium-ion batteries," *Small*, vol. 12, no. 5, pp. 588–594, 2016.
28. J. Wang, L. Wang, S. Zhang, S. Liang, X. Liang, H. Huang, W. Zhou, and J. Guo, "Facile synthesis of iron-doped SnO_2/reduced graphene oxide composite as high-performance anode material for lithium-ion batteries," *J. Alloys Compounds*, vol. 748, pp. 1013–1021, 2018.
29. H. Xu, L. Shi, Z. Wang, J. Liu, J. Zhu, Y. Zhao, M. Zhang, and S. Yuan, "Fluorine-doped tin oxide nanocrystal/reduced graphene oxide composites as lithium ion battery anode material with high capacity and cycling stability," *ACS Appl. Mater. Interfaces*, vol. 7, no. 49, pp. 27486–27493, Dec 16 2015.
30. Y. Liu, A. Palmieri, J. He, Y. Meng, N. Beauregard, S. L. Suib, and W. E. Mustain, "Highly conductive in-SnO_2/RGO nano-heterostructures with improved lithium-ion battery performance," *Sci. Rep.*, vol. 6, p. 25860, May 11 2016.
31. P. Zhao, W. Yue, Z. Xu, S. Sun, and H. Bao, "Graphene-based Pt/SnO_2 nanocomposite with superior electrochemical performance for lithium-ion batteries," *J. Alloys Compounds*, vol. 704, pp. 51–57, 2017.
32. S. Wang, L. Shi, G. Chen, C. Ba, Z. Wang, J. Zhu, Y. Zhao, M. Zhang, and S. Yuan, "In situ synthesis of tungsten-doped SnO_2 and graphene nanocomposites for high-performance anode materials of lithium-ion batteries," *ACS Appl. Mater. Interfaces*, vol. 9, no. 20, pp. 17163–17171, May 24 2017.
33. X. Ye, W. Zhang, Q. Liu, S. Wang, Y. Yang, and H. Wei, "One-step synthesis of Ni-doped SnO_2 nanospheres with enhanced lithium ion storage performance," *New J. Chem.*, vol. 39, no. 1, pp. 130–135, 2015.
34. N. I. Zaaba, K. L. Foo, U. Hashim, S. J. Tan, W.-W. Liu, and C. H. Voon, "Synthesis of graphene oxide using modified hummers method: solvent influence," *Procedia Eng.*, vol. 184, pp. 469–477, 2017.

11 Polymer Electrolytes Based on Ionic Liquid and Poly(ethylene glycol) via in-situ Photopolymerization of Lithium-Ion Batteries

Jeng-Shiung Jan, Ting-Yuan Lee, Yuan-Shuo Hsu, Song-Hao Zhang, and Chia-Chi Chang
Department of Chemical Engineering, National Cheng Kung University, Tainan City, Taiwan

CONTENTS

11.1 INTRODUCTION

The demand for secondary batteries with superior capacity, larger energy density, and better safety has sharply risen due to the recent development of electric/hybrid electric vehicles, energy storage devices, portable devices, and household appliances [1–3]. Among the recently developed secondary batteries, lithium-ion batteries (LIBs) have

DOI: 10.1201/9781003367215-11

received the most attention and, consequently, have been extensively studied in the past few decades. The overall performance and safety of LIBs are crucially determined by the battery electrolytes, carrying Li ions between the electrodes to complete a circuit of charges. Currently, the commercially available LIBs are mainly comprised of organic liquid electrolytes, giving fatal flaws including volatility, easy leakage, flammability, and safety mishap. To circumvent these problems, research has prompted to design alternatives such as polymer electrolytes (PEs) to replace liquid electrolytes [4]. PEs can be designed to meet the requirement of good thermal, mechanical, and electrochemical stability. However, some major problems inherited in PEs including low room-temperature ionic conductivity and poor contact at the interfaces between electrodes and electrolytes limit their utilization in LIBs.

PEs could be an effective alternative with expandability and good processability, [4]. since PEs would be easily processed to possess suitable thermal stability and high mechanical strength if properly designed. Moreover, PEs can effectively guarantee the intrinsic safety of the batteries by decreasing flammability, preventing leakage, and mechanically inhibiting the growth of lithium dendrites [5,6] Poly(ethylene glycol) (PEG) is by far the most employed polymer for PEs due to its good compatibility with lithium salts, large dielectric constants, and proper membrane processability [7,8]. The movement of PEG chains and complexation-dissociation behavior within PEG segments facilitate lithium-ion transport under an electric field [9]. Though its ionic conductivity at room temperature can be greatly improved by incorporating plasticizers or nanofillers to suppress the crystallization of PEG, the *as-prepared* PEs would exhibit impaired mechanical properties [10–12]. That is to say, it is challenging to prepare PEs featuring both superior ionic conductivity and excellent mechanical properties at room temperature.

Several approaches have been proposed to optimize both the ionic conductivity and mechanical strength via creating interconnected ion-channels and introducing cross-linked networks in the polymer matrices [13–15]. The introduction of poly (ionic liquid) (PILs) can render the *as-prepared* PEs displaying the intrinsic properties of ILs including negligible flammability, extremely low vapor pressure, and broad electrochemical window [16–19]. Several studies have demonstrated that the electrolytes comprised of cationic PILs are promising candidates for next-generation batteries [17,20–24]. It is even proposed that cationic PILs could stabilize the electric field near lithium anodes by providing local reservoirs, mitigating the disorder of ion intercalation [25,26]. Herein, we report the preparation of PEs based on imidazolium-based IL and flexible PEG via in-situ solvent-less photopolymerization by depositing precursors directly onto the lithium anode. It is expected that these *as-prepared* PEs can address the above-mentioned problems or challenges.

In order to achieve the in-situ solvent-less photopolymerization, thiol-ene click chemistry was chosen to perform reactions between thiol (-SH) and vinyl groups (-C=C) containing compounds due to the advantages including simplified polymerization kinetics, benign reaction conditions with the higher tolerance to oxygen, and easy to obtain a homogeneous cross-linked network with a high yield and low shrinkage [27–29]. The combination of thiol-ene click chemistry and in-situ solvent-less polymerization strategy could facilitate the formation of a seamless contact at the PE/electrode interface with low interfacial resistance [30–32].

It is also anticipated that a fully cross-linked network can be achieved by adjusting the PE composition ratio. The influences of cross-linked networks and composition in the *as-prepared* PEs on physical, thermal, and mechanical properties were evaluated, accordingly. With the conformal adhesion between the PE/electrode interfaces, the assembled battery cells are projected to exhibit outstanding electrochemical performance in terms of ionic conductivity and charge-discharge capacity. Moreover, a roll-to-roll manufacturing process can be adopted with this approach, rendering the scalability and processability of LIBs with high electrochemical performance.

11.2 EXPERIMENTS

11.2.1 Materials

Polyvinylidene difluoride (PVDF), 1-vinylimidazole (99%), and poly(ethylene glycol) (PEG400) (average Mn = 400 g·mol^{-1}) were purchased from Alfa Aesar. Magnesium sulfate (>99%, J.T. Baker) and a charcoal activated (TCI) were kept in the dry box. Chemicals from Sigma-Aldrich including poly(ethylene glycol) methyl ether methacrylate (PEGMEMA) (average Mn = 950 g·mol^{-1}), poly(ethylene glycol) dimethyl ether (PEGDME) (average Mn = 500 g·mol^{-1}), pentaerythritol tetrakis(3-mercaptopropionate) (>95%) (PETMP) and 2,2-dimethoxy-2-phenylacetophenone (DMPA) were stored under argon and used as received. Phosphorous tribromide (99%), lithium iron phosphate ($LiFePO_4$, LFP), and super P were the product of Acros Organics, Aleees, and Timcal, respectively. Lithium bis(trifluoromethanesulfonyl) imide (LiTFSI) was supplied from Solvay and stored in the glove box with filled argon. Lithium metal and aluminum foil were provided by UBIQ. All the solvents used in this research are of ACS reagent grade and used as received.

11.2.2 Preparation of Brominated PEG Polymer (Br-PEG400-Br)

Br-PEG400-Br was synthesized as described in the literature reported by Cecchini et al [33]. Briefly, PEG400 (50 g, 0.125 mol) and PBr_3 (45.1 g, 0.167 mol) were dissolved in chloroform with 160 mL and 40 mL in the round-bottom flask, respectively. The *as-prepared* solutions were mixed under liquid nitrogen, followed by halogenating in 0°C ice–water bath with an inert atmosphere for 24 hours. Subsequently, 50 mL of deionized water was added to the solution dropwise to terminate the reaction, further extracted with deionized water three times, and the organic layer was collected. After mixing with anhydrous $MgSO_4$ and charcoal activated, the *as-prepared* solution was filtered and condensed via vacuum distillation, therefore, Br-PEG400-Br in transparent light-yellow appearance was obtained.

11.2.3 Synthesis of PEG-Containing poly(ionic liquids) Cross-Linker (VIm-PEG400-VIm)

Initially, 1-vinylimidazole (21.5 g, 0.228 mol) and Br-PEG400-Br (48.1 g, 0.091 mol) were dissolved in anhydrous ethyl acetate (215 mL) in the round-bottom flask with reflux and stirring under 75°C for 12 hours. The yellow precipitates were dissolved in deionized water and extracted with ethyl acetate three times to remove excess reactant and byproducts before the ion exchange. Next, LiTFSI aqueous solution was slowly added to the [VIm-PEG400-VIm] [Br] solution with hydrophobic ion-exchanged white particles precipitating simultaneously. After dissolving ion-exchanged products in ethyl acetate, the measures to remove residue moisture in the last paragraph using $MgSO_4$ and charcoal activated were adopted here again. Accordingly, the poly(ionic liquids) were obtained after vacuum distillation for 24 hours.

11.2.4 Preparation of the Solid-State Polymer Electrolytes

The electrolytes abbreviated as O-X were prepared by in-situ UV-cured polymerization of thiol-branched crosslinker PETMP, vinyl-terminated crosslinker VIm-PEG400-VIm and PEO crosslinker PEGMEMA in polymer nature, PEGDME as a plasticizer additive, LiTFSI as a lithium salt, and DMPA as a photo initiator. First of all, the chemicals including PETMP, VIm-PEG400-VIm, PEGMEMA, PEGDME, and LiTFSI were mixed in vials on the basis of composition listed in Table 11.1 under continuous stirring in a glove box for 30 minutes. Subsequently, the solution was vacuumed for 2 hours, in which enabled the removal of moisture and bubbles trapped, followed by adding DMPA initiator. The as-prepared homogeneous solution was dripped on the lithium anode and the in-situ click polymerization was ignited by exposing to UV lamps at 365 nm for 20 minutes. Here, a series of polymer electrolytes with a fixed molar ratio of thiol group to vinyl group, combined with lithium anode were successfully synthesized.

11.2.5 Characterization

The ^1H NMR analyses were performed on a Bruker AVANCE III HD 600MHz NMR Spectrometer using a 5 mm NMR probe with dimethyl sulfoxide-d_6 (DMSO-d_6) and chloroform-d ($CDCl_3$) as solvents. Matrix-assisted laser desorption/ionization-time of flight (MALDI-TOF) mass spectrometer was recorded by the Shimadzu, MALDI-7090 TOF/TOF with nitrogen laser at 335 nm and 2,5-dihydroxybenzoic acid (2,5-DHB) as a matrix. Attenuated total reflectance-Fourier transform infrared

TABLE 11.1
Composition of the Prepared Electrolytes

Electrolyte	Prepolymer (eq.)			Additive (g)	Salt (g)
	PETMP	VIm-PEG-VIm	PEG MEMA 950	PEG DME	LiTFSI
O-33	1	2	2	0.56	0.403
O-25	1	2.25	1.5	0.54	0.385
O-16	1	2.5	1	0.51	0.367
O-0	1	3	0	0.46	0.331

(ATR-FTIR) spectroscopy was performed at room temperature over 32 scans using a Thermo Nicolet 5700 FT-IR with a scan range from 4000–400 cm^{-1}. Raman spectroscopy was characterized by the CCD sensor DV401A-BV and 532 nm excitation light. Rheology properties were measured on TA instruments HR-2 equipped with real-time UV curing kit, 20-mm parallel quartz plate, and Excelitas Omnicure S2000 mercury lamp source. All dynamic measurements were performed under oscillation fast sampling mode with a strain amplitude of 1%, angular frequency of 1.0 rad s^{-1}, and light amplitude of 1%. Mechanical property measurements were carried out by TA instruments ARES-G2 with a constant compression rate of 0.01 mm s^{-1} on the electrolyte shaped 10 mm of diameter and 5 mm of height cylinder. Crystallinity characterization was determined by XRD Rigaku Ultima IV-9407F701 with Cu-Kα radiation (λ = 1.54 Å) under 40 kV and 30 mA and DSC 200 F3 with a scan rate of 10°C min^{-1} under nitrogen purging and liquid nitrogen cooling system. Thermal stability was evaluated by thermogravimetric analyzer.

Perkin-Elmer TGA 4000, heating the electrolyte sample to 600°C with an increasing rate of 10°C·min^{-1} under nitrogen atmosphere. The morphology of the samples and cross-sectional thickness were characterized using a scanning electron microscope (FE-SEM, Hitachi SU8010, 10 kV).

11.2.6 Electrochemical Measurements

In this research work, all electrochemical tests were carried out using CR2032-type coin cells. Electrochemical measurements were carried out on an electrochemical analyzer (CH Instruments 6116E). The ionic conductivity of electrolytes at temperature ranging from 25°C to 80°C was measured by AC impedance spectroscopy with a frequency ranging from 0.1 Hz to 1 MHz. The thickness of a corresponding cell assembled as a stainless-steel/electrolyte/stainless-steel cell was measured by a digital micrometer. With the sample thickness, the ionic conductivity σ (S·cm^{-1}) was calculated by the following equation: σ (S·cm^{-1}) = D (cm)/[R_b (S^{-1})·A (cm^2)], where D (cm) is the thickness of the sample, A (cm^2) is the contact area of the stainless-steel and the electrolyte, and R_b (S^{-1}) is the bulk resistance of the cell. The electrochemical window was evaluated on a cell assembled as lithium metal/electrolyte/stainless-steel by linear sweep voltammetry (LSV) with a fixed scan rate of 5 mV·s^{-1} from 0 V to 6 V. To monitor dendrite suppression capability and stability, cycling performance of symmetric lithium batteries assembled as lithium metal/electrolyte/lithium metal was tested including a fixed current density of 50 μA·cm^{-2} and a variant current density including 50, 100, 150 and 200 μA cm^{-2}, respectively, and the polarization time in each cycle was 6 hours. Charge/discharge performance was carried out on a multichannel analyzer (BAT-750B). The LFP cathode composed of 80 wt% $LiFePO_4$ (LFP), 10 wt% Super-P and 10 wt% PVDF was coated on aluminum foils, and the loading density of $LiFePO_4$ was 2.6 mg cm^{-2}. The as-prepared cathode, lithium metal, and electrolyte were fabricated as a half-cell for the following tests. Variant rate capacity and cycle life test were cycled between 2.5 V to 4.0 V at 60°C. Furthermore, the AC impedance of the cells after the 0th, 5th, 10th, 20th, and 30th cycles was measured, respectively, to investigate the impact of SEI formation. After the end of the test, battery was disassembled for the characterization of surface morphology and interfacial adhesion.

11.3 RESULTS AND DISCUSSION

The proton NMR and MALDI-TOF analyses of PIL (VIm-PEG400-VIm) confirmed the successful synthesis of PIL (VIm-PEG400-VIm) (data not shown), consistent with the results in the previous report [34]. With the presence of UV light, the liquid precursor could be polymerized to PEs within seconds through thiol-ene click chemistry. Therefore, the detailed kinetics of a series of PEs and O-X, in which X is named after the molar percentage of vinyl groups, dominated by PEGMEMA in prepolymer contents should be examined further with rheometer equipped with the in-situ UV-curing chamber. In addition, the molar ratios between thiol group from PETMP and vinyl group from PIL and PEGMEMA were all kept constant at 1:1.5, while the mass ratios between prepolymers, additives, and lithium salts were also fixed as shown in Table 11.1. From the rheological response, the rate and the completion of polymerization could be determined by monitoring the crossover of storage modulus (G') and loss modulus (G'') and the slope of moduli as a function of time in Figure 11.1. As shown in the figure, all the loss moduli G'' were greater than the storage moduli G', indicating the inherent liquid nature in the precursor before the illumination of the UV light, while the drastic increase of storage moduli surpassed the corresponding loss moduli among all samples upon turning on the UV light. The moduli and gelation time were summarized in Table 11.2. The crossovers of G′ and G″, viewed as gelation points in this research, revealed that the photopolymerization rate was dominated by the percentage of VIm-PEG400-VIm for its cross-linking property. Compared with the O-0 sample, the O-33 one exhibited a much slower gelation trend because of the lower percentage of crosslinker, which also resulted in the lowest moduli in G' and G''. Besides, after being irradiated for 20 min, both moduli maintained stable, which can be viewed as the completion of photopolymerization, and determined as an exposure time for fabricating PEs.

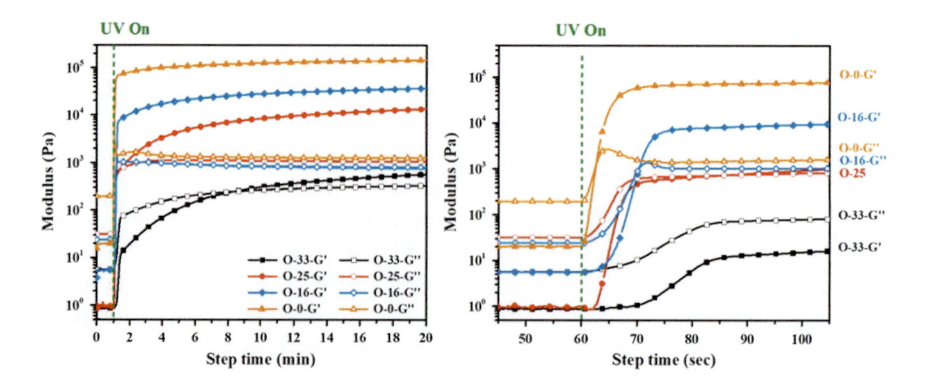

FIGURE 11.1 The storage modulus (G′) and loss modulus (G″) as a function of time for PEs with constant stress of 10 Pa at 55°C.

TABLE 11.2
Moduli and Gelation Time of PEs

PE	Modulus (Pa)				Gelation Time (s)
	G′ Before	G″ Before	G′ After	G″ After	
O-33	0.8	5.6	6.2×10^2	3.4×10^2	441
O-25	0.9	31	1.4×10^4	1.1×10^3	26.4
O-16	5.5	24	3.8×10^4	8.0×10^2	10.3
O-0	20	196	1.5×10^5	1.3×10^3	2.8

In order to observe the completion of the cross-linking, optical spectra including FTIR and Raman scanned the mixture before and after the UV irradiation. In Figures 11.2a and 11.2c, a characteristic absorption peak related to C=C stretching band was observed in both Raman and FTIR spectra in VIm-PEG400-VIm at 1,655 cm^{-1}. However, it disappeared in all electrolyte samples after photopolymerization because

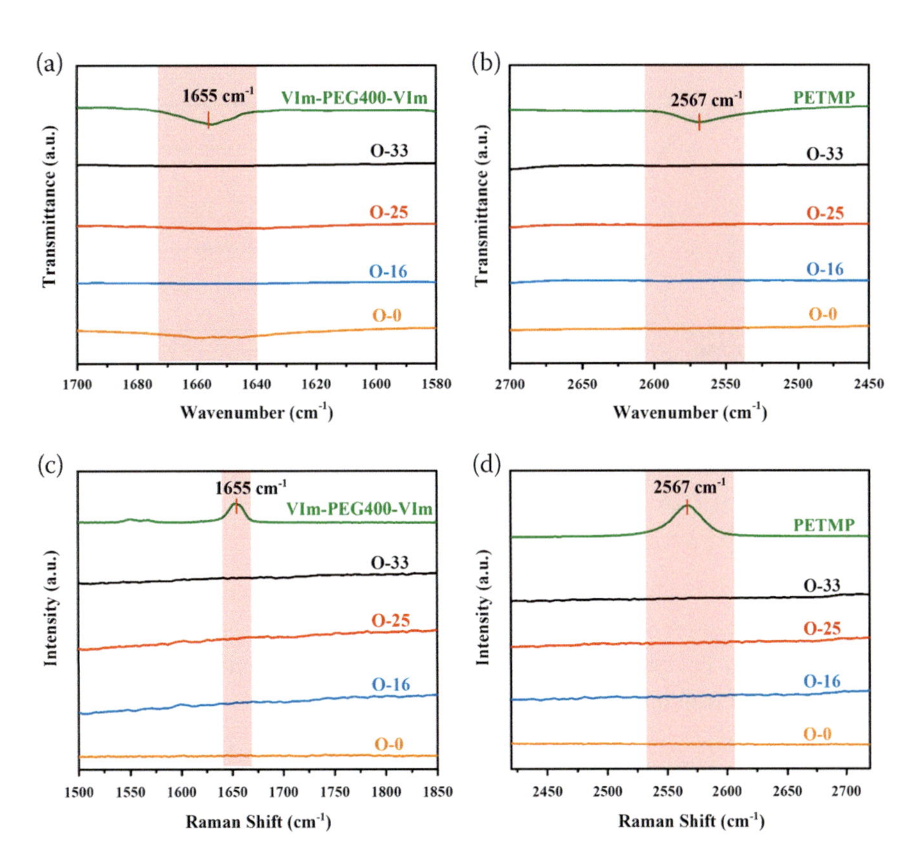

FIGURE 11.2 (a, b) FT-IR and (c, d) Raman spectra of PETMP, VIM-PEG400-VIm and PEs.

of the cross-linking not only between thiol and vinyl groups, but also between vinyl and vinyl groups [35]. The consumption of thiol groups was also observed in Figure 11.2b and 11.2d. A single characteristic peak at 2,567 cm^{-1} could be attributed to the stretching mode of thiol group in PETMP, and the smooth curves appeared in accordance with those of vinyl group [36]. Therefore, with the aid of Raman and FTIR analyses, the possession of vinyl and thiol groups was confirmed to be thoroughly consumed after the photopolymerization, which diminished the possibilities of side reaction triggered by the reactive groups in the following electrochemical tests.

After fabricating the PEs, their thermal properties were further tested for evaluating the stabilities and the polymer phase under different scenarios of operation temperatures by TGA and DSC (Figure 11.3). The decomposition temperature (T_d) defined as the temperature at which the sample lost 5% of its weight, as well as the glass transition temperature (T_g) are listed in Table 11.3. Overall, the PEs were thermally stable up to 300°C without significant decomposition profile owing to the cross-linking of polymer matrix and the inert nature of PIL as shown in Figure 11.3a, whereas pure PEGDME tends to decompose earlier than PIL and LiTFSI at 200°C resulted from its ether property. The thermogram implied that the interaction among the interpenetrating network slowed down the decomposition

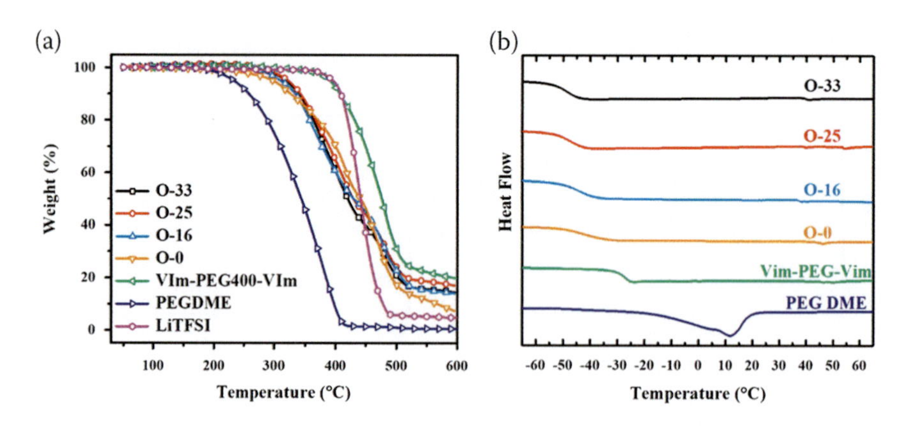

FIGURE 11.3 (a) TGA curves of PEs, salts, additives, and PIL (b) DSC curves of PEs, additives, and PIL.

TABLE 11.3
Decomposition Temperature (T_d) and Glass Transition Temperature (T_g) of PEs and Electrolyte Components

PE	T_d (°C)	T_g (°C)	Component	T_d (°C)	T_g (°C)
O-33	319	−48	LiTFSi	396	N/A
O-25	322	−46	PEG DME	233	N/A
O-16	310	−43	VIm-PEG_{400}-VIm	390	−29
O-0	298	−40			

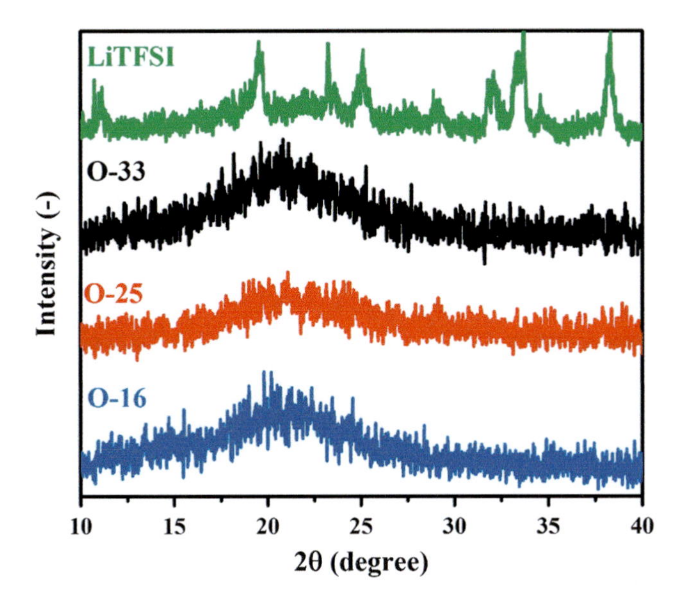

FIGURE 11.4 XRD patterns of the prepared PEs and lithium salt.

process of PEGDME. The DSC curves are shown in Figure 11.3b. All the curves of PEs in the observed region only exhibited a glass transition point between −50°C and −40°C, suggesting that the electrolytes were in an amorphous state. XRD patterns of the as-prepared PEs at room temperature revealed a broad peak at $2\theta =$ 16°~ 30° (Figure 11.4), revealing that there is no crystalline structure present in the PEs. The amorphous form of PEG was confirmed to be beneficial to the ionic conductivity as the increasing in segmental motion. Notably, a broad endothermic peak related to the melting point of pure PEGDME disappeared in that of electrolytes, which was mainly realized by the excellent miscibility between each ingredient, creating a wide range of the amorphous region.

Measurements of the electrochemical window via linear sweep voltammetry (LSV) would help determine the electrochemical stabilities and the charge/discharge operating voltage. From Figure 11.5a, the electrochemical windows, i.e., the onset of the oxidative potential (vs. Li/Li+), of the O-33, O-25, and O-16 samples are obviously more than 5.0 V. Notably, the O-33 sample can stable up to 5.3 V of onset oxidative potential (vs. Li/Li^+). When it comes to solid-state electrolytes, ionic conductivity is regarded as an important index to evaluate its feasible potential. Figure 11.5b and Table 11.4 show the ionic conductivities of PEs at different temperatures from 25°C to 80°C. The ionic conductivities of the three samples were all more than 10^{-4} $S{\cdot}cm^{-1}$ at 60°C. Nonetheless, only the O-33 sample can reach nearly 10^{-4} $S{\cdot}cm^{-1}$ at 25°C. A lower ratio of VIm-PEG400-Vim and PEGMEMA gives rise to a higher ionic conductivity for its larger segmental motion caused by single-side active sites, compared with PIL crosslinker. That is why the ionic conductivity raises as VIm-PEG400-VIm decreases. The excellent chemical and electrochemical stabilities exhibited on PEs can be attributable to the

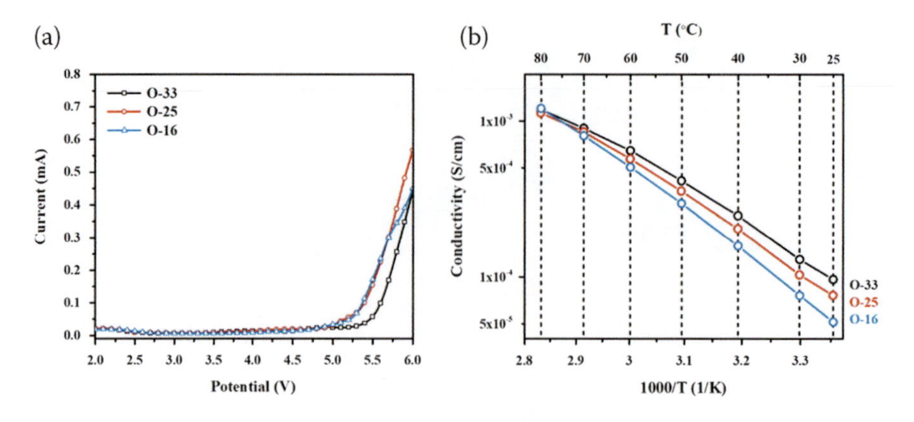

FIGURE 11.5 (a) LSV curves of PEs (b) ionic conductivities curves of PEs at different temperatures.

TABLE 11.4
Ionic Conductivity of PEs at Different Temperatures

PE	Ionic Conductivity (S cm^{-1})	
	25 °C	60 °C
O-33	9.7×10^{-5}	6.5×10^{-4}
O-25	7.6×10^{-5}	5.7×10^{-4}
O-16	5.1×10^{-5}	5.1×10^{-4}

properties of VIm-PEG400-VIm and the designed crosslinker, which enabled a series of PEs to be safe and reliable candidates. With the outcomes of the ionic conductivity measurements, the O-33 sample was confirmed to give the largest value among the three candidates. Therefore, the promising O-33 was chosen for further cycling tests.

The following cycling test was carried out on the Li/O-33/$LiFePO_4$ cell at 60°C to observe its charge-discharge performance. The test results under various C-rates are presented in Figure 11.6a and 11.6b. From the voltage profiles, the specific discharge capacity at 0.1 C, 0.2 C, and 0.3 C were all around 161 mAh·g^{-1}, near 95% of the theoretical capacity of $LiFePO_4$. Compared with the specific discharge capacity at 0.2 C, i.e., 162.2 mAh·g^{-1}, it still remained 41% of that at 1 C. The reversible cycling process from 0.1 C to 0.5 C was performed with a 99% of retention capacity at 0.1 C, attributed to the affordable ionic conductivity and cross-linking network. Figure 11.6c and 11.6d present the specific charge/discharge capacities, Coulombic efficiency, and EIS plot during long-term cycling test at 0.2 C under 60°C. The red datapoints representing discharge performance maintained a stable value at about 160 mAh·g^{-1} for 100 cycles, indicating the adhesion between the electrolyte and cathode was strong enough to afford the reversible process of deintercalation and intercalation of lithium ions and maintained a stable

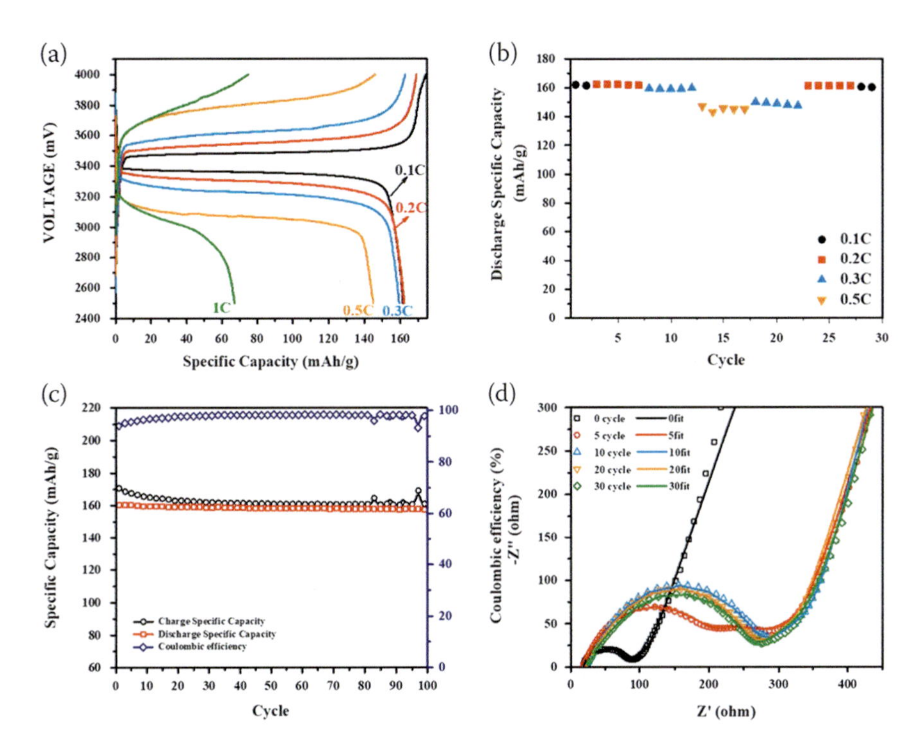

FIGURE 11.6 (a) charge-discharge curves (b) various current densities (c) cycling performance (d) EIS plot of Li/O-33/LFP cell under 60 °C.

ion transfer channel in the electrolyte. With the aid of EIS plot, the correlation between the formation of solid electrolyte interphase (SEI) and interfacial resistance could be simulated via a specific equivalent circuit illustrated in Figure 11.7. The values of fitting results including R_b, R_{sei}, and R_{ct} at specific cycles were listed in Table 11.5. Generally, R_b represents the bulk resistance corresponding to the diffusion of lithium ions in the electrolyte. R_{sei}, the impedance of ion diffusion in SEI layer, can be determined from the semicircle at high frequencies. R_{ct}, the charge transfer resistance between the electrolyte and active material layers can be inferred from the semicircle at the intermediate frequency. During the first 5 cycles, the resistance (R_{sei}) mainly increased from 42.4 Ω to 125.8 Ω, indicating the formation

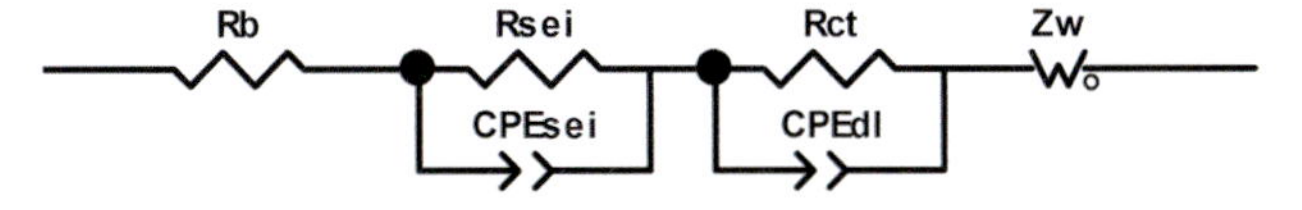

R_b: resistance of the liquid electrolyte
R_{sei}: resistance of solid electrolyte interphase (SEI)
R_{ct}: resistance of charge transfer at the interface between SEI and active materials
Z_w: Warburg impedance of lithium ions in electrode

FIGURE 11.7 The equivalent circuit model used for the electrochemical impedance spectroscopy (EIS) data fitting.

TABLE 11.5
The Corresponding Simulated Impedance Parameters of the Li/O-33/LFP Cell in an Equivalent Circuit

Sample	Cycle	R_b (Ω)	R_{sei} (Ω)	R_{ct} (Ω)
O-33	0	17.7	42.4	18.6
	5	15.4	125.8	216.8
	10	20.9	33.9	202.8
	20	20.4	44.9	181.6
	30	21.7	42.7	183.2

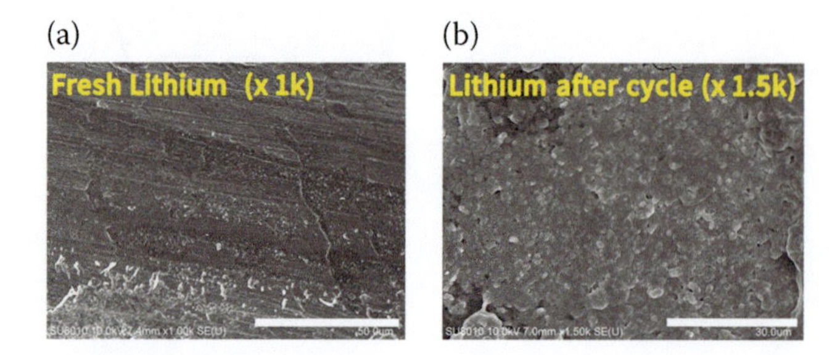

FIGURE 11.8 SEM images for the lithium anode of the cell based on O-33 (a) before and (b) after testing at 0.2 C for 100 cycles.

process of SEI. After 20 cycles, R_{sei} exhibited a stable value at about 40 Ω and the resistance was dominated by R_{ct}. The results show that the O-33 sample can inhibit the unlimited formation of the SEI layer and maintain a stable value of resistance during cycling tests.

The formation of SEI layer can also be observed from SEM images as shown in Figure 11.8a and 11.8b. The fresh lithium appearance exhibited a smooth surface. After cycling the Li/O-33/LFP cell for 100 cycles, the deposition of lithium rearranged as a dense layer without apparent fractures or cracks, which implied that the O-33 sample could effectively ease the uneven formation of lithium dendrite by its crosslinking network and feasible segmental motions.

11.4 CONCLUSIONS

In this work, a dicationic imidazolium-based crosslinker and PETMP with a plasticizer (PEGDME) have been successfully fabricated by an in-situ UV-curving thiol-ene click reaction. High ionic conductivity of 6.5×10^{-4} S cm^{-1} and excellent specific discharge capacity (161 mAh·g^{-1}) at 0.2 C under 60°C were obtained, which can be attributed to in-situ strategy and the interaction between imidazolium and anions of the Li salt. The prepared electrolyte shows a wide electrochemical window up to 5.3V (vs. Li/Li$^+$), demonstrating outstanding electrochemical stability. All these impressive properties give the synthesized electrolyte a promising potential for practical application in lithium-ion batteries.

REFERENCES

1. Sarma DD, Shukla AK. 2018 Building better batteries: A travel back in time. *ACS Energy Lett* 3:2841–2845.
2. Goodenough J, Chemical KP-J of the A. 2013 The Li-ion rechargeable battery: A perspective. *ACS Publ* 135:1167–1176.
3. Goodenough JB. 2018 How we made the Li-ion rechargeable battery. *Nat Electron* 1:204–204.
4. Goodenough JB, Kim Y. 2010 Challenges for rechargeable Li batteries. *Chem Mater* 22:587–603.
5. Choudhury S, Stalin S, Vu D, Warren A, Deng Y, Biswal P, et al. 2019 Solid-state polymer electrolytes for high-performance lithium metal batteries. *Nat Commun* 10:1–8.
6. Long L, Wang S, Xiao M, Meng Y. 2016 Polymer electrolytes for lithium polymer batteries. *J Mater Chem A* 4:10038–10069.
7. Arya A, Sharma AL. 2017 Insights into the use of polyethylene oxide in energy storage/conversion devices: a critical review. *J Phys D Appl Phys* 50:443002.
8. Xue Z, He D, Xie X. 2015 Poly(ethylene oxide)-based electrolytes for lithium-ion batteries. *J Mater Chem A* 3:19218–19253.
9. Zhao Y, Bai Y, Li W, An M, Bai Y, Chen G. 2020 Design strategies for polymer electrolytes with ether and carbonate groups for solid-state lithium metal batteries. *Chem Mater* 32:6811–6830.
10. Kobayashi K, Pagot G, Vezzù K, Bertasi F, DiNoto V, Tominaga Y. 2020 Effect of plasticizer on the ion-conductive and dielectric behavior of poly(ethylene carbonate)-based Li electrolytes. *Polym J* 53:149–155.
11. Hu J, Wang W, Zhou B, Feng Y, Xie X, Xue Z. 2019 Poly(ethylene oxide)-based composite polymer electrolytes embedding with ionic bond modified nanoparticles for all-solid-state lithium-ion battery. *J Memb Sci* 575:200–208.
12. Lin D, Yuen PY, Liu Y, Liu W, Liu N, Dauskardt RH, et al. 2018 A silica-aerogel-reinforced composite polymer electrolyte with high ionic conductivity and high modulus. *Adv Mater* 30:1802661.
13. Wang J, Yang J, Shen L, Guo Q, He H, Yao X. 2021 Synergistic effects of plasticizer and 3D framework toward high-performance solid polymer electrolyte for room-temperature solid-state lithium batteries. *ACS Appl Energy Mater* 4: 4129–4137.
14. Wang H, Wang Q, Cao X, He Y, Wu K, Yang J, et al. 2020 Thiol-branched solid polymer electrolyte featuring high strength, toughness, and lithium ionic conductivity for lithium-metal batteries. *Adv Mater* 32:2001259.
15. Zhu Y, Cao J, Chen H, Yu Q, Li B. 2019 High electrochemical stability of a 3D cross-linked network PEO@nano-SiO_2 composite polymer electrolyte for lithium metal batteries. *J Mater Chem A* 7:6832–6839.
16. Luo G, Yuan B, Guan T, Cheng F, Zhang W, Chen J. 2019 Synthesis of single lithium-ion conducting polymer electrolyte membrane for solid-state lithium metal batteries. *ACS Appl Energy Mater* 2:3028–3034.
17. Li Y, Sun Z, Shi L, Lu S, Sun Z, Shi Y, et al. 2019 Poly(ionic liquid)-polyethylene oxide semi-interpenetrating polymer network solid electrolyte for safe lithium metal batteries. *Chem Eng J* 375:121925.
18. Li X, Zhang Z, Li S, Yang L, Hirano SI. 2016 Polymeric ionic liquid-plastic crystal composite electrolytes for lithium ion batteries. *J Power Sources* 307:678–683.

19. Safa M, Chamaani A, Chawla N, El-Zahab B. 2016 Polymeric ionic liquid gel electrolyte for room temperature lithium battery applications. *Electrochim Acta* 213: 587–593.
20. Tejero R, López D, López-Fabal F, Gómez-Garcés JL, Fernández-García M. 2015 Antimicrobial polymethacrylates based on quaternized 1,3-thiazole and 1,2,3-triazole side-chain groups. *Polym Chem* 6:3449–3459.
21. Yi S, Leon W, Vezenov D, Regen SL. 2016 Tightening polyelectrolyte multilayers with oligo pendant ions. *ACS Macro Lett* 5:915–918.
22. Obadia MM, Jourdain A, Serghei A, Ikeda T, Drockenmuller E. 2017 Cationic and dicationic 1,2,3-triazolium-based poly(ethylene glycol ionic liquid)s. *Polym Chem* 8:910–917.
23. Tseng YC, Ramdhani FI, Hsiang SH, Lee TY, Teng H, Jan JS. 2026 Lithium battery enhanced by the combination of in-situ generated poly(ionic liquid) systems and TiO_2 nanoparticles. *J Memb Sci* 641:119891.
24. Tseng YC, Hsiang SH, Tsao CH, Teng H, Hou SS, Jan JS. 2021 In situ formation of polymer electrolytes using a dicationic imidazolium cross-linker for high-performance lithium ion batteries. *J Mater Chem A* 9:5796–5806.
25. Li X, Zheng Y, Pan Q, Li CY. 2019 Polymerized ionic liquid-containing interpenetrating network solid polymer electrolytes for all-solid-state lithium metal batteries. *ACS Appl Mater Interfaces* 11:34904–34912.
26. Zhou Y, Wang B, Yang Y, Li R, Wang Y, Zhou N, et al. 2019 Dicationic tetraalkylammonium-based polymeric ionic liquid with star and four-arm topologies as advanced solid-state electrolyte for lithium metal battery. *React Funct Polym* 145:104375.
27. Lowe AB. 2010 Thiol-ene "click" reactions and recent applications in polymer and materials synthesis. *Polym Chem* 1:17–36.
28. Hoyle CE, Lowe AB, Bowman CN. 2010 Thiol-click chemistry: A multifaceted toolbox for small molecule and polymer synthesis. *Chem Soc Rev* 39:1355–1387.
29. Kwisnek L, Heinz S, Wiggins JS, Nazarenko S. 2011 Multifunctional thiols as additives in UV-cured PEG-diacrylate membranes for CO_2 separation. *J Memb Sci* 369: 429–436.
30. Liu T, Zhang J, Han W, Zhang J, Ding G, Dong S, et al. 2020 Review—In situ polymerization for integration and interfacial protection towards solid state lithium batteries. *J Electrochem Soc* 167:070527.
31. Zhao Q, Liu X, Stalin S, Khan K, Archer LA. 2019 Solid-state polymer electrolytes with in-built fast interfacial transport for secondary lithium batteries. *Nat Energy* 4:365–373.
32. Chen X, He W, Ding LX, Wang S, Wang H. 2019 Enhancing interfacial contact in all solid state batteries with a cathode-supported solid electrolyte membrane framework. *Energy Environ Sci* 12:938–944.
33. Cecchini MM, Bendjeriou A, Mnasri N, Charnay C, Angelis FDe, Lamaty F, et al. 2014 Synthesis of novel multi-cationic PEG-based ionic liquids. *New J Chem* 38: 6133–6138.
34. Tseng YC, Hsiang SH, Lee TY, Teng H, Jan JS, Kyu T. 2021 In situ polymerized electrolytes with fully cross-linked networks boosting high ionic conductivity and capacity retention for lithium ion batteries. *ACS Appl Energy Mater* 4:14309–14322.
35. McManis GE, Gast LE. 1971 IR spectra of long chain vinyl derivatives. *J Am Oil Chem Soc* 48:668–673.
36. Miao JT, Yuan L, Guan Q, Liang G, Gu A. 2018 Water-phase synthesis of a biobased allyl compound for building UV-curable flexible thiol-ene polymer networks with high mechanical strength and transparency. *ACS Sustain Chem Eng* 6:7902–7909.

12 Synthesis of Multiporous Carbons with Biomaterials for Applications in Supercapacitors and Capacitive Deionization

Chun-Han Hsu
General Education Center, National Tainan Junior College of Nursing, Tainan City, Taiwan

Zheng-Bang Pan and Hong-Ping Lin
Department of Chemistry, National Cheng Kung University, Tainan City, Taiwan

CONTENTS

DOI: 10.1201/9781003367215-12

12.1 INTRODUCTION

Energy is required for all human activities. However, as the global population grows and energy demands increase accordingly, the world's remaining supplies of fossil fuels, such as oil, coal, and gas, are being rapidly depleted [1–3]. Moreover, extensive evidence now demonstrates that the combustion of such fuels results in numerous adverse environmental effects, including global warming, the greenhouse effect, extreme weather patterns, climate change, and pollution [4,5]. Consequently, the need for alternative, renewable, and green energy sources, such as solar energy, wind power, tidal power, and fuel cells, has become a pressing concern [5–9]. Although substantial advances have been made in increasing the power generation efficiency of these energy sources, the question of how best to store the produced energy has yet to be properly answered [10,11].

Supercapacitors are energy storage devices that have attracted extensive interest in recent years due to their advantages of high power density, long cycle lifetimes, and high reversibility [12–16]. Figure 12.1 presents a diagram of a supercapacitor device. The electrodes in supercapacitors are typically fabricated using porous carbon due to its good conductivity, high surface area, and low cost [17–21]. Although porous carbon is traditionally prepared using synthetic materials (e.g., sucrose or phenolic resin), the feasibility of using biomaterial or biomass waste as an alternative preparation material has received increasing attention. Furthermore, biomass waste is naturally rich in carbon and its use is environmentally friendly [22–25]. The advantage of using biomass as raw material comes from the fact that the net CO_2 emissions from the utilization process are considered to be zero. Thus, in the present study, water chestnut shell biochar (WCSB) from Guan-Tan district in Tainan was used as a carbon source for synthesizing multiporous carbon for use as the electrode materials of the supercapacitors.

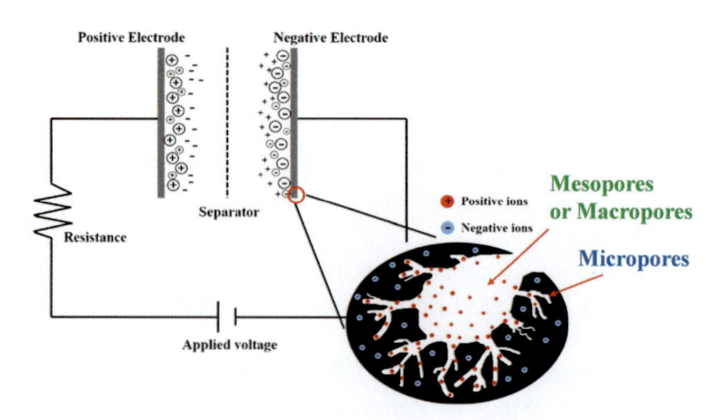

FIGURE 12.1 Diagram of a supercapacitor using multiporous carbon materials as electrode.

In addition to caring about energy issues, freshwater shortages are an increasingly serious problem in many parts of the world [26–28]. Many desalination techniques have been developed, including reverse osmosis, distillation, thermal separation, and electrodialysis [29–34]. However, most of these technologies are complex to operate and have high energy costs. Therefore, the potential for the large-scale application of these technologies is limited [35]. Thus, capacitive deionization (CDI) with porous carbon electrodes has emerged as a highly attractive method for water desalination due to its low energy consumption, minimal environmental impact, simple design, and straightforward operation [36–39]. Figure 12.2 presents a diagram of the adsorption procedure in CDI. However, the performance of CDI is critically dependent on the availability of carbon materials with a high specific surface area and distinctive pore structure [40–42]. Thus, the present study explores the feasibility of using the developed WCSB-synthesized multiporous carbon in supercapacitors and CDI applications.

The development of supercapacitors with high power density is gradually becoming a key topic in electrical energy research [43–45]. Previously, activated carbon was often chosen as the electrode material for supercapacitors due to its high stability and low price [18,46–48]. The microporous carbons with a pore size <2.0 nm can also be chosen as electrode materials to increase the capacitance; however, the capacitive behavior of activated carbon electrodes is no longer present due to the high resistance of ion transmission through micropores during fast charging and discharging processes [49–51]. To produce supercapacitors with high efficiency, a new carbon material with both microporous and mesoporous structures, denoted as multiporous carbon, must be manufactured. The synthesis of multiporous carbon is a critical research goal [52–57].

The advantages of multiporous carbon for use in supercapacitors are manifold. To achieve excellent high-speed charging and discharging performance, mesopores must be introduced into microporous carbon to act as charge transfer channels [58–63]. In our previous research, we used phenolic resin as a carbon source and mesoporous silica as a template to regulate the pore structure of a carbon electrode and form a three-dimensional network containing micropores (< 2 nm), mesopores

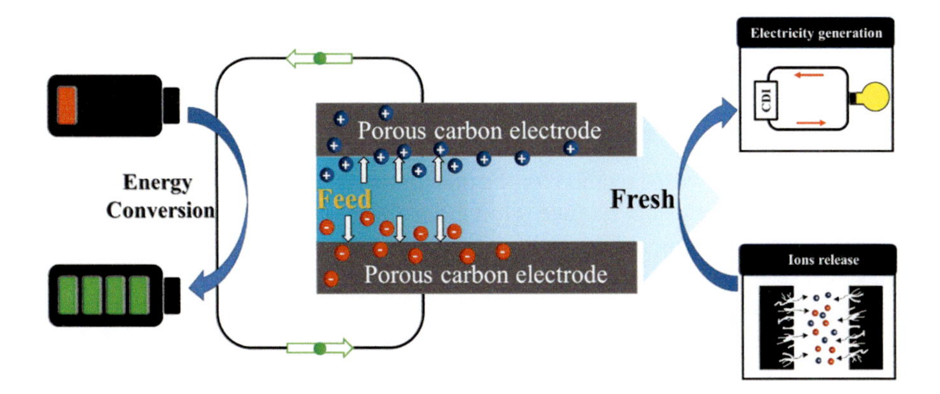

FIGURE 12.2 CDI adsorption procedure.

(2–50 nm), and macropores (>50 nm) [19,20]. The surface area of multiporous carbons can be increased by introducing micropores. The presence of the mesopores and macropores acting as fast channels for charge transfer can enable electrolytes to reach the surface of micropores. The surface area of the micropore is thus almost completely utilized, increasing the charge storage capacity (Figure 12.1). Because of fast charging and discharging capabilities, long cycle lifetimes, and low environmental pollution, supercapacitors may become the key novel green energy storage system of the 21st century. Therefore, the new type multiporous carbon of porous carbon materials with high surface area and high permeability mesopores and macropores has great potential for applications in energy storage and CDI.

12.2 WASTE WCS CARBON SOURCE FOR MULTIPOROUS CARBONS

12.2.1 Biomass Materials as a Carbon Source for Multiporous Carbon Preparation

With their high specific surface area and good electrical conductivity, porous carbons are promising electrode materials for supercapacitors, lithium-ion batteries, and fuel cells [64–67]. Agricultural waste materials are considered good alternative raw materials for producing porous carbon materials because they are abundant, inexpensive, and have a high carbon content [68–70]. If they can be converted into materials with high economic value, a circular economy of sustainable resources could be achieved [71,72].

In this study, waste WCS was used as a carbon source because tens of tons of waste WCS are produced in Guan-Tan district in Tainan, Taiwan every year, and this WCS has high lignin content [63]. Therefore, this study aimed to synthesize multiporous carbons from this primary carbonized WCS (water chestnut shell biochar, denoted as WCSB) through simple physical blending and chemical activation. The synthesized WCS multiporous carbons were used to assemble supercapacitors and CDI devices for desalination. Distinct from the microporous carbons, the existence of these mesopores and macropores can enhance the ion diffusion capacity of micropores when it used in supercapacitors or CDI devices [73–76].

12.2.2 Biomass Material as an Activating Agent for Preparing Porous Carbons

The most commonly used method for the preparation of high surface area carbons is chemical activation using alkaline substances such as KOH or K_2CO_3 [77–80]. However, after cracking at high temperature, the alkali metal is highly reactive and the synthesized product emits substantial heat in contact with water vapor, causing instantaneous combustion of the product; this chemical-activation procedure is not safe [81]. To overcome this drawback, many nanotemplates (e.g., MgO and ZnO) have been used for the synthesis of porous carbon materials [62–64]. In our laboratory, we have developed a new synthesis technique by mixing nano-$CaCO_3$ as hard templates with an appropriate amount of activating agent (K_2CO_3 or KOH) and carbon sources to synthesize multiporous carbons. The synthesis steps of this method are simple. And the nanotemplate can be directly removed with hydrochloric acid washing. In addition, the waste with Ca^{2+} ions can be recycled to reduce environmental pollution, thus achieving green production and a circular economy [62,63].

12.2.2.1 Eggshells

Eggshells have a primarily inorganic composition of 95% $CaCO_3$ and 3.5% organic matrix proteins [82]. Thus, eggshells were used in place of industrial-grade nano-$CaCO_3$ and nitrogen-containing compounds due to their advantages of being both an activating agent and a nitrogen source. In this study, the eggshells were used as a solid dispersion and activating agent also act as nitrogen-containing compounds (e.g. melamine, chitosan, and urea) to prepare the nitrogen-doped multiporous carbons during high-temperature carbonization [48,83,84].

12.3 PREPARATION AND CHARACTERIZATION OF MULTIPOROUS CARBONS

12.3.1 Multiporous Carbons Produced with a WCSB Carbon Source

Figure 12.3 demonstrates the experimental process for synthesizing multiporous carbons through physical mixing. A blender was used to mix 8.0 g of WCSB carbon precursor, 8.0–40.0 g of a nanosized $CaCO_3$ inorganic template, and 5.0–24.0 g of activating agent (KOH or K_2CO_3) for 3 minutes. The resultant powder was sealed in a stainless-steel container and pyrolyzed at a heating rate of 8°C min^{-1} to a carbonization temperature of 750°C for 2 h and subsequently to an activation temperature of 950°C for a further 3 hours. After being cooled to room temperature, the black powder was washed with water to remove the alkali salt and was then immersed in 37% HCl solution to remove the CaO template. After being stirred overnight, the solution was filtered, and the product was washed with deionized water. The resulting multiporous carbons were dried in an oven at 100°C.

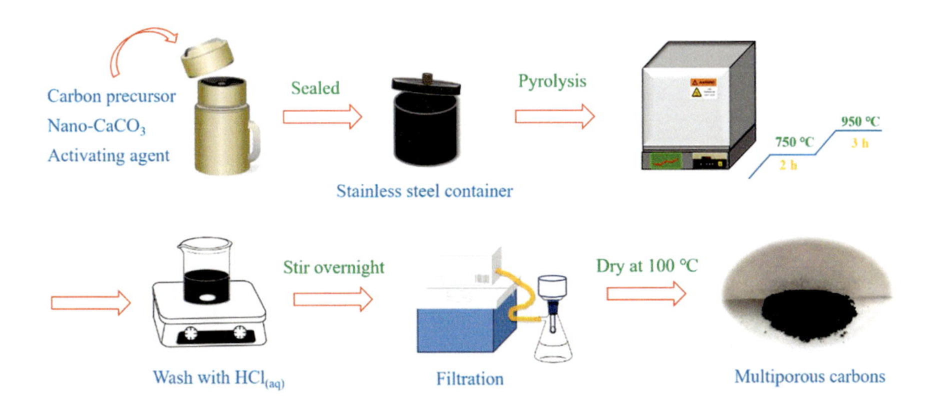

FIGURE 12.3 Synthesis of multiporous carbon by physical mixing.

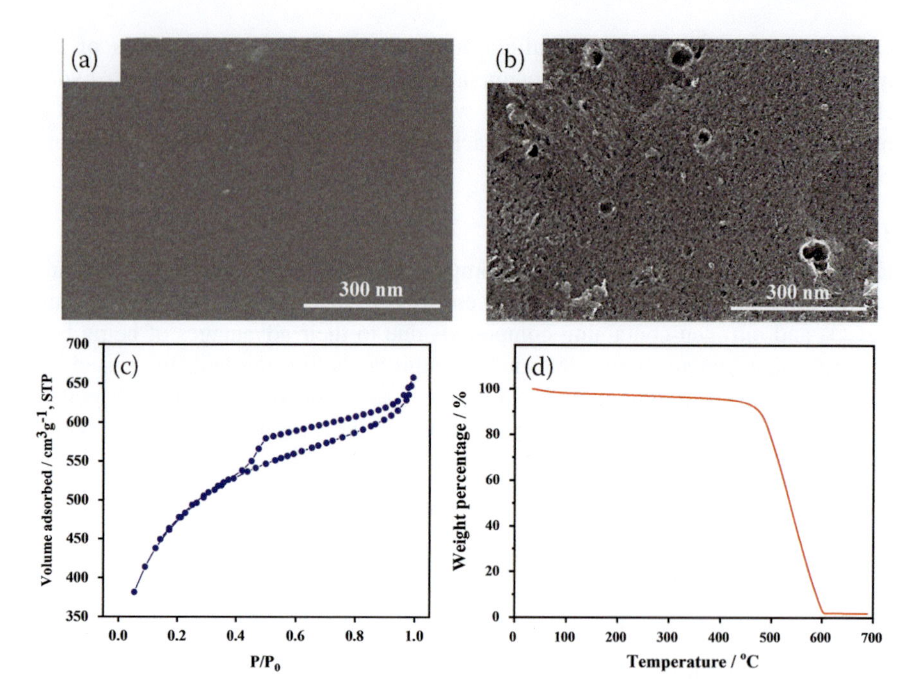

FIGURE 12.4 SEM images of (A) WCS biochar and (B) multiporous carbon. (C) N_2 adsorption–desorption isotherms and (D) TGA curve of multiporous carbon.

The scanning electron microscope (SEM) images in Figure 12.4A and B present the typical surface morphologies of the WCSB and multiporous carbon, respectively. The WCSB has a smooth surface without mesopores, even under high magnification. By contrast, multiporous carbon materials have numerous pores of different sizes. Furthermore, the N_2 adsorption–desorption isotherms of the multiporous carbon are of type IV and have a high specific surface area of up to 1584 m^2 g^{-1} (Figure 12.4C). And the low ash content (< 1%) and high thermal stability of multiporous carbon were shown in TGA curve (Figure 12.4D). Multiporous carbon materials with different specific surface areas and pore properties can be obtained by adjusting the relative content of the nano-$CaCO_3$ template and K_2CO_3 activating agent, respectively (Figure 12.5).

Because the WCSB carbon source is solid, the nano-$CaCO_3$ has different functions. First, it is used for solid dispersion of the carbonized material to avoid the hardening of the resulting product when only the activating agent KOH was added. Second, the carbon dioxide from $CaCO_3$ can be released at a high temperature (Equation 12.3.1) and then react with carbon structure to further erodes the micropores (Equation 12.3.2) [85–87]. Therefore, to synthesize multiporous carbons, a two-stage temperature program has been used for cracking and carbonization: at the temperature of 750°C, the $CaCO_3$ decomposes completely for CO_2-gas activation; at a temperature of 800–950°C, the reaction between the activating agent (KOH or K_2CO_3) and carbon dominated (Equation 12.3.3 to 12.3.10) [87]. The scheme of the synthetic procedure is presented in Figure 12.6.

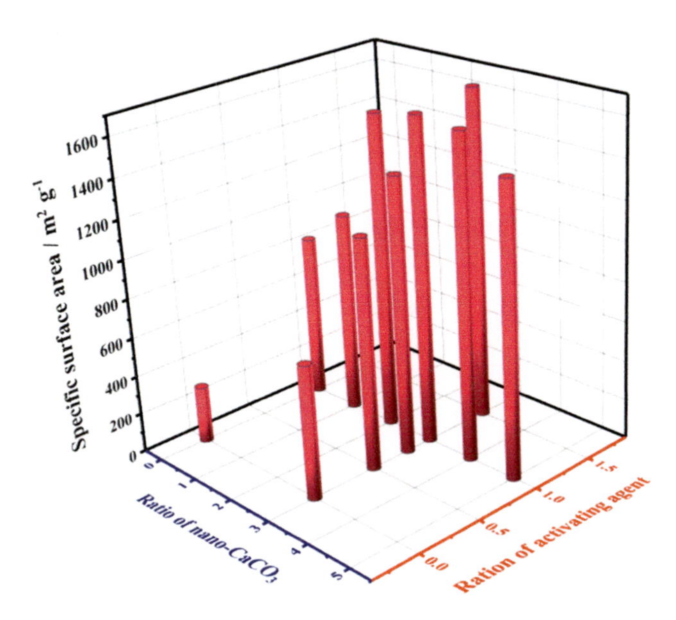

FIGURE 12.5 Specific surface areas of multiporous carbons synthesized with different templates and activating agent ratios.

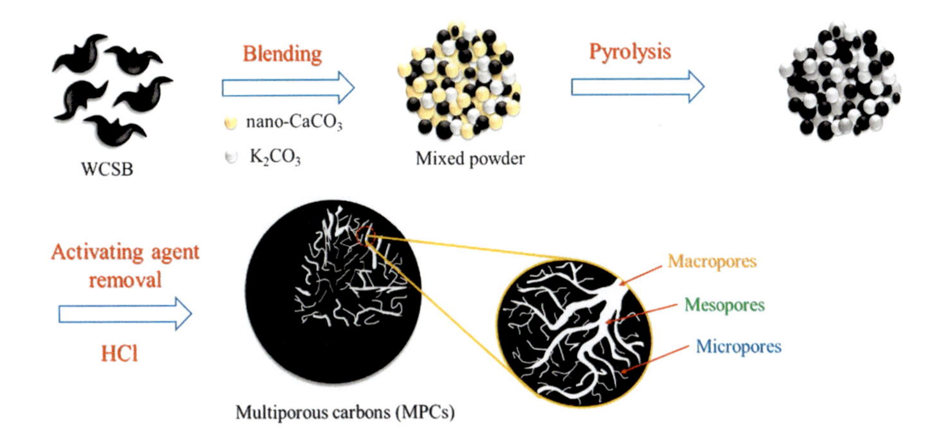

FIGURE 12.6 Synthesis procedure for multiporous carbon materials.

The $CaCO_3$ activation mechanism is presented in the following equations: [85]

$$CaCO_3(s) \rightarrow CaO\ (s) + CO_2(g) \tag{12.3.1}$$

$$CO_2(g) + C(s) \rightarrow 2CO(g) \tag{12.3.2}$$

The activation mechanism of the activator (KOH, K_2CO_3) is presented in the following equations: [86,87]

$$2KOH(s) \rightarrow K_2O(s) + H_2O(g) \quad (12.3.3)$$

$$C(s) + H_2O(g) \rightarrow CO(g) + H_2(g) \quad (12.3.4)$$

$$CO(g) + H_2O\ (g) \rightarrow CO_2(g) + H_2(g) \quad (12.3.5)$$

$$CO_2(g) + K_2O(s) \rightarrow K_2CO_3(s) \quad (12.3.6)$$

$$K_2CO_3(s) \rightarrow K_2O(s) + CO_2(g) \quad (12.3.7)$$

$$CO_2(g) + C(s) \rightarrow 2CO(g) \quad (12.3.8)$$

$$K_2CO_3(s) + 2C(s) \rightarrow 2K(s) + 3CO(g) \quad (12.3.9)$$

$$C(s) + K_2O(s) \rightarrow 2K(s) + CO(g) \quad (12.3.10)$$

12.3.2 Preparation of Nitrogen-Doped Multiporous Carbons Using Wasted Eggshells

To reduce the use of costly chemicals, we directly mixed the wasted eggshells consisted of $CaCO_3$ and proteins with the KOH activating agent for the preparation of the and N-doped multiporous carbons. The thermogravimetric analysis (TGA) curves displayed in Figure 12.7 reveal that the decomposition temperature of $CaCO_3$ in eggshells is between 700 and 800°C. The decomposition temperature of $CaCO_3$ in eggshells is higher than that of nano-$CaCO_3$. This is primarily because the $CaCO_3$ in eggshells has a larger particle size; thus, its decomposition temperature should be higher than that of nano-$CaCO_3$. Therefore, with the eggshell, the holding temperature at first stage was increased from 750°C to 800°C to induce the $CaCO_3$ decomposition.

Table 12.1 reveals the physical properties of the multiporous carbons synthesized using nano-$CaCO_3$ and eggshell as templates, respectively. From these data, the eggshell can replace the commercial nano-$CaCO_3$ as templates. Although the physical properties of the carbons are similar, the data of the elemental analysis demonstrate that the carbon produced with eggshell as a template had a higher nitrogen content of approximately 3.4 wt.%. The introduction of nitrogen with lone pairs of electrons into the carbon structure can increase the overall defect level and the electrical conductivity of the carbon matrix [88–90]. In addition, the nitrogen-containing functional groups increase the surface polarity of the carbon material and improve its wettability in

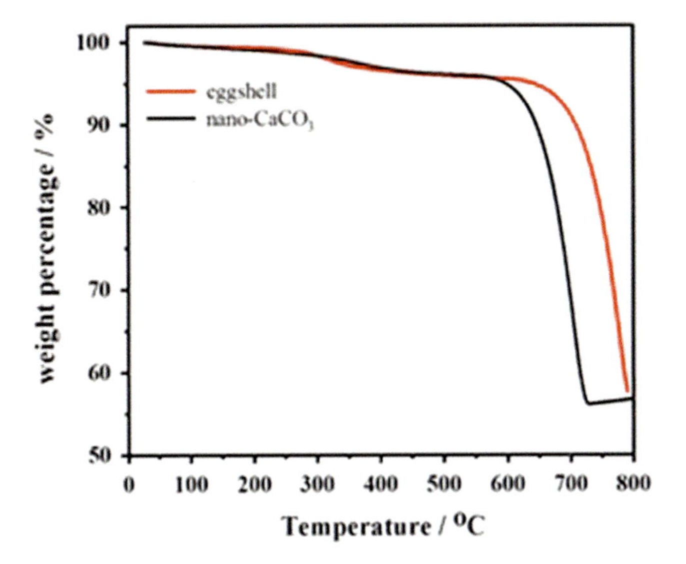

FIGURE 12.7 TGA curves of nano-$CaCO_3$ and eggshell under nitrogen.

TABLE 12.1
Specific Surface Area, Pore Volume, and Nitrogen Content of the Multiporous Carbons Synthesized with Different $CaCO_3$ Templates under the Same Conditions

Template Source	$S_{BET}(m^2g^{-1})$	$S_{mic}{}^{a}(m^2g^{-1})$	$S_{mes}{}^{b}(m^2g^{-1})$	Pore Volume (cm^3g^{-1})	N^{c}(wt%)
nano-$CaCO_3$	1509	540	969	0.94	1.0%
Eggshell	1549	590	959	1.00	3.4%

Notes
[a] Micropore surface area.
[b] Mesopore surface area.
[c] N content is from elemental analysis.

electrolytic solutions. These advantages significantly improve the capacitive performance of materials when used in supercapacitors. Besides, nitrogen-doped carbon materials have been widely studied as cathodes in fuel cells (such as air cells or methanol fuel cells) due to their active sites for oxygen reduction reactions [91]. Previous literatures have demonstrated that nitrogen-doped carbon materials have properties approaching commercial catalytic standards and thus could replace expensive commercial Pt–C catalysts, leading to fuel cells with reduced cost and high stability toward the CO deactivation [91–93].

12.3.3 Multiporous Carbon for Supercapacitor Applications

To improve the performance of the porous carbons, doping to produce functional groups (e.g., nitrogen, oxygen, or phosphorus groups) [53–56] can enhance the hydrophilicity of the carbon material and increase its electrical conductivity [18–20]. We added eggshells as a nitrogen-doping agent to prepare nitrogen-doped multiporous carbon materials and investigated the performance of these materials in supercapacitor applications. The nitrogen atoms in the material act as electron donors, and if the carbon material is enriched with nitrogen-containing functional groups, it can undergo rapid oxidation–reduction reactions with electrolyte ions to store charge, increasing the total capacitance of the material; this phenomenon is called *pseudocapacitance* [94].

Aqueous (6.0 M KOH) and organic (1.0 M $LiClO_4$/PC) electrolyte systems were used to investigate the performance of the synthesized nitrogen-doped multiporous carbon materials in supercapacitor applications. In the aqueous system (Figure 12.8A), the cyclic voltammetry (CV) curves of the nitrogen-doped multiporous carbons (with 3.4 wt.% nitrogen) have a large rectangular area and a specific capacitance value of 117 F g^{-1}, which is 30% greater than that of the multiporous

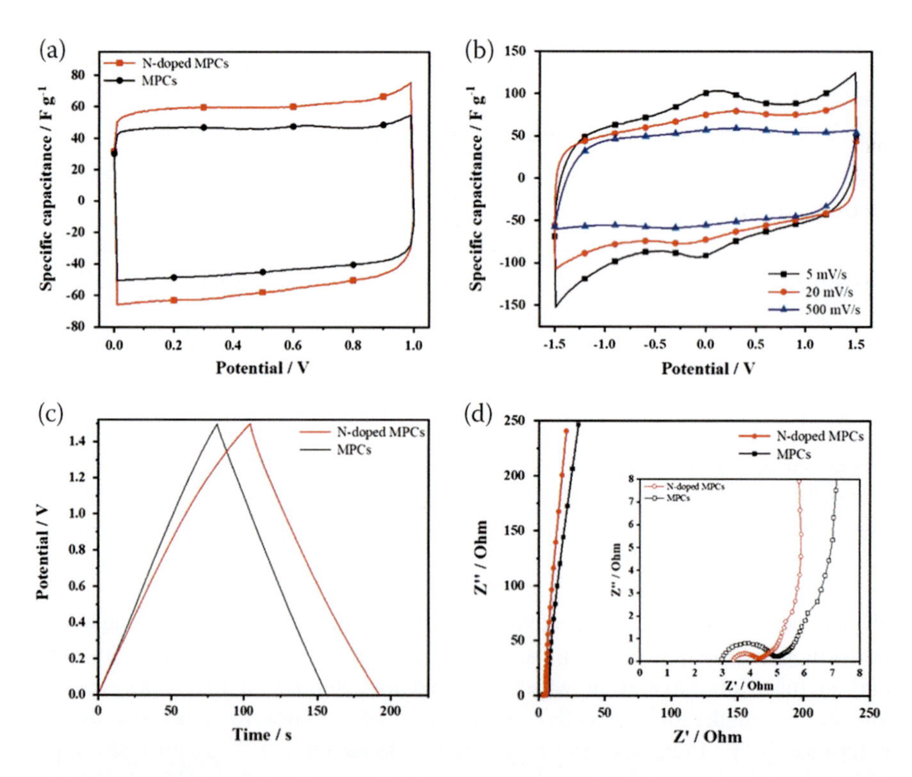

FIGURE 12.8 CV spectra of the nitrogen-doped and undoped multiporous carbons at 5 mV s^{-1} in (A) 6.0 M KOH and (B) 1.0 M $LiClO_4$/PC (for nitrogen-doped multiporous carbon). (C) Constant current charge–discharge plots for the presence of an organic electrolyte at a current density of 0.5 A g^{-1}. (D) AC impedance analysis.

carbons without nitrogen-doping (90 F g^{-1}). The electrochemical properties of the multiporous carbon electrodes were further evaluated by conducting CV and galvanostatic charge–discharge (GCD) measurements in 1.0 M $LiClO_4$/PC electrolyte as shown in Figure 12.8B and C, respectively. The CV curves of the two-electrode cell were measured at different scan rates over a potential range of 1.5 to −1.5 V. Notably, the cell assembled with the nitrogen-doped multiporous carbons had a high specific capacitance of up to 132 F g^{-1} at 5 mV s^{-1} and had a good retention rate (~70%) even at a scanning rate of 500 mV s^{-1}. In the organic electrolyte system, the CV curves have a clear redox peak and a shape slightly deviating from a rectangular shape that is primarily due to pseudocapacitance. The overall capacitance is mainly due to the double-layer capacitance, but also to the pseudocapacitance generated by the Faraday reaction. The overall specific capacitance can reach 132 F g^{-1} [94,95]. In addition, the nitrogen functional group on the porous carbons also improves the wettability of the carbons, facilitating the penetration of the electrolyte into the internal pores, increasing the accessibility of the surface areas from micropores in the carbon matrixes.

In Figure 12.8C, the charge–discharge curves of multiporous carbons are symmetric triangles at a current density of 0.5 A g^{-1}, and the specific capacitance can reach 160 F g^{-1}, indicating that the electrode material has high reversibility during charging and discharging. The charge–discharge curves of the nitrogen-doped multiporous carbons deviated slightly from symmetry due to the oxidation–reduction reaction during the charging process, further confirming the capacitance improvement contribution of pseudocapacitance [95]. Finally, an AC impedance analysis (Figure 12.8D) was used to compare the difference in conductivity between the materials. Semicircles with larger radii indicate poorer conductivity and larger charge transfer resistance. According to the literature, the introduction of nitrogen atoms can form local functional groups on the surface of carbon materials. The lone pair of electrons on nitrogen atoms can also donate a negative charge to the sp^2 structure of carbon, which can enhance the electrical conductivity of carbon materials [88–90].

12.3.3.1 Cycle Life and Safety of a Multiporous Carbon Supercapacitor

In practice, the long-time stability of a supercapacitor assembled with the multiporous carbon was tested by conducting 10,000 charge–discharge cycles of the electrode at a scan rate of 100 mV s^{-1}. After 10,000 cycles, 97% of the original capacitance value was retained (Figure 12.9). While at the first 20 cycles the specific capacitance decreased by approximately 7%, this slight decrease in capacitance was mainly ascribed to the changes in the chemical structure of unstable functional groups on the surface of the carbon and the electrolyte ions initially resisting entering the micropores. As the charging and discharging process continued, the electrolyte penetration into the pores of the carbon material gradually improved, and the specific capacitance value returned to 97% of its original value after 10,000 cycles. Thus, the supercapacitor has higher stability compared with most batteries, and its capacity does not deteriorate over long-term use.

To mimic a real application, the synthesized multiporous carbons were assembled into the pouch-cell supercapacitor with a size of 4 cm × 6 cm (Figure 12.10A), and the capacitance of approximately 25 F; these capacitors are

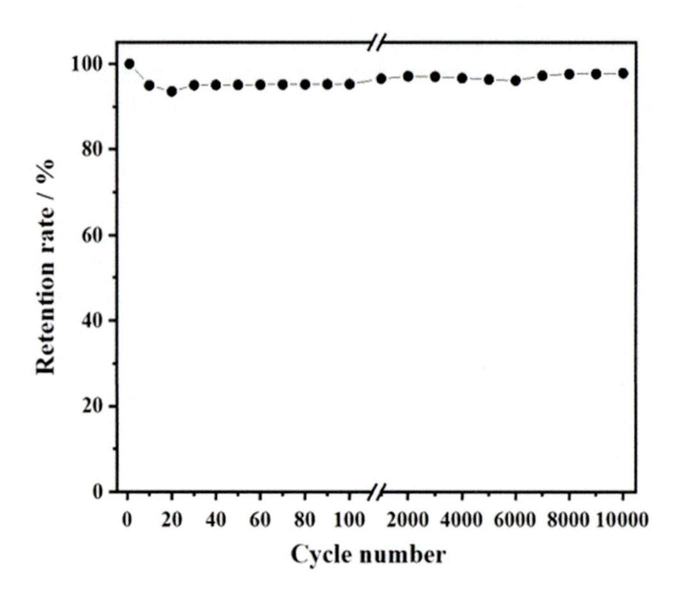

FIGURE 12.9 Stability test of the supercapacitor of the N-doped multiporous carbon.

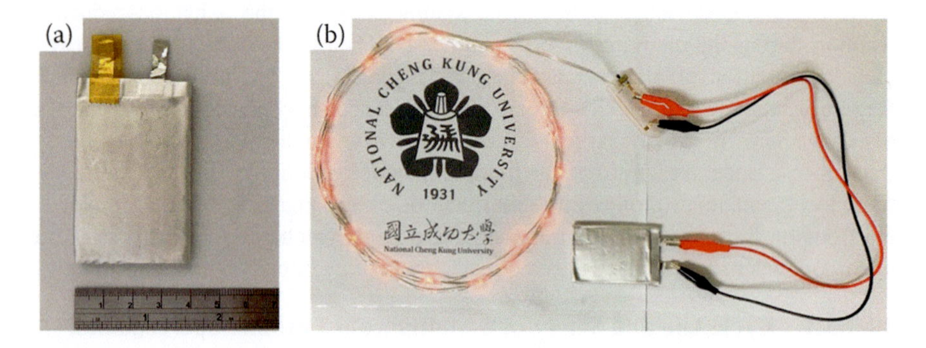

FIGURE 12.10 (A) Photograph of a pouch-cell supercapacitor (25 F) and (B) pouch-cell supercapacitor–driven LED.

sufficient to cause a set of red LEDs to glow for 5 minutes and can be fully charged within 30 seconds, as depicted in Figure 12.10B.

To test the safety of the supercapacitor, the multiporous carbons were assembled into pouch-cell supercapacitors (2.5 V and specific capacitance of 1500 F, Figure 12.11A and B), and a nail penetration test was performed. Before testing, the capacitors were fully charged, and a 3-mm nail was then inserted for 10 s before being removed; the voltage and temperature changes were observed (Figure 12.11C). When the electrode is damaged after nail penetration, the voltage instantaneously drops to approximately 2.1 V. After the removal of the nail, the voltage remains approximately constant but begins to decrease slowly. This result reveals that the energy storage device begins to lose power slowly

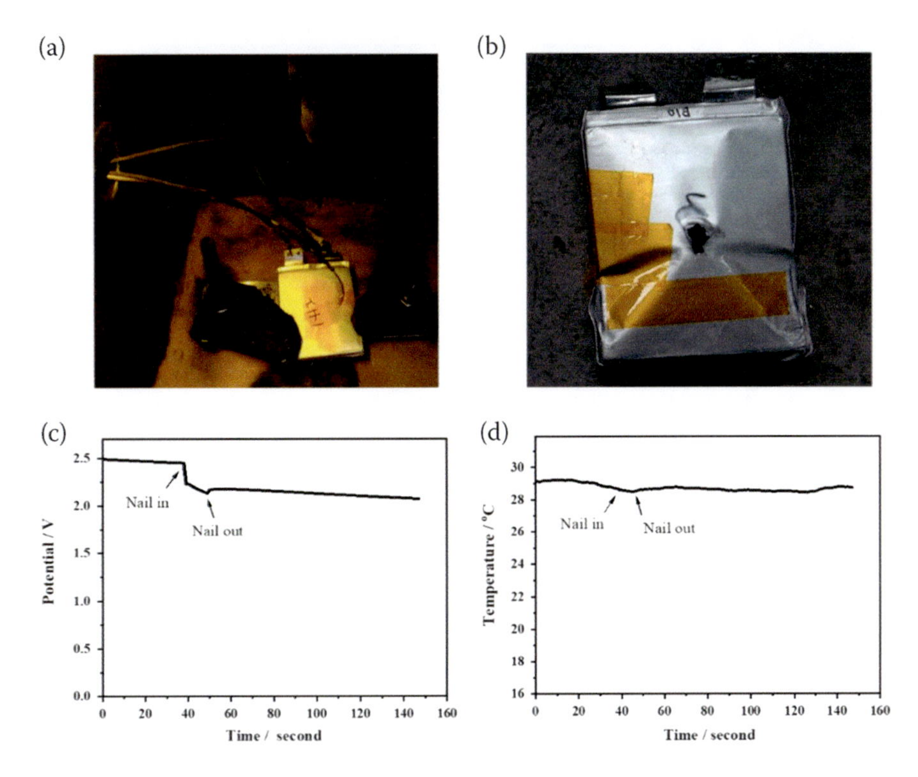

FIGURE 12.11 Photographs of the pouch cell supercapacitor (1500 F) (A) before and (B) after the nail penetration test. (C) Voltage and (D) temperature values during the test.

with time. Moreover, the temperature of the electrodes did not increase substantially during the nail penetration experiment (Figure 12.11D). Thus, external damage to the device will not cause a temperature increase resulting raise to induce combustion or explosion. The energy storage in supercapacitor is highly safe because of only the physical migration of the electrolyte ions in the electrode materials. By contrast, batteries store energy with a reversible chemical reaction; internal short circuits or high-temperature environments can cause violent, dangerous chemical reactions in batteries.

To further evaluate the energy storage performance of the multiporous-carbon supercapacitors, constant current charge and discharge data were measured, and the relationship between power density and energy density was calculated and expressed as a Ragone plot (Figure 12.12). The results revealed that the multiporous carbon supercapacitor has a high power density of 13,452 W kg^{-1} at an energy density of 6.35 W h kg^{-1} at fast charging and discharging rate.

In brief, the nitrogen-doped multiporous carbons were prepared by using eggshells as a hard template. The nitrogen functional groups underwent a rapid redox reaction with the electrolyte that increased the capacitance of the multiporous carbon electrodes; this phenomenon is termed pseudocapacitance. The experimental results revealed that the nitrogen-doped multiporous carbon electrodes had a high

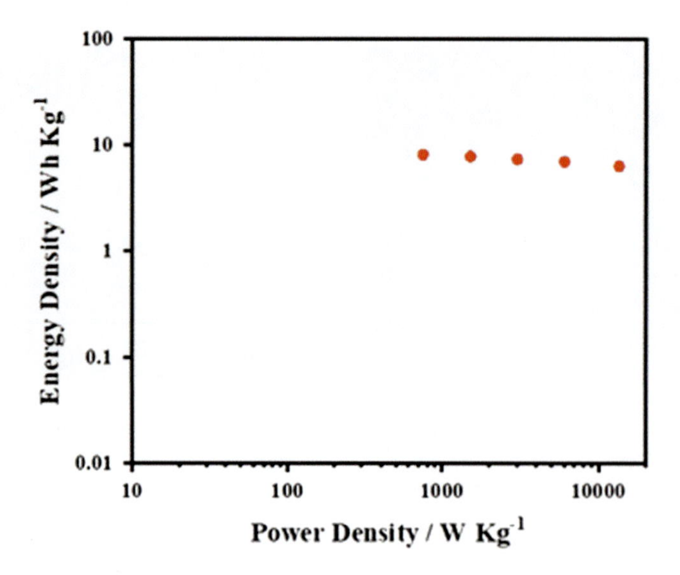

FIGURE 12.12 Ragone plot of the multiporous-carbon supercapacitor.

specific capacitance of 160 F g^{-1}. Overall, the multiporous carbon electrodes had a high power density of 13,452 W kg^{-1} at an energy density of 6.35 W h kg^{-1} and retained 97% of their original capacitance after 10,000 cycles.

12.3.4 Multiporous Carbons for CDI Applications

The CDI performance of a fabricated multiporous carbon electrode was investigated by using a potentiostat to support a CDI at 1.2 V in a 5 mM NaCl electrolyte solution. Figure 12.13A presents the conductivity changes of the CDI during the electrosorption process. The electrosorption capacity of the multiporous carbon electrode in the CDI system was 9.09 mg g^{-1} (Figure 12.13B). The adsorption capacity was measured at different times (Figure 12.13B). The CDI electrode had a high electrosorption rate at 5 minutes. As a result, the removal efficiency and charge efficiency are 11.5% and 43.6%, respectively, in the CDI system (Figure 12.13D).

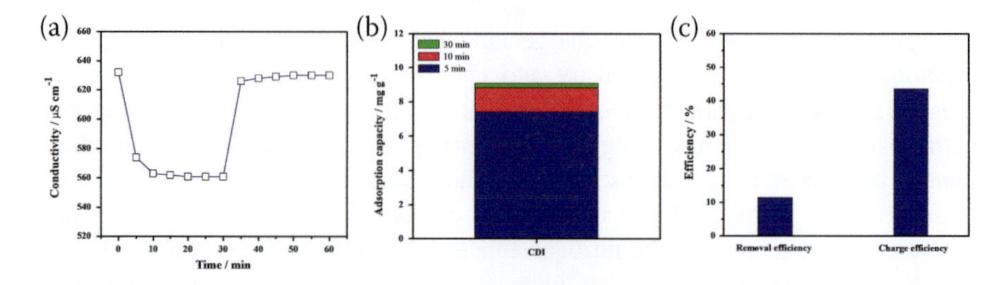

FIGURE 12.13 (A) Electrosorption profile, (B) adsorption capacity at different times, and (C) removal efficiency and charge efficiency of CDI systems.

12.3.4.1 Study of Porous Carbons with Different Pore Sizes for CDI Applications

Materials with high specific surface area have more adsorption sites, which enables the storage of more ions. If the carbons are more graphitized, the electrical conductivity is improved, which increases the energy efficiency of CDI. Theoretically, the higher the specific surface area of the carbon material, the higher the adsorption capacity of the carbon electrode for CDI applications. The multiporous carbons with microporous–mesoporous–macroporous hierarchical structure can cause ions in solution to have good diffusion and transfer efficiency among the pores, increasing the adsorption rate and desorption efficiency. In this section, the performance of multiporous carbons with different pore sizes on the performance of CDI is investigated.

In practice, an increase in the contents of the mesopores or macropores decreases the bulk density of the resulting carbons. Here, we use two porous carbons with different bulk densities: one is 0.08 g cm^{-3}, the other is 0.26 g cm^{-3}. Low bulk density can cause the electrode paste to seriously crack after coating which will cause the electrode material to detach from the electrode. (Figure 12.14A) Therefore, the low bulk-density carbon with large pore volume is unsuitable for use in CDI systems. When the carbon with high bulk density, the multiporous carbons can be smoothly coated on the electrodes without any crack (Figure 12.14B).

With the appropriate bulk density, the effects of the surface area and pore structure of the multiporous carbons synthesized under different conditions on the CDI performance were discussed. Table 12.2 presents the basic physical properties of the multiporous carbons. The materials are denoted as predominantly micropore MPCs-1 with a microporous/mesoporous ratio of 3.1, and predominantly mesopore MPCs-3 and MPCs-4 with a microporous/mesoporous ratio of 0.23-0.37. MPCs-2 has a similar ratio of mesopore to micropore composition with a microporous/mesoporous ratio of 1.25.

The desalination results are presented in Table 12.3. As expected, the larger the specific surface area is, the higher the salt adsorption capacity; however, the salt adsorption is affected by the pore structure at the microscopic scale. The pore size affects the diffusion rate of ions (Na^+ and Cl^- have hydration radii of 4 and 3 Å,

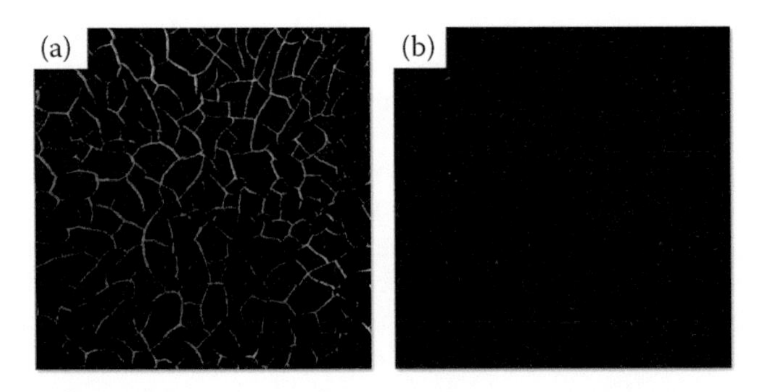

FIGURE 12.14 Photographs of (A) low bulk-density carbon (0.08 g cm^{-3}) and (B) high bulk-density carbon (0.26 g cm^{-3}) coated on electrodes.

TABLE 12.2
Surface Area and Pore Size of Different Multiporous Carbons

No.	$S_{BET}(m^2g^{-1})$	S_{mic}[a](m^2g^{-1})	S_{mic}[b](m^2g^{-1})	S_{mic}/S_{mes}	Pore size (nm)
MPCs-1	1029	779	250	3.1	2.32
MPCs-2	1397	778	619	1.25	2.44
MPCs-3	1538	281	1257	0.23	2.55
MPCs-4	1622	442	1180	0.37	2.59

Notes
[a] Micropore surface area.
[b] Mesopore surface area.

TABLE 12.3
Adsorption and Desorption Efficiency of Multiporous Carbon with Different Pore Sizes

No.	Carbon electrodes weight	SAC[a]	Removal efficiency (%)	Charge efficiency[b] (%)
MPCs-1	0.25 g	5.81 mg/g	9.6 %	66.6 %
MPCs-2	0.20 g	9.09 mg/g	11.5 %	54.5 %
MPCs-3	0.26 g	6.74 mg/g	11.9 %	55.0 %
MPCs-4	0.20 g	7.44 mg/g	9.7 %	45.3 %

Notes
[a] Salt adsorption capacity.
[b] Charge efficiency measured at 10 minutes.

respectively) into the pores, and CDI is a fast physical adsorption process. Therefore, if the micropores are more dominant in the electrode material, and the pore size is sufficiently large for the ions to fill the pores. Thus, micropores can be used more efficiently than can mesopores and macropores. By contrast, if mesopores and macropores are dominant on the electrode surface, ions are more efficiently adsorbed into the pore wall. However, for these mesoporous and macroporous carbons, when ions begin to fill the pores after the inner pore walls are saturated, these ions may be excessively far from the electrodes. The electrostatic force acting on these ions is low; thus, the ions cannot fill the entire pore, resulting in low pore utilization efficiency and a decrease in adsorption capacity. By contrast, an increase in the ratio of mesopores to macropores may reduce the density of the material (increase its porosity), resulting in a decrease in electrical conductivity and overall adsorption efficiency. Selecting a material with an appropriate pore size for capacitive desalination can enhance the adsorption efficiency.

12.4 CONCLUSION

In this study, multiporous carbons with high specific surface area were successfully synthesized. The biomass materials used for synthesis are relatively environmentally friendly, and the method is simple and inexpensive. These characteristics are beneficial for future economical large-scale production. Moreover, a novel method of replacing nano-$CaCO_3$ and nitrogen-containing compounds in the synthesis by using eggshells as both $CaCO_3$ templates and nitrogen dopants was attempted and demonstrated to be effective. The method substantially reduces the cost of the manufacturing process and is consistent with a circular economy due to the reuse of otherwise wasted eggshells. The synthesized multiporous carbons have not only a high specific surface area but also both microporous and mesoporous structures; thus, these materials can be used in supercapacitors to provide appropriate structural permeability and buffer zones for ions and enhance the retention rate of supercapacitors during rapid charging and discharging. Future research could apply multiporous carbons for CDI with the goal of adsorbing heavy metal ions in aqueous solutions to reduce water pollution and improve water recyclability; CDI is thus a promising technology for next-generation water treatments.

REFERENCES

1. Tarascon, JM; Armand, M. Issues and challenges facing rechargeable lithium batteries. *Nature* 2001, 414, 359–367.
2. Kim, MH; Kim, D; Heo, J; Lee, DW. Energy performance investigation of net plus energy town: Energy balance of the Jincheon Eco-Friendly energy town. *Renewable Energy* 2020, 147, 1784–1800.
3. Sharma, A; Tyagi, VV; Chen, CR; Buddhi, D. Review on thermal energy storage with phase change materials and applications. *Renewable & Sustainable Energy Reviews* 2009, 13, 318–345.
4. Agarwal, AK. Biofuels (alcohols and biodiesel) applications as fuels for internal combustion engines. *Progress in Energy and Combustion Science* 2007, 33, 233–271.
5. Olabi, AG; Wilberforce, T; Abdelkareem, MA. Fuel cell application in the automotive industry and future perspective. *Energy* 2021, 214.
6. Abdelkareem, MA; Elsaid, K; Wilberforce, T; Kamil, M; Sayed, ET; Olabi, A. Environmental aspects of fuel cells: A review. *Science of The Total Environment* 2021, 752,141803.
7. Magazzino, C; Mele, M; Schneider, N. A machine learning approach on the relationship among solar and wind energy production, coal consumption, GDP, and CO_2 emissions. *Renewable Energy* 2021, 167, 99–115.
8. Nasab, NM; Kilby, J; Bakhtiaryfard, L. Case study of a hybrid wind and tidal turbines system with a microgrid for power supply to a remote off-grid community in New Zealand. *Energies* 2021, 14, 12.
9. Rabaia, MKH; Abdelkareem, MA; Sayed, ET; Elsaid, K; Chae, KJ; Wilberforce, T; Olabi, AG. Environmental impacts of solar energy systems: A review. *Science of The Total Environment* 2021, 754, 141989.
10. Kalogirou, SA. Solar thermal collectors and applications. *Progress in Energy and Combustion Science* 2004, 30, 231–295.
11. Wang, ZL. Triboelectric nanogenerators as new energy technology for self-powered systems and as active mechanical and chemical sensors. *ACS Nano* 2013, 7, 9533–9557.

12. Wang, GP; Zhang, L; Zhang, JJ. A review of electrode materials for electrochemical supercapacitors. *Chemical Society Reviews* 2012, 41, 797–828.
13. Kuila, T; Bose, S; Mishra, AK; Khanra, P; Kim, NH; Lee, JH. Chemical functionalization of graphene and its applications. *Progress in Materials* 2012, 57, 1061–1105.
14. Lai, XY; Halpert, JE; Wang, D. Recent advances in micro-/nano-structured hollow spheres for energy applications: From simple to complex systems. *Energy & Environmental Science* 2012, 5, 5604–5618.
15. Brownson, DAC; Kampouris, DK; Banks, CE. An overview of graphene in energy production and storage applications. *Journal of Power Sources* 2011, 196, 4873–4885.
16. Sharma, P; Bhatti, TS. A review on electrochemical double-layer capacitors. *Energy Conversion and Management* 2010, 51, 2901–2912.
17. Wang, CC; Su, WL. Ultrathin artificial solid electrolyte interface layer-coated biomass-derived hard carbon as an anode for sodium-ion batteries. *ACS Applied Energy Materials* 2022, 5, 1052–1064.
18. Zhang, LL; Zhao, XS. Carbon-based materials as supercapacitor electrodes. *Chemical Society Reviews* 2009, 38, 2520–2531.
19. Chang-Chien, CY; Hsu, CH; Lee, TY; Liu, CW; Wu, SH; Lin, HP; Tang, CY; Lin, CY. Synthesis of carbon and silica hollow spheres with mesoporous shells using polyethylene oxide/phenol formaldehyde polymer blend. *European Journal of Inorganic Chemistry* 2007, 24, 3798–3804.
20. Hsu, CH; Lin, HP; Tang, CY; Lin CY. Synthesis of mesoporous silicas with different pore sizes using PEO polymers via hydrothermal treatment: A direct template for mesoporous carbon. *Materials Chemistry and Physics*, 2006, 100, 112–116.
21. Li, BW; Xiong, H; Xiao, Y. Progress on synthesis and applications of porous carbon materials. *International Journal of Electrochemical Science* 2020, 15, 1363–1377.
22. Yana, S; Nizar, M; Irhamni; Mulyati, D. Biomass waste as a renewable energy in developing bio-based economies in Indonesia: A review. *Renewable & Sustainable Energy Reviews* 2022, 160, 112268.
23. Wang, XY; Wang, HQ; Dai, Q; Li, QY; Yang, JH; Zhang, AN; Yan, ZX. Preparation of novel porous carbon spheres from corn starch. *Colloids and Surfaces A-Physicochemical and Engineering Aspects* 2009, 346, 213–215.
24. Atkinson, CJ; Fitzgerald, JD; Hipps, NA. Potential mechanisms for achieving agricultural benefits from biochar application to temperate soils: A review. *Plant and Soil* 2010, 337, 1–18.
25. Hsu, CH; Chung, CH; Hsieh, TH; Lin, HP. Green and highly-efficient microwave synthesis route for sulfur/carbon composite for Li-S Battery. *International Journal of Molecular Sciences* 2022, 23, 39.
26. Ullah, I; Rasul, MG. Recent developments in solar thermal desalination technologies: A review. *Energies* 2019, 12, 1.
27. Li, CN; Goswami, Y; Stefanakos, E. Solar assisted sea water desalination: A review. *Renewable & Sustainable Energy Reviews* 2013, 19, 136–163.
28. Ali, MT; Fath, HES; Armstrong, PR. A comprehensive techno-economical review of indirect solar desalination. *Renewable & Sustainable Energy Reviews* 2011, 15, 4187–4199.
29. Yang, H; Fu, MX; Zhan, ZL; Wang, R; Jiang, YF. Study on combined freezing-based desalination processes with microwave treatment. *Desalination* 2020, 475, 114201.
30. Sharon, H; Reddy, KS. A review of solar energy driven desalination technologies. *Renewable & Sustainable Energy Reviews* 2015, 41, 1080–1118.
31. Ghalavand, Y; Hatamipour, MS; Rahimi, A. A review on energy consumption of desalination processes. *Desalination and Water Treatment* 2015, 54, 1526–1541.

32. Amiri, A; Brewer, CE. Biomass as a renewable energy source for water desalination: A review. *Desalination and Water Treatment* 2020, 181, 113–122.
33. Mohammadi, K; McGowan, JG. A survey of hybrid water desalination systems driven by renewable energy based components. *Desalination and Water Treatment* 2019, 150, 9–37.
34. Al-Mutaz, IS; Wazeer, I. Current status and future directions of MED-TVC desalination technology. *Desalination and Water Treatment* 2015, 55, 1–9.
35. Shrivastava, A; Rosenberg, S; Peery, M. Energy efficiency breakdown of reverse osmosis and its implications on future innovation roadmap for desalination. *Desalination* 2015, 368, 181–192.
36. Bundschuh, J; Kaczmarczyk, M; Ghaffour, N; Tomaszewska, B. State-of-the-art of renewable energy sources used in water desalination: Present and future prospects. *Desalination* 2021, 508, 115035.
37. Zaib, Q; Fath, H. Application of carbon nano-materials in desalination processes. *Desalination and Water Treatment* 2013, 51, 627–636.
38. Subielaa, VJ; Penateb, B; Garcia-Rodriguez, L. Design recommendations and cost assessment for off-grid wind-powered - seawater reverse osmosis desalination with medium-size capacity. *Desalination and Water Treatment* 2020, 180, 16–36.
39. Tan, VT; Chien, HT; Vinh, LT. Expanded graphite-based membrane for water desalination. *Desalination and Water Treatment* 2021, 234, 324–332.
40. Jiang, YX; Alhassan, SI; Wei, D; Wang, HY. A review of battery materials as CDI electrodes for desalination. *Water* 2020, 12, 3030.
41. Oren, Y. Capacitive delonization (CDI) for desalination and water treatment - past, present and future (a review). *Desalination* 2008, 228, 10–29.
42. Anderson, MA; Cudero, AL; Palma, J. Capacitive deionization as an electrochemical means of saving energy and delivering clean water. Comparison to present desalination practices: Will it compete? *Electrochimica ACTA* 2010, 55, 3845–3856.
43. Kotz, R; Carlen, M. Principles and applications of electrochemical capacitors. *Electrochimica ACTA* 2000, 45, 2483–2498.
44. Frackowiak, E; Beguin, F. Carbon materials for the electrochemical storage of energy in capacitors. *Carbon* 2001, 39, 937–950.
45. Pandolfo, AG; Hollenkamp, AF. Carbon properties and their role in supercapacitors. *Journal of Power Sources* 2006, 157, 11–27.
46. Zhai, YP; Dou, Y; Zhao, DY; Fulvio, PF; Mayes, RT; Dai, S. Carbon materials for chemical capacitive energy storage. *Advanced Materials* 2011, 23, 4828–4850.
47. Xie, P; Yuan, W; Liu, XB; Peng, YM; Yin, YH; Li, YS; Wu, ZP (Wu, Ziping) [1] Advanced carbon nanomaterials for state-of-the-art flexible supercapacitors. *Energy Storage Materials* 2021, 36, 56–76.
48. Liang, XD; Liu, RN; Wu, XL. Biomass waste derived functionalized hierarchical porous carbon with high gravimetric and volumetric capacitances for supercapacitors. *Microporous and Mesoporous Materials* 2021, 310.
49. Yu, ZN; Tetard, L; Zhai, L; Thomas, J. Supercapacitor electrode materials: Nanostructures from 0 to 3 dimensions. *Energy & Environmental Science* 2015, 8, 702–730.
50. Muzaffar, A; Ahamed, MB; Deshmukh, K; Thirumalai, J. *Renewable & Sustainable Energy Reviews* 2019, 101, 123–145.
51. Bo, XK; Xiang, K; Zhang, Y; Shen, Y; Chen, SY; Wang, YZ; Xie, MJ; Guo, XF. Microwave-assisted conversion of biomass wastes to pseudocapacitive mesoporous carbon for high-performance supercapacitor. *Journal of Energy Chemistry* 2019, 39, 1–7.
52. Jiang, WC; Pan, JQ; Liu, XG. A novel rod-like porous carbon with ordered hierarchical pore structure prepared from Al-based metal-organic framework without

template as greatly enhanced performance for supercapacitor. *Journal of Power Sources* 2019, 409, 13–23.
53. Chinnadurai, D; Karuppiah, P; Chen, SM; Kim, HJ; Prabakar, K. Metal-free multi-porous carbon for electrochemical energy storage and electrocatalysis applications. *New Journal of Chemistry* 2019, 43, 11653–11659.
54. Li, XC; Zhang, L; He, GH. Fe_3O_4 doped double-shelled hollow carbon spheres with hierarchical pore network for durable high-performance supercapacitor. *Carbon* 20166, 99, 514–522.
55. Cao, JH; Zhu, CY; Aoki, Y; Habazaki, H. Starch-derived hierarchical porous carbon with controlled porosity for high performance supercapacitors. *ACS Sustainable Chemistry & Engineering* 2018, 6, 7292–7303.
56. Wang, HY; Deng, J; Chen, YQ; Xu, F; Wei, ZZ; Wang, Y. Hydrothermal synthesis of manganese oxide encapsulated multiporous carbon nanofibers for supercapacitors. *Nano Research* 2016, 9, 2672–2680.
57. Li, YD; Xu, XZ; He, YZ; Jiang, YQ; Lin, KF. Nitrogen doped macroporous carbon as electrode materials for high capacity of supercapacitor. *Polymers* 2017, 9, 2.
58. Chinnadurai, D; Karuppiah, P; Chen, SM; Kim, HJ; Prabakar, K. Metal-free multi-porous carbon for electrochemical energy storage and electrocatalysis applications. *New Journal of Chemistry* 2019, 43, 11653–44659.
59. Wu, CG; Zhang, H; Hu, MJ; Shan, GC; Gao, JF; Liu, JZ; Zhou, XL; Yang, J. In situ nitrogen-doped covalent triazine-based multiporous cross-linking framework for high-performance energy storage. *Advanced Electronic Materials* 2020, 6, 2000253.
60. Huang, JS; Sumpter, BG; Meunier, V. A universal model for nanoporous carbon supercapacitors applicable to diverse pore regimes, carbon materials, and electrolytes. *Chemistry-A European Journal* 2008, 14, 6614–6626.
61. Han, Y; Dong, XT; Zhang, C; Liu, SX. Hierarchical porous carbon hollow-spheres as a high performance electrical double-layer capacitor material. *Journal of Power Sources* 2012, 211, 92–96.
62. Liu, GW; Chen, TY; Chung, CH; Lin, HP; Hsu, CH. Hierarchical micro/mesoporous carbons synthesized with a ZnO template and petroleum pitch via a solvent-free process for a high-performance supercapacitor. *ACS Omega* 2017, 2, 2106–2113.
63. Hsu, CH; Pan, ZB; Chen, CR; Wei, MX; Chen, CA; Lin, HP; Hsu, CH. Synthesis of multiporous carbons from the water caltrop shell for high-performance supercapacitors. *ACS Omega* 2020, 5, 10626–10632.
64. Chen, SK; Chang, KH; Hsu, CH; Lim, ZY; Du, FY; Chang, KW; Chang, MC; Lin, HP; Hu, CC; Tang, CY; Lin, CY. Synthesis of mesoporous carbon platelets of high surface area and large porosity from polymer blends-calcium phosphate nanocomposites for high-power supercapacitor. *Journal of The Chinese Chemical Society* 2021, 68, 462–468.
65. Wang, Q; Astruc, D. State of the art and prospects in metal-organic framework (MOF)-based and MOF-derived nanocatalysis. *Chemical Reviews* 2020, 120, 1438–1511.
66. Wang, CY; Xia, KL; Wang, HM; Liang, XP; Yin, Z; Zhang, YY. Advanced carbon for flexible and wearable electronics. *Advanced Materials* 2019, 31, 1801072.
67. Kuo, PL; Hsu, CH. Stabilization of embedded Pt nanoparticles in the novel nanostructure carbon materials. *ACS Applied Materials & Interfaces* 2011, 3, 115–118.
68. Tsai, WT; Hsu, CH; Lin, YQ. Highly porous and nutrients-rich biochar derived from dairy cattle manure and its potential for removal of cationic compound from water. *Agriculture-Basel* 2019, 9, 6.
69. Keiluweit, M; Nico, PS; Johnson, MG; Kleber, M. Dynamic molecular structure of plant biomass-derived black carbon (biochar). *Environmental Science & Technology* 2010, 44, 1247–1253.

70. Hendriks, ATWM; Zeeman, G. Pretreatments to enhance the digestibility of lignocellulosic biomass. *Bioresource Technology* 2009, 100, 10–18.
71. Wang, JS; Zhang, X; Li, Z; Ma, YQ; Ma, L. Recent progress of biomass-derived carbon materials for supercapacitors. *Journal of Power Sources* 2020, 451, 227794.
72. Gopalakrishnan, A; Badhulika, S. Effect of self-doped heteroatoms on the performance of biomass-derived carbon for supercapacitor applications. *Journal of Power Sources* 2020, 480.
73. Nadakatti, S; Tendulkar, M; Kadam, M. Use of mesoporous conductive carbon black to enhance performance of activated carbon electrodes in capacitive deionization technology. *Desalination* 2011, 268, 182–188.
74. Noonan, O; Liu, Y; Huang, XD; Yu, CZ. Layered graphene/mesoporous carbon heterostructures with improved mesopore accessibility for high performance capacitive deionization. *Journal of Materials Chemistry A* 2018, 6, 14272–14280.
75. Baroud, TN; Giannelis, EP. High salt capacity and high removal rate capacitive deionization enabled by hierarchical porous carbons. *Carbon* 2018, 139, 614–625.
76. Quan, GX; Wang, H; Zhu, F; Yan, JL. Porous biomass carbon coated with SiO_2 as high performance electrodes for capacitive deionization. *Bioresources* 2018, 13, 437–449.
77. Wang, JC; Kaskel, S. KOH activation of carbon-based materials for energy storage. *Journal of Materials Chemistry* 2012, 22, 23710–23725.
78. Heidarinejad, Z; Dehghani, MH; Heidari, M; Javedan, G; Ali, I; Sillanpaa, M. Methods for preparation and activation of activated carbon: a review. *Environmental Chemistry Letters* 2020, 18, 393–415.
79. Kalyani, P; Anitha, A. Biomass carbon & its prospects in electrochemical energy systems. *International Journal of Hydrogen Energy* 2013, 38, 4034–4045.
80. Linares-Solano, A; Lozano-Castello, D; Lillo-Rodenas, MA; Cazorla-Amoros, D. Carbon activation by alkaline hydroxides preparation and reactions, porosity and performance. *Chemistry and Physics of Carbon* 2008, 30, 1–62.
81. Yoon, SH; Lim, S; Song, Y; Ota, Y; Qiao, WM; Tanaka, A; Mochida, I. KOH activation of carbon nanofibers. *Carbon* 2004, 42, 1723–1729.
82. Wei, ZK; Xu, CL; Li, BX. Application of waste eggshell as low-cost solid catalyst for biodiesel production. *Bioresource Technology* 2009, 100, 2883–2885.
83. Wu, YL; Li, XF; Wei, YS; Fu, ZM; Wei, WB; Wu, XT; Zhu, QL; Xu, Q. Ordered Macroporous superstructure of nitrogen-doped nanoporous carbon implanted with ultrafine Ru nanoclusters for efficient pH-Universal hydrogen evolution reaction. *Advanced Materials* 2021, 33, 2006965.
84. Rehman, A; Heo, YJ; Nazir, G; Park, SJ. Solvent-free, one-pot synthesis of nitrogen-tailored alkali-activated microporous carbons with an efficient CO_2 adsorption. *Carbon* 2021, 172, 71–82.
85. Yang, GW; Han, HY; Li, TT; Du, CY. Synthesis of nitrogen-doped porous graphitic carbons using nano-$CaCO_3$ as template graphitization catalyst, and activating agent. *Carbon* 2012, 50, 3753–3765.
86. Zhao, CR; Wang, WK; Yu, ZB; Zhang, H; Wang, AB; Yang, YS. Nano-$CaCO_3$ as template for preparation of disordered large mesoporous carbon with hierarchical porosities. *Journal of Materials Chemistry* 2010, 20, 976–980.
87. Zhang, Y; Sun, Q; Xia, KS; Han, B; Zhou, CG; Gao, Q; Wang, HQ; Pu, S; Wu, JP. Facile synthesis of hierarchically porous N/P codoped carbon with simultaneously high-level heteroatom-doping and moderate porosity for high-performance supercapacitor electrodes. *ACS Sustainable Chemistry & Engineering* 2019, 7, 5717–5726.
88. Lian, J; Xiong, LS; Cheng, R; Pang, DQ; Tian, XQ; Lei, J; He, R; Yu, XF; Duan, T; Zhu, WK. Ultra-high nitrogen content biomass carbon supercapacitors and nitrogen forms analysis. *Journal of Alloys and Compounds* 2019, 809, 151664.

89. Chen, JJ; Mao, ZY; Zhang, LX; Wang, DJ; Xu, R; Bie, LJ; Fahlman, BD. Nitrogen-deficient graphitic carbon nitride with enhanced performance for lithium ion battery anodes. *ACS Nano* 2017, 11, 12650–12657.
90. Wu, FM; Gao, JP; Zhai, XG; Xie, MH; Sun, Y; Kang, HY; Tian, Q; Qiu, HX. Hierarchical porous carbon microrods derived from albizia flowers for high performance supercapacitors. *Carbon* 2019, 147, 242–251.
91. Yang, LJ; Shui, JL; Du, L; Shao, YY; Liu, J; Dai, LM; Hu, Z. Carbon-based metal-free ORR electrocatalysts for fuel cells: Past, present, and future. *Advanced Materials* 2019, 31, 1804799.
92. Singh, SK; Takeyasu, K; Nakamura, J. Active sites and mechanism of oxygen reduction reaction electrocatalysis on nitrogen-doped carbon materials. *Advanced Materials* 2019, 31, 13.
93. Zhu, XF; Amal, R; Lu, XY. N, P Co-coordinated manganese atoms in mesoporous carbon for electrochemical oxygen reduction. *Small* 2019, 15, 29.
94. Wang, G; Oswald, S; Loffler, M; Mullen, K; Feng, XL. Beyond activated carbon: Graphite-cathode-derived li-ion pseudocapacitors with high energy and high power densities. *Advanced Materials* 2019, 31, 1807712.
95. Xie, YM; Yin, J; Zheng, JJ; Wang, LJ; Wu, JH; Dresselhaus, M; Zhang, XC. Synergistic cobalt sulfide/eggshell membrane carbon electrode. *ACS Applied Materials & Interfaces* 2019, 11, 32244–32250.

13 Low-Dimensional Heterostructure-Based Solar Cells

Shih-Hsiu Chen, Tsung-Yen Wu, and Chia-Yun Chen
Department of Materials Science and Engineering, National Cheng Kung University, Tainan City, Taiwan

CONTENTS

13.1 INTRODUCTION

More and more attention has been paid to developing renewable energies such as wind, tidal, geothermal, and solar power owing to the gradual depletion of natural reserve gasoline and natural gas. Among those reversible green energies, solar power has been recognized as one of promising alternatives that allow for providing unlimited energy in a rather clean way (1,2). However, there remain technical challenges to be solved in order to practical employment of solar energy for electricity generation. To seek high conversion efficiency and low-cost solar cells, several critical issues should be addressed. It has been theoretically and experimentally identified that efficient charge separation, carrier transportation, and collection processes are vital for the improvement of power conversion efficiency (3–5). Based on these requirements, employment of nanomaterials is considered to be highly potential due to superior characteristics including tunability of band

DOI: 10.1201/9781003367215-13

gap energy, creation of high junction area, and strong light-trapping effect overwhelming the conventional bulk materials. Besides, in the material aspect, silicon (Si) is the second most abundant element in the earth, and the history of commercialized solar cells made by crystalline Si has been widely employed for decades. Taking together, nanostructures fabricated from silicon, such as nanowires, nanorods and nanotubes have been tested as antireflection layers in order to absorb more sunlight power. These one-dimensional materials are able to facilitate cost reduction and improve efficiency at the same time (1,6,7).

A general perspective to enhance the performances of one-dimensional configurational solar cells can be achieved through maintaining the balance between transparency and electrical conductivity which are two important factors that usually play a decisive role in determining device performance. Thinner transparent conductive layer represents less loss of light utilization while inevitably causing increased sheet resistance and lower conversion efficiency of the solar cells. In addition, rapid charge transport highways and great transparency for penetration of incident light provided by one-dimensional metallic nanowire networks are comparable to conventional conductive coating. For instance, copper nanowire (CuNW)-based transparent electrodes as one-dimensional nanowire networks were fabricated in polymer-based solar cells, with Cu NW-TiO_2-polyacrylate composite electrode reached 3.11% of conversion efficiency showing great flexibility after 500 times bending with a sheet resistance of 35 Ω/sq, compared the PET/ITO with 15 Ω/sq (8,9).

Except for one-dimensional structures, there are still other kinds of structures that have attracted attention worldwide. With the scaling trends in photovoltaics, there is no doubt that the atomically thin bodies and high flexibility 2D materials are high potential for the employment of the next generation of solar cells and may revolutionize the conventional solar community (10–13). 2D materials have emerged as potential candidates for replacing traditional Si-based devices, on account of their layered-depended tunable bandgap, flexibility, and lightweight, whereas currently, the synthesis of these monolayer 2D materials is still confronted with the high cost and complexity of the processed method, some researchers had realized innovative approaches, solution-processed method, to overcome the hinders while producing with the smaller scale of area and enlarge the scale with another process like spin coating, suction filtration, or spray coating (14–16). Herein, in this chapter, we aim to provide comprehensive discussions on the synthesis and application of low-D materials for photovoltaic applications. These promising materials along with novel strategies for cell design indeed impact the advanced development of solar cells, while the issues regarding the operation stability and reliability of device operation are yet to be improved.

13.2 BACKGROUND AND SYNTHETIC METHODS

13.2.1 History

Since the first ever monolayered graphene was mechanically exfoliated by Andre Geim and Konstantin Novoselov (17), this kind of novel 2d material turns out to be realistic and comes with plenty of stunning features, which quickly drew people's eye. Recently, low D semiconductors, including 0D, 1D, and 2D features have been used to demonstrate proof-of-concept devices reporting the basic feasibility of photovoltage generation in such material systems (18–20). Indeed, one can find that these novel materials are promising for photovoltaic applications and have the potential to not only match but also surpass the performance and complement the conventional photovoltaic technologies based on Si and GaAs.

13.2.2 Synthesis of 1D Materials with Top-Down Approach

Chemical etching process is one of the important techniques to fabricate controlled 1D structures, and it is eligible for various kinds of materials. Before the chemical etching process is applied, the involvement of surface structures or patterns is required in order to proceed with selective etching. In the patterning process, optical or electro-beam lithography techniques have been considered the reliable and robust ways that can accomplish regular and ordered patterns (21,22), whereas drawbacks including low productivity, high cost, and lengthy process stem against their practical manufacturing (23,24). Besides, the colloid solution of polystyrene spheres for the realization of template-assisted lithography was proposed to ease the fabrication burden, The templates with nanoscale can be realized by self-assembling of polystyrene spheres into a hexagonal structure between the interface of gas and liquid, enabling to form uniform nanopatterns distributed in a large area with no need for expensive instruments nor complicated processes. Once the nanosphere-based patterns are formed, either electron beam evaporation or sputtering can be used to deposit metal layers, such as gold and silver, as the catalyst for conducting chemical etching selectively occurred at these defined catalyst sites. The formed 1D nanomaterials with adjustable size, direction, and spacing have been found to act as important for the employment of high-performance solar cells (25).

13.2.3 Synthesis of 1D Materials with Bottom-Up Approach

Aside from the top-down method, another concept to fabricate 1D structures correlates with "bottom-up" approach. It is essentially based on the growth kinetics via molecular recognition and self-assembly strategies (26,27), where the synthetic structures are allowed with a wide range of sizes from nanometers to micrometers. Many approaches are involved with bottom-up growth including vapor-phase growth (28,29), electrospinning (30–32), liquid phase growth (33), and so on. Compared with top-down method, the involvement of bottom-up growth particularly enables to fabricated functional multi-component devices by the controlled assembly of atoms and molecules, without wasting or additional requirement for introducing or eliminating parts of the final components, which benefits their practical employment of solar cells with fabrication compatibility and simplicity. There still exist several drawbacks remaining to be solved, such as surface preparation, control of site impurities, uniformity, etc. Overall, by adjusting different growing seeds and sources, various materials, such as sole elements, compounds, alloys, and heterostructures can be directly synthesized via bottom-up method with the need for a sophisticated transfer technique.

13.2.4 Synthesis of 2D Materials with Top-Down Approach

Basically, synthesis of 2D semiconductor materials can also be categorized by top-down and button-up approaches. After completing the synthesis of monolayer 2D materials, a transfer method is required for integration with substrates for the construction of solar cells. In this approach, few-layer thin 2D materials are exfoliated from crystal bulk, either by the traditional Scotch Tape method or solution-processable method (34,35), where the externally applied force is required in the form of mechanical tape, sonication in liquids or chemical reactant involved for transfer requirement while maintaining the structural regularity of monolayer features. In Scotch Tape method, a small piece of tape is pasted and pressed on the demanded bulk materials, by gently peeling off the tape where the cleaved flakes or debris would be glued, the demanded flake can then transfer to the substrate. However, the size, thickness, and morphology of flakes cannot be precisely controlled, which usually comes with random distribution (36).

Among various 2D materials, Graphene has been considered one of the extensively employed materials for photovoltaic applications. The synthesized graphene in monolayer form can be either exfoliated by liquid-phase sonication, chemical intercalation, or electrical exfoliation (37–41). There are three basic steps in liquid-phase sonication. First, disperse the bulk carbon-based materials into specific liquid media; second, sonicate the dispersions dipped in the bath; finally, purify the resultant nanoflakes by density gradient centrifugation. It should be noticed that the longer the sonication process, surface or structural defects might be created that hinder their electronic and photovoltaic properties. In chemical intercalation, the layer spacing of bulk graphite materials can be increased by introducing acids such as sulfuric or nitric acids into the gap between each graphite layer (42,43). The as-fabricated graphene layer conjugated with functional groups like epoxied, hydroxyl and carboxylic groups turn the structures into graphene oxides (GO) (43). GO is a water-soluble material with a high density of defects resulting from strong oxidation that breaks down the conjugated bonds of carbon atoms. Thus, further modification stays essential to eliminate oxygen-contained groups. In electrochemical exfoliation, the employment of external voltage on graphite materials enables to produce gram-scale of graphene structures in a few minutes. Subsequently, radical scavengers are involved to suppress the defects of graphene layer formed by exfoliation treatment.

13.3 BACKGROUND OF LOW-D MATERIALS IN PVS

Low-D materials, including 0D, 1D, and 2D materials, with sound flexibility and integration compatibility have been considered the potential design for future-generation photovoltaic technology. These materials provide the controllable energy bandgap, light trapping in nanometer-thickness absorber layers, and efficient carrier separation and transport, which turn out to be highly competitive with and

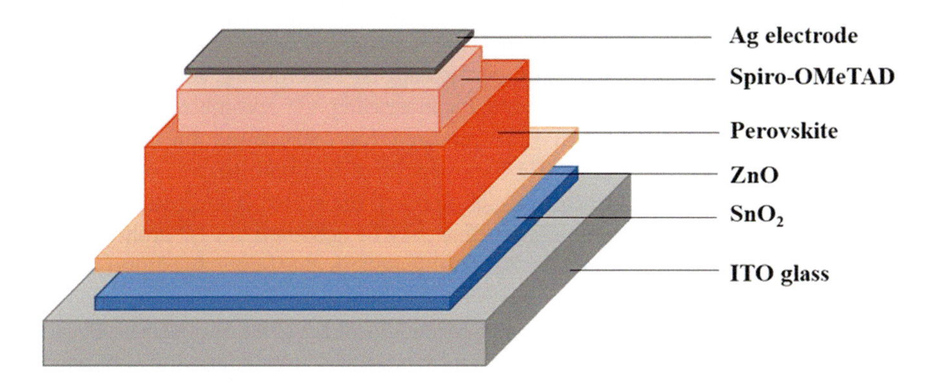

FIGURE 13.1 Structure of 0D/3D heterostructure all-inorganic halide perovskite solar cells device.

complementary to commercial solar cells and correlated optoelectronic devices. In the practical aspect, the long-term stability of device operation remains to be a critical issue that seems to be determinate for their applications. Recently, more reports about perovskite solar cells aim to raise their structural stability where perovskite materials are not stable in ambient condition, which often suffers from the phase transition owing to lower thermodynamics stability. Fujin Bai et al. found that by tuning the stoichiometry of the precursor between PbI and CsI, the zero-dimensional particles Cs_4PbI_6 consequently appear and attach on the grain boundary of the three-dimensional matrix inorganic perovskite material of $CsPbI_3$ (44). As shown in Figure 13.1, adopting the conventional n-i-p solar cells structure, also with the electron transport layer of SnO_2 which is modified by joining the nanoparticle of ZnO aiming to strengthen the charge transfer efficiency in the interface of SnO_2/perovskite, this final device with the 0D Cs_4PbI_6 hybrid with 3D $CsPbI_3$ form the matrix of $Cs_{1.2}PbI_{3.2}$ turns out the highest PCE of 16.39%, V_{OC} of 1.09 V, J_{SC} of 18.84 mA cm^{-2} and FF of 79.95. Under 500 hours of stability test, the normalized PCE of this device still holds nearly 100% of its beginning energy conversion ability, the modified perovskite matrix shows great stability and durability among all the other perovskite device.

13.4 APPLICATIONS OF 1D MATERIALS IN PVS

1D materials, such as nanowires or nanotubes, have been extensively investigated due to their compelling light-tapping effects. Compared with bulk type or thin-film counterparts, periodic 1D nanostructure arrays enable to efficiently reduce the light reflection covering the broadband visible regions. These features can be attributed to not only the involvement of multiple scattering within 1D architectures that trap the incoming light (45,46), but further exhibit the antireflection layer contributing to the reduction of light reflection from structure surfaces (47,48). In addition to the photonic benefits of 1D features, another important aspect correlates with carrier transport and collection. It has been reported that the recombination time of photogenerated electrons and holes and the collection time of photocarriers from the materials absorber materials to create the photocurrents, play important roles in photovoltaic performances. Specifically, improvement of conversion efficiency can be achieved by increasing the recombination time as well as reducing the collection time required for considering the kinetics of carrier diffusion. In this aspect, 1D nanostructures facilitate to create a reliable pathway for photoexcited electrons/holes to reach the electrolyte/electrode faster. On the other hand, long recombination time may be achieved via 1D architectures because radial heterojunction can be established that effectively suppress the carrier recombination (49).

The PV device is usually a semiconductor diode, which is made of sandwich-like stack of p-type and n-type materials joined by a p-n junction. Among the abundant variety of semiconductor materials, silicon-based cells are still dominating the PV market because of their theoretical limit efficiency of 29.4%. However, the drawbacks such as exorbitant cell design, and complicated and long-period manufacturing are still hindering the extensive replacement of the current petroleum-related energy. In order to break through the dilemma, the development of organic/inorganic hybrid solar cells has been reported to be a promising method. Considering the improvement of photovoltaic performances, the employment of 1D semiconductors acts as the decisive role that both greatly enhances the utilization of broadband solar lights and suppresses the charge recombination due to the establishment of heterojunction. While the strategies to boost the conversion efficiency of hybrid solar cells were evidenced with the involvement of 1D features, there are still some remaining issues related to the eventual conversion efficiency, such as carrier lifetime, charge transfer resistance, carrier recombination, and charge selectivity at metal contact. In this aspect, several potential solutions have been reported in this review to improve the efficiency, including perovskite-passivated layer on silicon nanowires, rearrangement of conductive organic films, and employment of transparent carrier-selective layer. These new strategies indeed provide the promising and reliable solutions toward realizing the high-performance and low-cost solar cells, and may further offer the high potential for other functional applications based on hybrid material designs.

In the passivation design, the strategy was proposed for minimizing the possible charge recombination existed at nanowire surfaces. The excess PbI_2 as well as antisolvent treatment are designed to function as the perovskite-passivated layer for compensating dangling states at the surfaces of silicon nanowires. Compared with the reference cell, the optimized procedure is associated with 5% excess PbI_2 based on CB (chlorobenzene) treatment, resulting in an increased carrier lifetime of up to 8.84 ns. In addition, EIS analysis was employed to understand the charge transfer resistance (R_{ct}) and analytically monitored the transport of photogenerated carriers at interfaces. The estimated R_{ct} was found to be significantly decreased for the perovskite-passivated devices prepared through antisolvent treatments in comparison with the perovskite-free solar cells, particularly reaching 13.5 Ω while using CB as the antisolvent. Such features facilitate the charge extraction from photoactive regions and enable quenching the possible radiative recombination of excited carriers. Finally, the noticeable improvement of cell performance, open circuit voltage (V_{oc}) of 0.512 V and fill factor (FF) of 55.5% is found in the passivated hybrid solar cells with optimal 5% excess PbI_2, in comparison with the reference cells (J_{sc} = 37.1 mA/cm^2, V_{oc} = 0.507 V and FF = 53.9).

In addition, the facile employment of conductive polymers has been considered for behaving as hole-transport layer that assists the carrier transport and collection. Among them, the most compelling material is poly(3,4-ethylene dioxythiophene): poly(styrene sulfonate) (PEDOT:PSS) due to its high transmittance, high temperature stability, and tunable conductivity compared to other organic materials. However, in order to further meet the practical requirement, its intrinsic low electrical conductivity due to the existence of insulating PSS molecules should be properly improved. In photovoltaic fields, the application of PEDOT:PSS in immerging organic/inorganic hybrid solar cells has been extensively investigated owing to several advantages such as facile fabrication, low-temperature process, and high device performance. In the synthesis of PEDOT:PSS film with improved electric conductivity, Zonyl-FS300 surfactants along with EG as col-solvent were added, and the prepared solution was magnetically stirred under several different temperatures. The pristine film essentially displays the weak phase separation between coiled-like PEDOT chains and PSS molecules, where the grains composing the PEDOT chains embedded in the PSS matrix are randomly oriented. Such disconnected features of PEDOT chains result in the low electric conductivity. By introducing EG as a co-solvent for polymer preparation, the involvement of the screening effect reduces the Coulombic interaction of positively charged PEDOT chains and negatively charged PSS molecules. These results are found to critical for promoting the charge transfer characteristics, and the improved photovoltaic performances have been experimentally evidenced. In addition, the conversion efficiency of hybrid solar cells via the incorporation of highly conductive PEDOT:PSS films prepared by EG addition and a heated-stir process were further explored (50). By employing heated-stir treatment, the cell conversion efficiency is drastically increased. The optimal process condition was found to be 90^{0}C, where the photovoltaic parameters including fill factor and open circuit voltage are improved with a conversion efficiency of up to 12.2%.

Besides, the involvement of carrier-selective layers is found to enhance the conversion efficiency of solar cells. It is found that the intrinsic bandgap of PEDOT:PSS is narrow (1.5 eV) as well as the involved HOMO state (5.0~5.2 eV) turns to inefficient to block the photogenerated electrons, which might both cause the carrier recombination, particularly at interfaces of hole-transport layer and contact electrode. Several transition metal oxides (TMOs) such as V_2O_5, WO_3, NiO, and MoO_3 have been demonstrated and employed as a hole-transport layer in PVs (51–54). For instance, highly transparent vanadium oxide (VO_x), synthesized by all solution-processed methods, was found to boost the performances of hybrid silicon-based solar cells. Not only did the VO_x act as the transparent and ohmic-like contact, but it also offered the hole-selectivity for blocking electrons that might reduce the carrier recombination at the interfaces. The vanadium oxide layer is applied between the ITO electrode and the heterojunction, which is formed by PEDOT:PSS and silicon nanowire. The low LUMO level (3.5 eV) of p-type polymeric layer might allow the transport of photoexcited electrons reaching the

TABLE 13.1
Photovoltaic Parameters of Nanowire-Based Hybrid Solar Cells

Design Strategy	Cell Type	J_{SC} (mA/cm^2)	V_{OC} (V)	FF (%)	PCE (%)
Conductivity Enhancement	Reference Cell	30.8	0.565	53.6	9.3
	Hot Stir PEDOT: PSS	34.9	0.615	56.7	12.2
Passivation	Reference Cell	37.1	0.507	53.9	10.2
	Passivation Layer	44.8	0.512	55.5	14.1
Carrier Selective Layer	Reference Cell	35.6	0.499	50.91	9.1
	VO_x Layer	45.4	0.541	58.3	14.4

ITO/polymer interfaces according to the band diagram, resulting in significant charge recombination at ITO surfaces. With the addition of VO_x layers exhibiting a relatively high LUMO level (2.4 eV) and band gap energy (>3 eV), the electrons that may move toward ITO electrodes will be greatly decreased due to the created large potential barrier. Furthermore, the offered comparable HOMO energy of VO_x layer (4.7 eV) with PEDOT:PSS counterpart (5.1 eV) energetically stimulated the hole extraction through interfaces due to the formation of ohmic-like contact. Such an induced V_2O_5 layer acted as a hole-selective contact that could efficiently benefit the current gain by both favoring the collection of holes and on the other hand minimizing the surface recombination at ITO electrodes.

Combined with these modifications, the result of cell performances is greatly improved. The ohmic-like contact, which is introduced by the VO_x layer, facilitates the hole extraction generated by the light excitations. Furthermore, the VO_x layer can be further regarded as the electron-blocking component that effectively reduces the charge recombination encountered at ITO/polymer interfaces. Compared to the reference cell with an efficiency of 9.1%, sound improvement is found in VO_x-introduced hybrid solar cells, where the highest conversion efficiency with a value of 14.4% is realized. This result is found to reach above 36% enhancement of conversion efficiency compared with the reference cells, where the results are summarized in Table 13.1.

13.5 APPLICATIONS OF 2D MATERIALS IN PVS

2D materials, owing sound structural quality with almost defect-free crystallization, enable to provide effective charge transport along planar direction and controllable light absorption characteristics due to adjustable band-gap nature, which have been regarded as promising materials for the new-generation solar cells with advantages of thin architecture and high performance.

Among various 2D materials, mono- or few-layers of MoS_2 have been evidenced to act a significant role in PV applications due to their integration compatibility, high absorption coefficient, and tunable bandgap energy. It has been examined that the employment of MoS_2 monolayer was highly dependent on the transfer process. Wet transfer could be conducted to transfer the monolayer of 2D material through the spin coating of PMMA on top of it to strengthen mechanical properties, and then

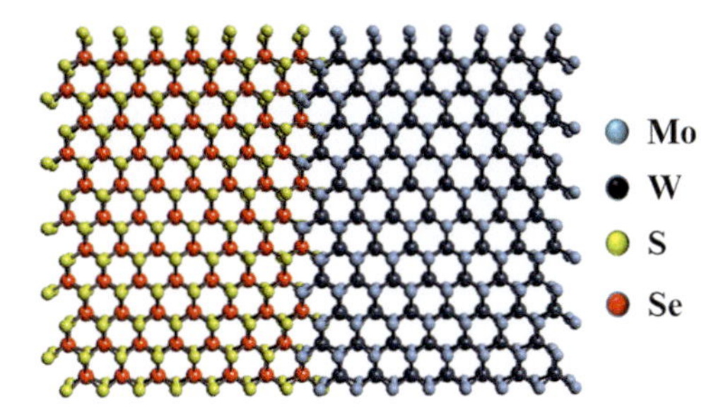

FIGURE 13.2 Schematic illustration of lateral junction formation by combining two kinds of 2D materials.

soaking it in HF solution to exfoliate PMMA/MoS_2 floating within the solutions. After transferring it to the desired substrate, the post-treatment of eliminating PMMA layer by soaking the fabricated in acetone under 80°C (36).

In addition to the employment of sole 2D materials in PV devices, heterostructures formed by two types of 2D materials have been extensively explored recently. These heterostructures allow the formation of atomic-scale interfaces that can address the fascinating management of photogenerated carriers and enable the construction of ultrathin solar devices (36,55). For instance, the vertically stacking heterostructure comprised of vertical growth of MoS_2/WS_2 exhibits the single atomically sharp lateral p-n heterojunction, as seen in Figure 13.2 below, which brings about 2.56% of conversion efficiency with fascinating omnidirectional light harvesting characteristics. Nevertheless, the remaining concern lies in the operation stability of solar cells made with 2D heterostructures as the ultrathin layered structures are highly sensitive to surrounding environment and may be contaminated due to chemisorption of environmental oxygen, and thus may lower the performance and stability of solar cells.

13.6 APPLICATIONS OF MIXED-DIMENSIONAL MATERIALS IN PVS

Mixed-dimensional materials, in terms of combining low-D materials in a hybrid form, have been extensively studied recently due to their remarkable light-absorption and charge-separation capabilities (56–58). These effects stand for the improvement of cell performance while maintaining facile and inexpensive solution-based synthesis. One promising example is shown in Figure 13.3. Different concentrations of MoS_2 flakes made by solution-processed method were mixed into PEDOT:PSS layer serving as p-type part. By combing these hybrid films with silicon nanowires as n-type component, it is found that adding 0.5 wt% of MoS_2 flakes in p-type polymeric layer achieves the highest power conversion efficiency (PCE) of 9.04%, with open circuit voltage (V_{oc}) of 0.527 V, short circuit density (J_{sc}) of 25.01 mA/cm^2 and fill factor (FF) of 68.6%. Compared with the reference without adding MoS_2 flakes in the preparation of p-type component (4.85 % of cell efficiency), the efficiency enhancement can reach 19.5%. The underlying reasons can be attributed to two possible effects. One is the efficient light absorption covering the visible-wavelength regions due to the

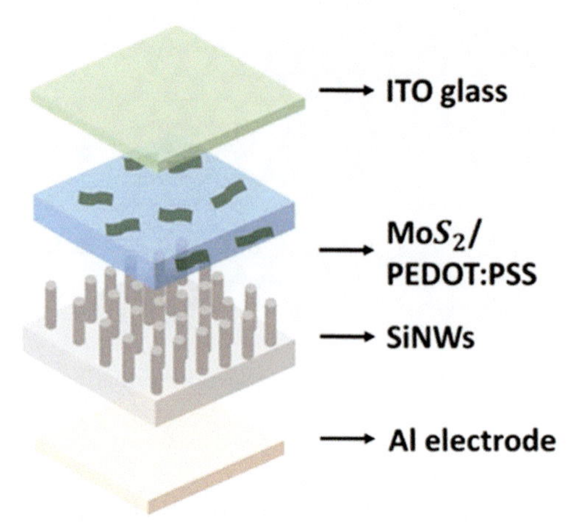

FIGURE 13.3 Schematic illustration of mixed-dimensional materials for the construction of hybrid solar cells: From top to the rear side: ITO glass/MoS_2@PEDOT:PSS/ Si nanowires/Al electrode.

introduction of MoS_2 flakes in the transparent PEDOT:PSS film, which facilitates the capture of incoming lights and in turn generated photoexcited carriers. Another reason might be originated from the short-diffusion length of photo-generated carriers due to hybrid design of p-type layer, thus effectively reducing the possible recombination of carriers.

13.7 CONCLUSION AND OUTLOOK

Current technological advances and developments require the use of renewable sources of energy that guarantee the sustainability for the next generations. Over the past few decades, increasingly evolved systems have been demonstrated that allow the effective use of energy resources on this planet. Solar cell is one of the promising types of unprecedented development towards a more technologically advanced and sustainable world. While silicon has dominated the commercial solar cell markets for decades, one-dimensional silicon structures are of great importance that enable to enhance solar energy harvesting efficiencies because of several unique optical and electric properties, and further sustain the minimum material usage and fabrication simplification. To further assess the viability of one-dimensional semiconductor photovoltaics, it is worthy evaluating them by several impacting criteria: (1) Charge separation, recombination, and collection. (2) The flexibility of conducting efficiency improvement methods, such as perovskite-passivated layer on silicon nanowires, rearrangement of conductive organic films and employment of transparent carrier-selective layer. In addition, the mixed-dimensional materials in solar cells seem to provide the balance between charge generation and collection by reducing the diffusion length of photoexcited carriers. Low-dimensional nanostructured materials have shown promising prospects for improving efficiency and will play a vital role to impact next-generation photovoltaic applications.

REFERENCES

1. Sun H, Deng J, Qiu L, Fang X and Peng H 2015 Recent progress in solar cells based on one-dimensional nanomaterials. *Energy Environ. Sci.* **8(4)** 1139–1159.
2. Devabhaktuni V, Alam M, Depuru SSSR, Green II RC, Nims D and Near C 2013 Solar energy: Trends and enabling technologies. *Renew. Sustain. Energy Rev.* **19** 555–564.
3. Garnett EC, Brongersma ML, Cui Y and McGehee MD 2011 Nanowire solar cells. *Annu. Rev. Mater. Res.* **41(1)** 269–295.
4. Martinson AB, McGarrah JE, Parpia MO and Hupp JT 2006 Dynamics of charge transport and recombination in ZnO nanorod array dye-sensitized solar cells. *Phys. Chem. Chem. Phys.* 2006 **8(40)** 4655–4659.
5. Würfel U, Cuevas A and Würfel P 2014 Charge carrier separation in solar cells. *IEEE J. Photovolt.* **5(1)** 461–469.
6. Han N, Wang F and Ho JC 2012 One-dimensional nanostructured materials for solar energy harvesting. *Nanomaterials and Energy*. **1(1)** 4–17.
7. Machín A, Fontánez K, Arango JC, Ortiz D, De León J and Pinilla S 2021 One-Dimensional (1D) Nanostructured Materials for Energy Applications. *Materials.* **14(10)** 2609.
8. Hu L, Kim HS, Lee J-Y, Peumans P and Cui Y 2010 Scalable coating and properties of transparent, flexible, silver nanowire electrodes. *ACS Nano*. **4(5)** 2955–2963.
9. Zhai H, Li Y, Chen L, Wang X, Shi L and Wang R 2018 Copper nanowire-TiO_2-polyacrylate composite electrodes with high conductivity and smoothness for flexible polymer solar cells. *Nano Res.* **11(4)** 1895–1904.
10. Das S, Pandey D, Thomas J and Roy T 2019 The role of graphene and other 2D materials in solar photovoltaics. *Adv. Mater.* **31(1)** 1802722.
11. Kostarelos K, Vincent M, Hebert C and Garrido JA 2017 Graphene in the design and engineering of next-generation neural interfaces. *Adv. Mater.* **29(42)** 1700909.
12. Palacios T 2011 Thinking outside the silicon box. *Nat. Nanotechnol.* **6(8)** 464–465
13. Yu X, Cheng H, Zhang M, Zhao Y, Qu L and Shi G 2017 Graphene-based smart materials. *Nat. Rev. Mater.* **2(9)** 1–13.
14. Zan W, Zhang Q, Xu H, Liao F, Guo Z, Deng J and Zhang, DW 2018 Large capacitance and fast polarization response of thin electrolyte dielectrics by spin coating for two-dimensional MoS_2 devices. *Nano Res.* **11(7)** 3739–3745.
15. Kwon O, Choi Y, Choi E, Kim M, Woo YC and Kim DW 2021 Fabrication techniques for graphene oxide-based molecular separation membranes: Towards industrial application. *Nanomaterials.* **11(3)** 757.
16. Dutta S, Banerjee I and Jaiswal KK 2020 Graphene and graphene-based nanomaterials for biological and environmental applications for sustainability. *J. Biosci.* **8(2)** 106–1012.
17. Kohlschütter V and Haenni P 1919 Zur Kenntnis des graphitischen Kohlenstoffs und der Graphitsäure. *Z. Anorg. Allg. Chem.* **105(1)** 121–1244.
18. Ali N, Hussain A, Ahmed R, Wang M, Zhao C and Haq BU 2016 Advances in nanostructured thin film materials for solar cell applications. *Renew. Sustain. Energy Rev.* **59** 726–737.
19. Soga T 2006 Nanostructured materials for solar energy conversion: Elsevier.
20. Afshar EN, Xosrovashvili G, Rouhi R and Gorji NE 2015 Review on the application of nanostructure materials in solar cells. *Mod. Phys. Lett. B* **29(21)** 1550118.
21. Chen Y 2015 Nanofabrication by electron beam lithography and its applications: A review. *Microelectron. Eng.* **135** 57–72.
22. Gangnaik AS, Georgiev YM and Holmes JD 2017 New generation electron beam resists: a review. *Chem. Mater.* **29(5)** 1898–1917.

23. Okazaki S 2015 High resolution optical lithography or high throughput electron beam lithography: The technical struggle from the micro to the nano-fabrication evolution. *Microelectron. Eng.* **133** 23–35.
24. Zhang R, Chen T, Bunting A and Cheung R 2016 Optical lithography technique for the fabrication of devices from mechanically exfoliated two-dimensional materials. *Microelectron. Eng.* **54** 62–68.
25. Chen SH, Kuo KY, Tsai KH and Chen CY 2022 Light trapping of inclined Si nanowires for efficient inorganic/organic hybrid solar cells. *Nanomaterials.* **12(11)** 1821.
26. Denkov N, Velev O, Kralchevski P, Ivanov I, Yoshimura H and Nagayama K 1992 Mechanism of formation of two-dimensional crystals from latex particles on substrates. *Langmuir.* **8(12)** 3183–3190.
27. Lash M, Fedorchak M, McCarthy J and Little S 2015 Scaling up self-assembly: Bottom-up approaches to macroscopic particle organization. *Soft Matter.* **11(28)** 5597–5609.
28. Barth S, Hernandez-Ramirez F, Holmes JD and Romano-Rodriguez A 2010 Synthesis and applications of one-dimensional semiconductors. *Prog. Mater. Sci.* **55(6)** 563–627.
29. Güniat L, Caroff P and Fontcuberta i Morral A 2019 Vapor phase growth of semiconductor nanowires: key developments and open questions. *Chem. Rev.* **119(15)** 8958–8971.
30. Hou Z, Li G, Lian H and Lin J 2012 One-dimensional luminescent materials derived from the electrospinning process: Preparation, characteristics and application *J. Mater. Chem.* **22** 5254.
31. Patil JV, Mali SS, Kamble AS, Hong CK, Kim JH and Patil PS 2017 Electrospinning: A versatile technique for making of 1D growth of nanostructured nanofibers and its applications: An experimental approach. *Appl. Surf. Sci.* **423** 641.
32. Shi X, Zhou W, Ma D, Ma Q, Bridges D and Ma Y 2015 Electrospinning of nanofibers and their applications for energy devices. *J. Nanomater.* **2015** 2015.
33. Kwak WC, Kim TG, Lee W, Han SH and Sung YM 2009 Template-free liquid-phase synthesis of high-density CdS nanowire arrays on conductive glass. *J. Phys. Chem. C.* **113** 1615.
34. Cai X, Luo Y, Liu B and Cheng HM 2018 Preparation of 2D material dispersions and their applications. *Chem. Soc. Rev.* **47** 6224.
35. Ricciardulli AG and Blom PW 2020 Solution-processable 2D materials applied in light-emitting diodes and solar cells. *Adv. Mater. Technol-US.* **5** 1900972.
36. Frisenda R, Molina-Mendoza AJ, Mueller T, Castellanos-Gomez A and Van Der Zant HS 2018 Atomically thin p–n junctions based on two-dimensional materials. *Chem. Soc. Rev.* **47** 3339.
37. Li L, Zhang D, Gao Y, Deng J, Gou Y and Fang J 2021 Electric field driven exfoliation of MoS_2. *J. Alloy. Compd.* **862** 158551.
38. Ciesielski A and Samorì P 2014 Graphene via sonication assisted liquid-phase exfoliation. *Chem. Soc. Rev.* **43** 381.
39. Zhang Q, Mei L, Cao X, Tang Y and Zeng Z 2020 Intercalation and exfoliation chemistries of transition metal dichalcogenides. *J. Mater. Chem. A.* **8** 15417.
40. Xu Y, Cao H, Xue Y, Li B and Cai W 2018 Liquid-phase exfoliation of graphene: An overview on exfoliation media, techniques, and challenges. *Nanomaterials.* **8** 942.
41. Zhou M, Tian T, Li X, Sun X, Zhang J and Cui P 2014 Production of graphene by liquid-phase exfoliation of intercalated graphite. *Int. J. Electrochem. Sci.* **9** 810.
42. Wang X and Dou W 2012 Preparation of graphite oxide (GO) and the thermal stability of silicone rubber/GO nanocomposites. *Thermochimica. acta.* **529** 25.
43. Hummers Jr WS and Offeman RE 1958 Preparation of graphitic oxide. *J. Am. Chem. Soc.* **80** 1339.

44. Bai F, Zhang J, Yuan Y, Liu H, Li X and Chueh CC 2019 A 0D/3D Heterostructured all-inorganic halide perovskite solar cell with high performance and enhanced phase stability. *Adv. Mater.* **31** 1904735.
45. Burresi M, Pratesi F, Riboli F and Wiersma DS 2015 Complex photonic structures for light harvesting *Adv. Opt. Mater.* **3** 722.
46. Fazio B, Artoni P, Antonia Iatì M, D'andrea C, Lo Faro MJ and Del Sorbo S 2016 Strongly enhanced light trapping in a two-dimensional silicon nanowire random fractal array. *Light-Sci. Appl.* **5** 16062.
47. Chen JZ, Ko WY, Yen YC, Chen PH and Lin KJ 2012 Hydrothermally processed TiO2 nanowire electrodes with antireflective and electrochromic properties. *ACS nano*. **6** 6633.
48. Sivakov V, Bronstrup G, Pecz B, Berger A, Radnoczi G and Krause M 2010 Realization of vertical and zigzag single crystalline silicon nanowire architectures. *J. Phys. Chem. C.* **114** 3798.
49. Zhang J, Zhong W, Liu Y, Huang J, Deng S and Zhang M 2021 A high-performance photodetector based on 1D perovskite radial heterostructure. *Adv. Opt. Mater.* **9** 2101504.
50. Wei TC, Chen SH and Chen CY 2020 Highly conductive PEDOT: PSS film made with ethylene-glycol addition and heated-stir treatment for enhanced photovoltaic performances. *Mater. Chem. Front.* **4** 3302.
51. Al Mamun A, Ava TT, Abdel-Fattah TM, Jeong HJ, Jeong MS and Han S 2019 Effect of hot-casted NiO hole transport layer on the performance of perovskite solar cells. *Sol. Energy.* **188** 609.
52. Chen CY, Wei TC, Hsiao PH and Hung CH 2019 Vanadium oxide as transparent carrier-selective layer in silicon hybrid solar cells promoting photovoltaic performances. *ACS Appl. Energ. Mater.* **2** 4873.
53. Hu X, Chen L and Chen Y 2014 Universal and versatile MoO_3-based hole transport layers for efficient and stable polymer solar cells. *J. Phys. Chem. C.* **118** 9930.
54. Li Z 2015 Stable perovskite solar cells based on WO_3 nanocrystals as hole transport layer. *Chem Lett.* **44** 1140.
55. Gong Y, Lin J, Wang X, Shi G, Lei S and Lin Z 2014 Vertical and in-plane heterostructures from WS_2/MoS_2 monolayers. *Nature. Materials.* **13** 1135.
56. Wang HP, Li S, Liu X, Shi Z, Fang X and He JH 2021 Low-dimensional metal halide perovskite photodetectors. *Adv. Mater.* **33** 2003309.
57. Mahmud MA, Duong T, Peng J, Wu Y, Shen H and Walter D 2009 Origin of efficiency and stability enhancement in high-performing mixed dimensional 2D-3D perovskite solar cells: A review. *Adv. Funct. Mater.* **32** 2009164.
58. Jana S, Mukherjee S, Bhaktha BNS and Ray SK 2021 Plasmonic silver nanoparticle-mediated enhanced broadband photoresponse of few-layer phosphorene/Si vertical heterojunctions. *ACS Appl. Mater. Inter.* **14** 1699.

14 Towards High Performance Indoor Dye-Sensitized Photovoltaics

A Review of Electrodes and Electrolytes Development

Shanmuganathan Venkatesan
Department of Chemical Engineering, National Cheng Kung University, Tainan, Taiwan

Yuh-Lang Lee
Department of Chemical Engineering, and Engineering and Hierarchical Green-Energy Materials (Hi-GEM) Research Center, National Cheng Kung University, Tainan City, Taiwan

CONTENTS

DOI: 10.1201/9781003367215-14

14.1 INTRODUCTION

As the world population continues to grow, the resulting energy needs have increased, together with concerns about carbon dioxide emissions and global warming. These have inspired scientists around the world to discover alternative, renewable and green technologies for energy generation. Among the renewable energy sources, solar energy provides adequate and environmentally benign energy that has huge potential for meeting world energy consumption needs [1,2]. Photovoltaic technologies (PV) including first, second, and third-generation solar cells offer a chance to easily convert the solar energy into electrical energy [3]. In the last few years, the usage of PV technologies has gradually moved to room light conditions from sunlight environments. For the applications of ambient-light harvesting, the internet of things (IoT) has been considered an important one [4,5]. IoT received ever-increasing technological and scientific attention because it describes networks of various small devices, such as wireless sensor nodes, wearable devices, consumer electronic devices, and smart meters. Since various portable devices are interconnected with communication systems, a reliable, stable power supply is vital for allowing these applications, owing to the demand for constant performance over a period of several months to years [5]. Energy storage devices such as batteries may be utilized as energy sources for IoT devices. However, maintenance and device replacement are considered to be very expensive. Therefore, PV technologies are considered to be the most appropriate means by which to supply stable power to IoT devices [5].

It is well known that conventional silicon solar cells have ruled the world PV market over the past 50 years [6]. This is because of their important features such as efficient power generation under sunlight and high long-term stability in all atmospheric conditions. A high-power conversion efficiency (PCE) of 26.7% is reached for the silicon solar cells under one-sun conditions [7]. However, there are various disadvantages related to silicon solar cells, including their high cost, complicated design, as well as the fact that they are heavy and have low PCE in low light intensities. In addition, these solar cells have restricted their use in indoor applications, portable electronic devices, and building integrated photovoltaic devices.

To overcome the problems of silicon solar cells, third-generation solar cells such as dye-sensitized solar cells (DSSCs), organic solar cells, and perovskite solar cells have been proposed [8–10]. These cells are fabricated with a simple, compact, and colorful design using readily available materials. So far perovskite solar cells and DSSCs have PCEs well above 25% and 15.2%, respectively under one-sun conditions [11,12]. However, these cells have lower PCE and lower stability under one-sun conditions compared to silicon solar cells. These characteristics limit their commercial use, especially mass electricity generation outdoors. However, their high PCEs under room light, which surpass the other solar cells under typical ambient light conditions. A PCE of 36.2% is reached for the cells using a triple-anion perovskite layer under room light conditions [13]. These characteristics make them suitable for IoT applications.

In recent years, DSSCs are considered to be efficient and stable power supply cells for IoT devices [14]. One of the reasons for this is the PCEs are much higher than those of DSSCs under sunlight illumination [15–42]. This is attributed to the narrower wavelength region of the artificial lights. Furthermore, the energy required for operating IoT devices is very small which can be easily provided by the room light DSSCs. Recently, by using the copper liquid electrolytes(LEs) and quasi-solid-state (QS) electrolytes, DSSCs have obtained the PCEs of 34.0% and 25%, respectively under 1000-lux fluorescent lighting [37,42].

This chapter summarizes classic and new materials in view of their physical and electrochemical characteristics for application in room light DSSCs. The influence of the thickness of the photoelectrodes (PEs), counter electrodes (CEs) materials, concentrations of the electrolyte components, nature of the solvents, and characteristics of the dyes on the performance of the cells are reviewed. The working mechanism of the cells based on the interaction of the polymers with the electrolyte components is presented. The long-term performance of the cells using different components of the DSSCs is presented.

14.2 OPTIMIZATION OF THE DSSC COMPONENTS

A classical DSSCs consists of a dye-sensitized PE, a platinum CE, and a redox electrolyte. The working function of DSSC involves photoexcitation of the dye, followed by electron injection into the conduction band of the titanium dioxide (TiO_2) anode [9,10]. The oxidized dye molecule is regenerated by accepting electrons from the redox couples, which itself is reduced at the CEs. In the literature, cells using iodide electrolytes and novel dyes are mostly utilized for room light applications. The optimal conditions utilized for these cells are different from those operating under one-sun conditions [34–36] In contrast to one-sun conditions, only a tiny number of electrons are excited under T5 light illumination. These electrons are recombined easily with the redox couple, which seriously affects the open circuit voltage (V_{oc}) of the cells. Likewise, the formation of excited holes is decreased under room light illumination, in which a small quantity of iodide is sufficient to reduce the number of holes formed. As a result of these two facts, the overall performance of the cell is decreased. By regulating the amount of iodide/triiodide (I^-/I_3^-) in the electrolytes, the efficiency of the cell can be somewhat improved [34,35]. The light intensity is also affected due to the amount of iodine in the electrolyte. This issue is realized in the incident photon-to-current conversion efficiency (IPCE) and electrochemical impendence spectroscopy (EIS) analysis of DSSCs. In recent years, cobalt and copper electrolytes have been considered to be suitable for the preparation of high-performance room light DSSCs [31–42]. However, different amounts of cobalt and copper redox couples, various kinds of solvents, and different concentrations of additives are required for the room light DSSCs compared to the DSSCs working under one-sun illumination. Similarly, different thicknesses of the TiO_2 layers, and different amounts of CE materials are required for room light DSSCs to achieve high cell efficiencies. Therefore, optimization of the DSSCs components is generally performed in the literature to obtain the maximum efficiency under room light conditions.

14.3 PHOTOELECTRODE MATERIALS

14.3.1 Compact TiO_2 Layer

In DSSCs, a blocking layer is generally formed in between the fluorine-doped tin oxide (FTO) and the TiO_2 film [33]. A compact blocking layer covers the rough surface of FTO and decreases the charge transfer resistance at the FTO substrate/electrolyte interface. The cells using titanium tetrachloride ($TiCl_4$) -based blocking layer works efficiently in one-sun conditions whereas severe charge transfer from the FTO to electrolyte under room light conditions decreases the efficiency of such cells. To overcome these issues, compact TiO_2 blocking layers were prepared using the spray pyrolysis method. In this method, an alcohol solution containing Titanium diisopropoxide bis(acetylacetonate) (TiAcAc, 75% in isopropanol) was sprayed manually 24 times onto the FTO surface to prepare the compact blocking TiO_2 layer. The thickness of the compact layer was proportional to the employed spraying cycles. The morphology of the plain FTO surface and FTO substrates covered with blocking layers were analyzed using scanning electron microscopy (SEM) (Figure 14.1a-c). Uniform coverage and a smoother morphology were observed for the blocking layers prepared based on the spray pyrolysis method. (Figure 14.1c). On the contrary, several crystalline grains and sharp facets of tin oxide (SnO_2) were observed for the plain FTO surface (Figure 14.1a). When the classical $TiCl_4$ treatment was utilized to prepare the blocking layer, several small TiO_2 particles were formed on the surface of the FTO (Figure 14.1b). However, the surface remained the same as that of the plain FTO surface, which was due to the low surface coverage and non-uniformity of $TiCl_4$ blocking layers. Therefore, $TiCl_4$-based blocking layers were unable to reduce such electron recombination, thus affecting the performance of cobalt-mediated room light DSSCs. The cells using compact blocking layer obtained a PCE of 15.26% under the room light illumination of 1000 lux. This PCE is much higher than the efficiency (8.56%) of the cells using the classical $TiCl_4$-based blocking layer (Table 14.1). This is attributed to the high V_{oc} of the related DSSCs.

FIGURE 14.1 Top view: SEM images with a scale bar of 200 nm: (a) pristine FTO surface and FTO substrates coated with blocking layers fabricated by (b) $TiCl_4$ treatment and (c) 12 spraying cycles.

TABLE 14.1
Room Light DSSCs Using Various Components of PEs

DSSC Components	Illuminance [lux]	J_{sc} (μAcm^{-2})	V_{oc} (V)	FF	P_{in} (μW cm^{-2})	P_{max} (μW cm^{-2})	η [%]
$TiCl_4$ [(a)]	251	21.90	0.472	0.472	78.8	4.87	6.19
Compact blocking layer[(a)]	251	22.76	0.652	0.803	78.8	11.96	15.12
Compact blocking layer[(b)]	251	25.55	0.712	0.771	79.0	14.02	17.75
2/4 μm[(c)]	200	17.61	0.684	0.747	64.0	9.00	14.05
4/4 μm[(c)]	200	21.03	0.651	0.719	64.0	9.74	15.39
6/4 μm[(c)]	200	21.74	0.631	0.706	64.0	9.56	15.13
8/4 μm[(c)]	200	23.05	0.595	0.700	64.0	9.50	15.03
2/4 μm[(d)]	201.8	23.05	0.764	0.736	63.84	12.98	20.33
4/4 μm[(d)]	201.8	23.20	0.757	0.741	63.84	13.04	20.42
6/4 μm[(d)]	201.8	24.33	0.756	0.732	63.84	13.47	21.09
8/4 μm[(d)]	201.8	24.35	0.745	0.743	63.84	13.49	21.13
10/4 μm[(d)]	201.8	23.86	0.694	0.775	63.84	12.86	20.14
4.16/3.63 /4.16 μm-Front side[(e)]	1,000	139.39	830.7	0.73	337.57	85.13	25.04
Rear side	1,000	193.56	843.21	0.76	337.57	80.59	23.70
Z907[(f)]	200	20.48	0.600	0.740	64.0	9.28	14.50
N719[(f)]	200	21.03	0.651	0.720	64.0	9.74	15.40
MK-2[(f)]	200	21.25	0.627	0.725	64.0	9.67	15.09
Yd2-O-C8[(f)]	200	15.54	0.467	0.701	64.0	5.09	7.94
N719[(g)]	201.8	4.54	0.495	0.774	63.84	1.74	2.72
Z907[(g)]	201.8	14.18	0.591	0.796	63.84	6.68	10.46
MK-2[(g)]	201.8	19.83	0.607	0.775	63.84	9.34	14.63
Y123[(g)]	201.8	21.75	0.774	0.774	63.84	12.63	19.78
D35/XY1B[(h)]	200	27.2	0.732	0.790	61.3	15.6	25.5
XY1/L1[(h)]	200	29.0	0.840	0.780	60.6	19.0	31.4

(*Continued*)

TABLE 14.1 (Continued)
Room Light DSSCs Using Various Components of PEs

DSSC Components	Illuminance [lux]	J_{sc} (μAcm^{-2})	V_{oc} (V)	FF	P_{in} (μW cm^{-2})	P_{max} (μW cm^{-2})	η [%]
XY1+5T[h]	1,000	131.2	0.860	0.780	303.1	88.5	29.20
MM6+MM3[i]	1,000	137	0.598	0.649			28.74
AN-11[j]	1,000	52.90	1.038	0.647	316	35.58	11.26
TY6+CDCA[k]	1,200	164	0.671	0.778	345		24.43
YL4+CDCA[l]	1,025	224	0.576	0.630	330		25.0
SK-6+CW10[m]	1,200	148	0.582	0.737	380		16.75
Y1A1+CDCA[n]	350	56.6	0.476	0.755	104.3	20.3	19.50
Y123+XY1[o]	200	29.6	0.808	0.722	63.6	49	27.5

Notes

[a] 2.3 μm (PST-18NR)/2.6 μm TiO_2 layer (PST-400C), MK-2 dye, 12.1 nm Pt layer, 0.20 M Co (II)/, 0.07 M Co (III), and 1.0 M *t*BP, ACN.
[b] 5.5 μm (30NRD)/5.2 μm (PST-400C), Y123 dye, 12.1 nm Pt layer,0.22 M Co (II)/0.05 M Co (III), and 0.5 M *t*BP, ACN.
[c] ML(PST-18NR)/SL(PST-400C), N719 dye, 12.1 nm Pt layer, 0.01 M I_2, MPN
[d] ML (30NR-D)/ SL (PST-400C), Y123 dye, 12.1 nm Pt layer, 0.11 M Co (II)/0.025 M Co (III), and 1.2 M *t*BP, MPN.
[e] ML (30NR-D)/ SL (PST-400C), D35/XY1b =7:3, 0.55 nm Pt layer, 0.11 M Co (II)/0.025 M Co (III) and1.2 M *t*BP, MPN.
[f] 4 μm (PST-18NR)/4 μm TiO_2 layer (PST-400C), 12.1 nm Pt layer, 0.01 M I_2, MPN.
[g] 2 (30NR-D) + 2 μm (PST-400C), TiO_2 layer, 0.11 M Co (II)/0.025 M Co (III), 1.2 M *t*BP, 12.1 nm Pt layer.
[h] 4 μm (30 NRD)/4 μm TiO_2 layer, 0.2 M Cu(I)/0.04M Cu(II), 0.1 M LiTFSI,0.6 M *t*BP, ACN or PPN, PEDOT CE.
[i] 12+6 μm TiO_2 layer,0.05 M I_2, MPN, Pt CE.
[j] F-17 module cell, 9+6 μm TiO_2 layer, 0.05 M I_2, ACN/VN, Pt CE.
[k] 12+4 μm TiO_2 layer, 0.05M I_2, 0.05M *t*BP, PVP-Pt layer, ACN/VCN(85:15).
[l] 10+5 μm TiO_2 layer,0.04 M I_2, 0.5 M *t*BP, Pt/Carbon cloth, ACN/MPN.
[m] 12+2 μm TiO_2 layer, 0.04 M I_2, 0.5 M *t*BP, ACN/VN, PVP-Pt CE.
[n] 12+3 μm TiO_2 layer,0.165 Co(II)/0.045 Co (III), 0.8 M tBP, PVP-Pt CE, ACN/VN.
[o] 4+3 μm TiO_2 layer,0.04 M Cu(II)/0.2 M Cu(II), 0.1 M LiTFSi, and 0.6M MBI, ACN.

14.3.2 TiO_2 Layer Thickness

It has been reported in the literature that the performance of DSSCs is strongly affected by the thickness of the TiO_2 layers in the PEs. For the classical DSSCs, generally 10–20 μm thick layers of TiO_2 have been utilized with ruthenium dyes (N719) to achieve high current density (J_{sc}) under one-sun conditions. However, cobalt electrolytes show slow diffusion with thick TiO_2 layers and reduce the chance of obtaining higher J_{sc} than obtained with iodide electrolytes. These problems have been solved using thin TiO_2 layers and dyes with high absorption coefficients. However, these very thick and too thin TiO_2 layers are not suitable for DSSCs to achieve high cell efficiencies under room light conditions. To overcome this problem, PEs with various main (2, 4, 6, and 8 μm) and scattering layer (SL) (2 or 4 μm) thicknesses were prepared by our group [31,35]. The performance of the cells using these PEs under the room light illumination of 200 lux is studied as shown in Table 14.1. An improvement in the J_{sc} of the iodide cells using N719 dye was observed, which was mainly ascribed to the high IPCE value of the corresponding cells (Figure 14.2a) [31]. The IPCE increased with increases in the thickness of the main layers (MLs). This is because the dye loading and scattering of light improved when the thickness of the TiO_2 layers increased. The cobalt DSSC using a 10 μm thick TiO_2 layer and Y123 dye caused red-shifts in the IPCE spectra by 23 nm (Figure 14.2b) [35]. This value was higher than those obtained for the other cells. Thus, this cell can absorb more light in a longer wavelength region and increase the J_{sc} of the related DSSC. Although the J_{sc} of the iodide and cobalt cell increased, the V_{oc} reduced with increases in the TiO_2 layer thickness. It is well known that thick TiO_2 layers have a large number of recombination sites that trigger a greater recombination reaction and therefore decrease the cell V_{oc}.

The chemical capacitance measured from the EIS analysis for the cobalt DSSCs with a thick layer (10 μm) was higher than those for the DSSCs with thin layers (4 and 6 μm), implying that the Fermi level of the electrode downshifted in the presence of the thick layer (Figure 14.3a). The high J_{sc} was consistent with the downshift of the Fermi level. The V_{oc} of the related DSSC decreased slowly, which was also consistent with what was expected from the downshift of the Fermi level.

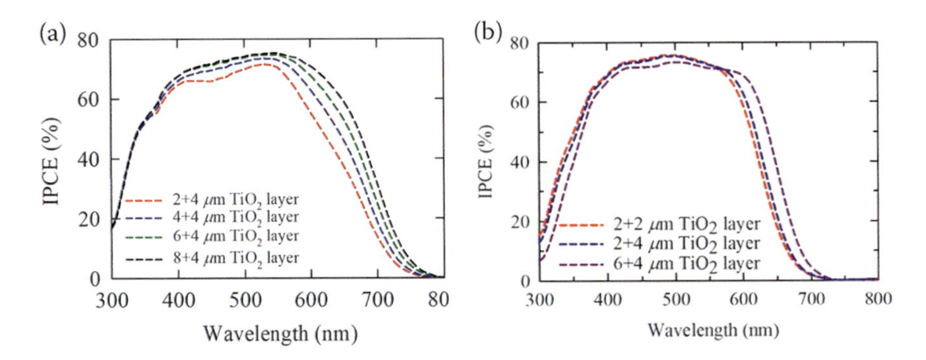

FIGURE 14.2 IPCE spectra of the (a) iodide (N719) and (b) cobalt DSSCs (Y123) containing PEs with various thicknesses of the main layer.

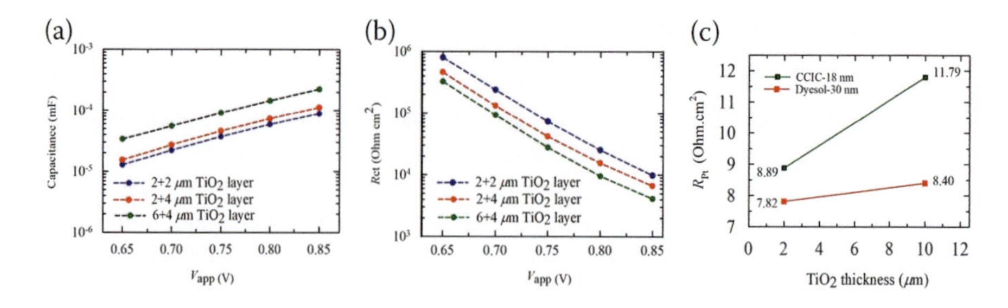

FIGURE 14.3 (a) Chemical capacitance, and (b) recombination resistance (R_{ct}) measured for the cobalt liquid cells using various TiO_2 layer thicknesses. (c) A comparison between the electrolyte penetration through the TiO_2 main layers prepared using two different particle sizes of 18 and 30 nm based on the R_{Pt}.

For the recombination resistance (R_{ct}) shown in Figure 14.3b, the DSSC using a thick TiO_2 layer had a lower value as compared to the DSSCs using thin layers. The lower R_{ct} value indicated a higher recombination reaction at the photoanode/electrolyte interface. According to these EIS analyses, the lower V_{oc} value of the cell using the thick layer was due to both the downshift of the Fermi level and higher electron recombination at the photoanode/electrolyte interface. Therefore, TiO_2 PEs with the thicknesses of (4+4) 8 μm and (6+4) 10 μm TiO_2 layers, respectively were considered optimal for the iodide and cobalt LEs--based DSSCs [31,35]. These iodide and cobalt cells using N719 and Y123 dyes achieved high PCEs of 15.39% and 21.08%, respectively under 200 lux T5 light illumination. With TiO_2 electrode thicknesses of 8 μm, copper electrolyte-based DSSCs using D35+ XY1, XY1+5T, and XY1+L1 dyes also obtained high PCEs of 28.9%, 29.2%, and 34.0%, respectively under 1000 lux of fluorescent light [39,41,42].

14.3.3 TiO_2 Particle Size

The penetration of the electrolytes into the pores of the TiO_2 can be improved by the use of large particle sizes [36]. To prove this, electrolyte penetration experiments were conducted and the results of the study are shown in Figure 14.3c. The 2 and 10 μm thick TiO_2 MLs with particles size of both 30 and 20 nm were printed separately onto platinum-coated FTO glasses. The penetration of the electrolytes was assessed by calculating the charge transfer resistance at the CE/PE interface (R_{Pt}). The 2 μm thick TiO_2 MLs using particle sizes of 30 and 20 nm had R_{Pt} values of 7.82 and 8.89 Ω cm^2, respectively (Figure 14.3c). These two values increased with increases in the thickness of the TiO_2 MLs from 2 to 10 μm. However, the increase was not significant for the ML with a particle size of 20 nm, which exhibited poor electrolyte penetration.

In comparison, the R_{Pt} decreased significantly when the ML with a particle size of 30 nm was used, revealing much higher electrolyte penetration. Thus, the obtained R_{Pt} values were 11.79 Ω cm^2 and 8.40 Ω cm^2, respectively, for the 10 μm thick TiO_2 ML containing particle sizes of 20 and 30 nm. These studies confirmed

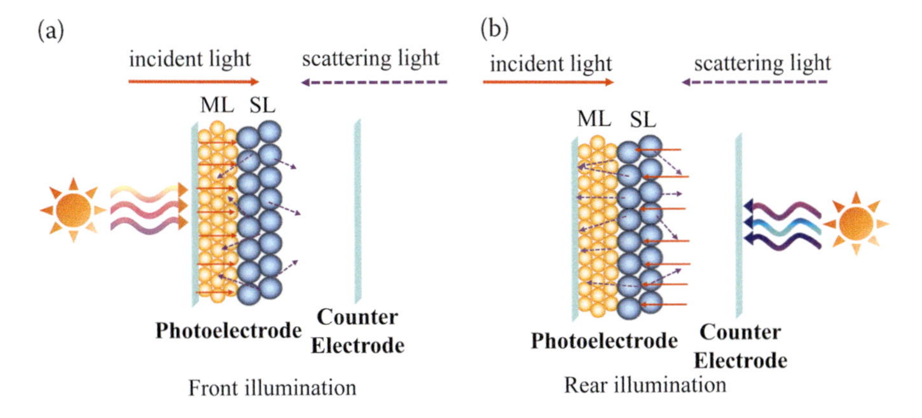

FIGURE 14.4 The classic structure of the bifacial DSSCs fabricated using main and scattering TiO_2 layers under: (a) front and (b) rear illumination.

the advantages of using a main TiO_2 layer with a particle size of 30 nm in terms of the contact of the electrolyte into the pores of the PE.

14.3.4 TiO_2 Layer Architecture

Very recently, novel TiO_2 architecture was developed for indoor bifacial DSSCs [38]. The structure of classical DSSCs always contains an ML and a light SL. For the front-side illumination, the function of the SL is to reflect the unabsorbed light back to the ML; and the corresponding mechanism is shown in Figure 14.4a.

However, this structure is unfavorable for the back-side efficiency since the rear-illuminated light irradiates to the SL first and, due to the scattering effect, less light can be harvested by the ML (Figure 14.4b). To solve this problem, the SL of the PE was removed (Figure 14.5a) in the fabrication of bi-facial cells, which can definitely increase the transmittance of the back-illuminated light to the ML and the rear-side conversion efficiency (Figure 14.5b). However, the

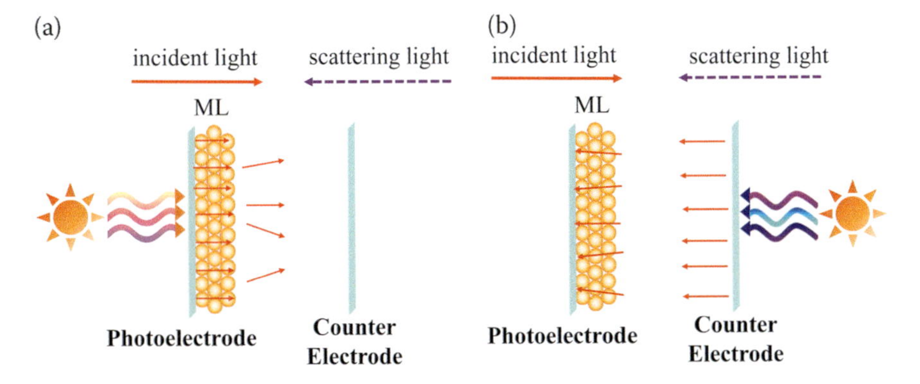

FIGURE 14.5 The bifacial DSSCs fabricated using only ML under: (a) front and (b) rear illumination.

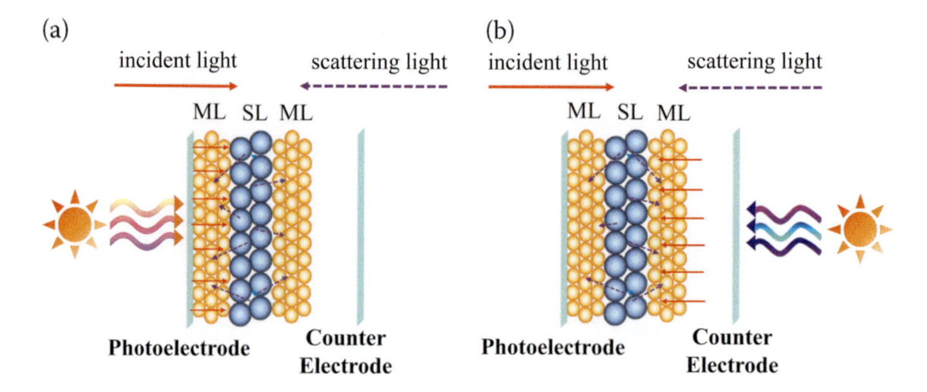

FIGURE 14.6 The novel architecture of the bifacial DSSCs fabricated using main, scattering, and main TiO_2 layers under: (a) front and (b) rear illumination.

front-side efficiency becomes smaller without an SL (Figure 14.5a). Our novel architecture PE was developed by introducing an additional ML behind the light SL of conventional PEs, as shown in Figure 14.6a. By employing this PE, the rear-side incident light will irradiate sequentially to the rear ML, the SL, and the front ML (Figure 14.6b). Except for the light absorbed by the electrolyte, the front-side and rear-side incident lights were harvested according to the same mechanism as the classic DSSCs. Since an SL was sandwiched between two MLs, this new architecture of the TiO_2 film is termed a "sandwich PE". By using D35 and XY1b as co-sensitizers of the PE, the front-side and rear-side efficiencies of the bifacial cells achieved 25.04 and 23.7%, respectively, under 1000-lux fluorescent lighting, which had a rear-to-front efficiency ratio of 95%. The cells also showed high PCEs under one-sun conditions. These efficiencies were higher than the PCEs of the cells using the PE with the ML3 and ML2/SL1 architecture.

To figure out the structure effect on the performance of the bifacial cells, the spectra of light absorption, transmittance, and diffuse reflectance were measured for various electrodes (ML3, ML2/SL1, and ML1/SL1/ML1) sensitized by Y123 dye (Figure 14.7). The ML1/SL1/ML1 electrode has the highest light absorption ability,

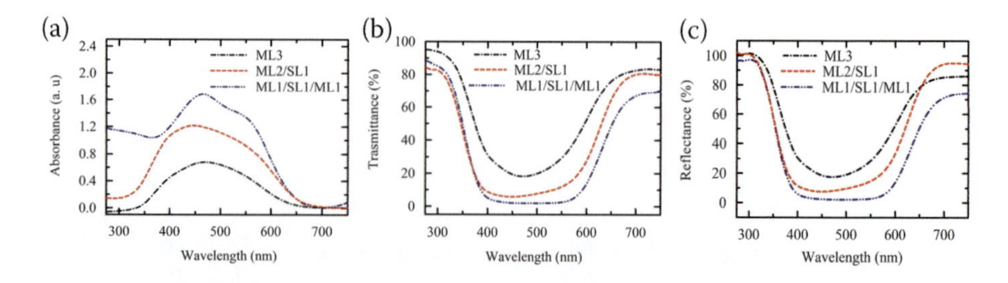

FIGURE 14.7 UV-visible absorption (a), transmittance (b) and reflectance (c) spectra of the Y123 dye sensitized photoelectrode with the ML1/SL1/ML1, ML2/SL1, and ML3 structures.

while the ML3 electrode has the lowest one (Figure 14.7a). For the transmittance and reflection properties shown in Figures 14.7(b) and 14.7(c), it decreases in the order of ML3, ML2/SL1, and ML1/SL1/ML1, which is also conformity with the tendency of light absorption and cell performance. The external quantum efficiency (EQE) spectra show that the values of the front-side illumination are much higher (about 70% for M1/S1/M1) than those of the rear-side ones (below 60%), which can be ascribed to the absorption, scattering, and reflection of the back-incident light by the electrolyte. Regarding the rear-side spectra, the EQE of the ML2/SL1 is the lowest in comparison to the others, indicating an unfavorable effect of the SL on the back-illuminated light, consistent with its lowest rear-side efficiency.

14.3.5 Photosensitizers

Apart from the thickness of the PEs, the performance of the DSSCs was found to be strongly affected by the structure of the dyes (Table 14.1). The DSSCs sensitized using N719 dye exhibited the best efficiency of 15.40% in Iodide LEs under the room light illumination of 200 lux [31]. However, the same dye exhibited a very low efficiency of 2% in Cobalt LEs [35]. This is because the N719 dye shows the shortest and longest electron lifetime in cobalt and Iodide LEs, respectively. For the cells using cobalt LEs-, MK2, Y123, and D35/XY1b = 7/3 were utilized. Among these dyes, the cells using D35/XY1b = 7/3 achieved high efficiencies of 25.3% under room light illuminance of 1000 lux [36]. The cells using D35/XY1b = 7/3 had the widest light absorption range and higher EQE values compared to other dyes, which contributed to the J_{sc} and efficiency of the related cells. For the cells using a copper electrolyte, D35/XY1b, XY1b/Y123, and XY1/L1 dyes were utilized, which achieved the PCEs of 28.9%, 32%, and 34%, respectively under room light illumination of 1000 lux [39,40,42]

14.4 POLYMER GEL AND PRINTABLE ELECTROLYTES

14.4.1 Electrolyte Solvents

Electrolyte, an important part of the DSSC, is responsible for the movement of charge carriers between electrodes [14,43,44]. LEs have been widely used in DSSCs owing to their high ionic conductivity and strong interfacial contact ability. The role of solvents in the electrolytes is very important and solvents with different physical properties such as viscosity, dielectric constant, and donor number directly influence the performance of the DSSCs [34]. From Table 14.2, it can be seen that by increasing the donor number (DN) of the solvents acetonitrile (ACN) (DN = 14.1), 3-methoxypropionitrile (MPN) (DN=15.4), propylene carbonate (PC) (DN=15.1), and gamma-butyrolactone (gBL) (DN =18)) the values of J_{sc} increased. The cell using the gBL-based electrolyte obtained a J_{sc} value of 26.82 μA cm^{-2}. This value was higher than those obtained for the cells using other solvent-based electrolytes. Therefore, this cell achieved the highest PCE of 20.18% under the room light illumination of 200 lux. The TiO_2 conduction band (CB) edge in the gBL-based electrolyte was relatively lower than that in the ACN, PC, and MPN-

TABLE 14.2
Room Light DSSCs Using Various Components of Electrolytes

DSSC Components	Illuminance [lux]	J_{sc} (μAcm^{-2})	V_{oc} (V)	FF	P_{in} (μW cm^{-2})	Pmax (μW cm^{-2})	η [%]
Acetonitrile[(a)]	200	24.28	0.627	0.777	64.0	11.83	17.82
3-methoxy propionitrile[(a)]	200	25.43	0.639	0.760	64.0	12.34	18.91
Propylene carbonate[(a)]	200	25.85	0.633	0.758	64.0	12.41	19.02
γ-butyrolactone[(a)]	200	26.82	0.631	0.778	64.0	13.17	20.18
0.07 M Co(II)/0.017 M Co (III)[(b)]	201.8	23.62	0.732	0.630	63.84	10.91	17.06
0.11 M Co(II)/0.025 M Co(III)[(b)]	201.8	24.03	0.726	0.660	63.84	11.50	18.03
0.15 M Co(II)/0.033 M Co(III)[(b)]	201.8	24.12	0.712	0.660	63.84	11.33	17.75
0.22 M Co(II)/0.05 M Co(III)[(b)]	201.8	24.43	0.666	0.687	63.84	11.17	17.50
0.6 M *t*BP[(c)]	201.8	21.81	0.750	0.730	63.84	11.94	18.70
1.2 M *t*BP[(c)]	201.8	21.75	0.774	0.774	63.84	12.63	19.78
PVDF-HFP PGE[(d)]	200	24.66	0.587	0.763	65.2	11.05	16.93
PVDF-HFP PGE[(e)]	200	25.54	0.607	0.759	65.2	11.78	18.05
PVDF-HFP PGE+ ZnO[(f)]	200	24.79	0.669	0.791	65.2	13.12	20.11
PVDF-HFP/PMMA[(g)]	200	27.12	0.710	0.778	68.2		22.0
	1000	137.6	0.770	0.786	329		25.3
PVDF-HFP/PMMA (Front)[(h)]	1000	127.7	0.786	0.765	329		23.3
PVDF-HFP/PMMA (Rear)	1000	115.4	0.785	0.780	329		21.5
0.01 M I_2[(i)]	200	23.90	0.587	0.695	64	9.62	15.23

0.05 M I_2[i]	200	23.03	0.549	0.729	64	9.23	14.40
0.2 M DMII[j]	200	23.48	0.608	0.720	64	10.28	16.06
0.8 M DMII[j]	200	24.71	0.581	0.700	64	10.01	15.70
0.2 M *t*BP[k]	200	23.74	0.588	0.720	64	10.19	15.70
0.8 M *t*BP [k]	200	23.16	0.618	0.730	64	10.45	16.32
9 wt.% PEO[l]	200	23.70	0.604	0.730	64	10.45	16.33
9 wt.% PEO[m]	200	25.10	0.727	0.779	67.56	14.23	21.06
PEO+PMMA[n]	200	24.14	0.664	0.787	67.56	12.61	18.67

Notes

[a] 2 μm (30NR-D) + 2 μm (PST-400C), TiO_2 layer, MK-2 dye, 0.15 M Co (II)/0.035 M Co (III), 0.5 M *t*BP, MPN, 12.1 nm Pt layer.

[b] 2 μm (30NR-D) + 2 μm (PST-400C), TiO_2 layer, Y123 dye, 0.2 M *t*BP, 12.1 nm Pt layer.

[c] 2 μm (30NR-D) + 2 μm (PST-400C), TiO_2 layer, Y123 dye, 0.11 M Co (II)/0.025 M Co (III), MPN, 12.1 nm Pt layer.

[d] 2 μm (30NR-D) + 2 μm (PST-400C), TiO_2 layer, MK-2 dye, 0.15 M Co (II)/0.035 M Co (III), 0.5 M T *t*BP, ACN,12.1 nm Pt layer.

[e] 2 μm (30NR-D) + 2 μm (PST-400C), TiO_2 layer, MK-2 dye, 0.15 M Co (II)/0.035 M Co (III), 0.5 M *t*BP, MPN,12.1 nm Pt layer.

[f] 2 μm (30NR-D) + 2 μm (PST-400C), TiO_2 layer, MK-2 dye, 0.15 M Co (II)/0.035 M Co (III), 0.5 M *t*BP, MPN,12.1 nm Pt layer.

[g] 9.4 μm (30NR-D) +4.9 μm (PST-400C) TiO_2 layer, D35/XY1B = 7:3, 0.11 M Co (II)/0.025 M Co (III) and1.2 M *t*BP in ACN.

[h] 9.4 μm main TiO_2 layer, D35/XY1B = 7:3, 0.11 M Co (II)/0.025 M Co (III) and1.2 M *t*BP in ACN.

[i] 8 μm ML (PST-18NR)/ 4 μm SL (PST-400C), N719 dye, 12.1 nm Pt layer, MPN.

[j] 6 μm (30NR-D) + 4 μm (PST-400C) TiO_2 layer, N719 dye, 12.1 nm Pt layer, 0.01 M I_2, 0.8 M *t*BP in MPN.

[k] 6 μm (30NR-D) + 4 μm (PST-400C) TiO_2 layer, N719 dye, 12.1 nm Pt layer, 0.01 M I_2, 0.2 M DMII in MPN.

[l] 6 μm (30NR-D) + 4 μm (PST-400C) TiO_2 layer, N719 dye, 12.1 nm Pt layer, 0.01 M I_2, 0.2 M DMII,MPN.

[m] 6 μm (30NR-D) + 4 μm (PST-400C) TiO_2 layer, Y123 dye, 12.1 nm Pt layer, 0.11 M Co (II)/0.025 M Co (III), 1.2 M *t*Bp, MPN.

[n] 6 μm (30NR-D) + 4 μm (PST-400C) TiO_2 layer, Y123 dye, 12.1 nm Pt layer, 0.11 M Co (II)/0.025 M Co (III), 1.2 M *t*Bp, MPN

based LEs, which was thus responsible for the higher J_{sc} [34]. The V_{oc} of the cell using the MPN-based electrolyte was higher than those of cells using other solvent-based electrolytes. This is because of a higher ability of the MPN to increase the recombination resistance in the LE. Although the V_{oc} was higher for the cell using the MPN-based electrolyte, the efficiency of the cell was lower than that of the cell using the gBL-based LE. This implies that J_{sc} is an important factor determining the efficiencies of the cells using ACN, MPN, PC, and gBL-based LEs. The choice of solvents to prepare LEs for indoor DSSCs should be in the following order: gBL>PC or MPN>ACN. However, gBL and PC-induced dye desorption affected the long-term performance of the DSSCs. This problem was solved by using ACN or MPN based- LEs.

14.4.2 Redox Couples

The iodide/triiodide ions (I^-/I_3-), cobalt complexes, and copper complexes are mostly utilized as redox couples for the DSSCs applications [14,45]. The cells using I^-/I_3- redox couples show good efficiency. However, the I^-/I_3- couple causes corrosion at the metal CEs and they also have strong light absorption and low redox potential. These disadvantages affect the overall performance of the DSSCs [14,45]. To solve these problems, cobalt and copper complexes have been utilized to replace the I^-/I_3- redox couples [31–42,45]. These redox couples have low light adsorption and good compatibility with the metal and polymer-based-CEs. The cells using these redox couples achieved high V_{oc}(1V), which allows them to achieve high PCEs under room light conditions.

The ratio of the redox couples in electrolytes greatly influences the performance of the cells under both AM1.5 G and indoor light conditions [35]. The J_{sc} increased, and the V_{oc} decreased when the Co(II)/Co(III) ratio was increased from 0.07/0.017 M to 0.22/0.05 M (Table 14.2). The cells using Co(II)/Co(III) ratios of 0.07/0.017 M and 0.22/0.05 M, respectively, obtained a higher V_{oc} (0.732 V) and J_{sc} (24.42 V) than the other cells. However, higher PCE (18.03%) was obtained for the cell using a Co(II)/Co(III) ratio of 0.11/0.025 M. Therefore, this ratio was considered to be optimal for the cobalt-DSSCs working under room light conditions. For the DSSCs working under one-sun conditions, the electrolyte was prepared by introducing $Co^{II}(bpy)_3(PF_6)_2$ (0.22 M), Co^{III} $(bpy)_3(PF_6)_3$ (0.05 M), 4-tert-butyl pyridine (*t*BP) (0.2 M), and lithium perchlorate ($LiClO_4$) (0.1 M) in ACN solvent [35,38].

To increase the PCE of the cells, copper LEs have been utilized for DSSCs. The cells using ACN-based copper LEs and XY1+L1 dyes obtained the PCEs of 31.4%, 32.7%, and 34.0%, respectively under 200, 500, and 1,000 lux illuminations [42]. These cells with active area of 16 cm^2 were used to power machine learning on wireless nodes. In another study, D35+XY1-based cells using ACN-Cu electrolytes achieved the PCEs of 25.5% and 28.9%, respectively under 200 and 1000 lux illumination [39]. These efficiencies were higher than the efficiencies of the cells using iodide and cobalt LEs. Solid-state DSSCs-based on the copper redox couples were also fabricated for room light applications. For the preparation of this cell, at first, copper redox couples were introduced into the cells and then electrolyte solvents were allowed to evaporate for several hours. This procedure forms a solid mass of copper complexes in between the PE and -CE for working as a solid-state hole transporter. The cells using this solid-state electrolyte achieved an power conversion efficiency of 30% under the illumination of 1000 lux [42].

14.4.3 Polymer Gel Electrolytes

Almost all of the high-efficiency DSSCs reported in the literature used LEs. However, the cells with LEs frequently suffer from some practical issues, such as leakage and evaporation of organic solvents, and desorption of dye materials, which affects the long-term performance of DSSCs [45]. A good solution to these problems is polymer gel electrolytes (PGEs) and printable electrolytes (PELs) [43]. To prepare the PGEs, various concentrations of poly(vinylidene fluoride-hexafluoropropylene (PVDF-HFP) were also introduced into the ACN -based cobalt LEs, and the performance of the DSSCs using these PGEs was investigated [34]. PCEs of ~16.93% were obtained for the QS-DSSCs prepared using these PGEs. This study suggests that the performance of DSSCs under room light conditions is independent of the polymer concentrations. These efficiencies were also lower than that of the cell using the LE (17.82%). The large decrease in the V_{oc} caused low PCEs in these QS-DSSCs. To increase the efficiency and stability, various amounts zinc oxide (ZnO) -nanofillers were added into the MPN-based cobalt PGEs [43]. A high PCE of 20.11% was attained at 4 wt. % ZnO nanofillers [34]. This efficiency was higher than the efficiencies of the corresponding liquid version (18.91%) and the ACN-based QS-DSSCs using metal oxides such as silicon dioxide -, SnO_2, aluminium oxide --, ZnO, and zirconium oxide ZrO_2 nanofillers.

In another study, various compositions PVDF-HFP and poly (methyl methacrylate) (PMMA) polymers were introduced into the ACN-cobalt LEs [37] (Figure 14.8a & 14.8b). The use of these polymers caused slow diffusion of cobalt redox couples. However, the efficiency of the related QS-DSSCs using Y123 dye

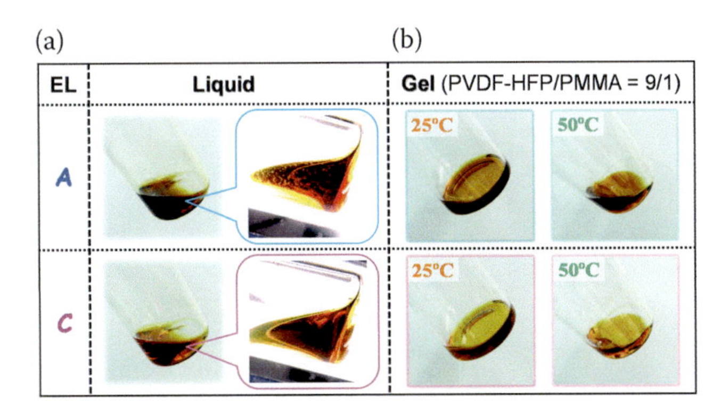

FIGURE 14.8 Photographs of the cobalt LEs(a) and PVDF-HFP - PGEs (b) Electrolyte A consists of 0.22 M [Co (II) $(bpy)_3](PF_6)_2]$, 0.05 M [Co (III) $(bpy)_3](PF_6)_3]$, 0.1 M $LiClO_4$, 0.2 M *t*Bp in ACN. Electrolyte C consists of 0.11 M [Co (II) $(bpy)_3](PF_6)_2]$, 0.025 M [Co (III) $(bpy)_3](PF_6)_3]$, 0.1 M $LiClO_4$, 0.2 M *t*Bp in ACN.

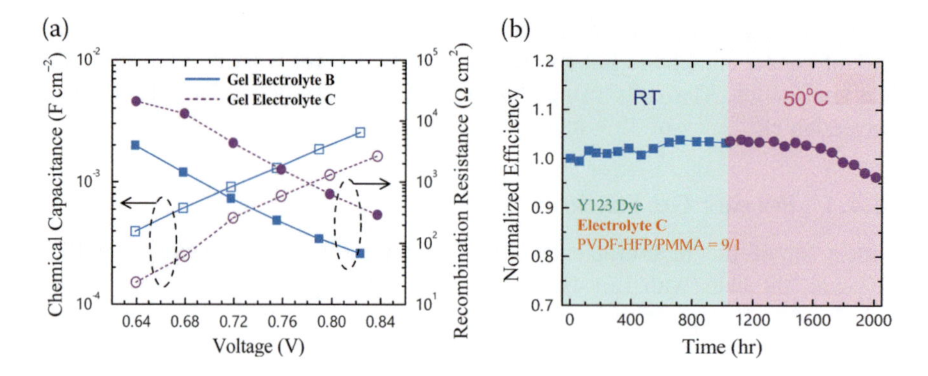

FIGURE 14.9 (a) PVDF-HFP PGE B consists of 0.11 M [Co (II) $(bpy)_3](PF_6)_2$], 0.025 M [Co (III) $(bpy)_3](PF_6)_3$], 0.1 M $LiClO_4$, 1.2 M *t*Bp in ACN.(b) trace on the performance stability based on three cells using electrolyte.

surpassed the corresponding liquid cells under room light conditions, which was due to the suppressed electron recombination at the PE/PGE interface. By optimizing the PVDF-HFP/PMMA ratio and the electrolyte composition, the charge recombination was inhibited, thereby creating high Voc (Figure 14.9a). Furthermore to increase the J_{sc}, a sensitization system (D35/XY1b = 7/3) and a polymeric catalyst was introduced, which respectively broaden the light-harvesting region and facilitate interfacial charge transfer at the CE. Consequently, the resultant QS-DSSCs not only showed long-term performance during a 2000 h test (Figure 14.9b), but also achieved a PCE beyond 25% under 1000-lux fluorescent lighting. The bifacial DSSCs using these PGEs obtained high front and rear side efficiencies which had an efficiency ratio of 96%. The module cells were prepared using the polymer blend electrolytes which showed the ability to power small electronics under room light conditions (Figure 14.10).

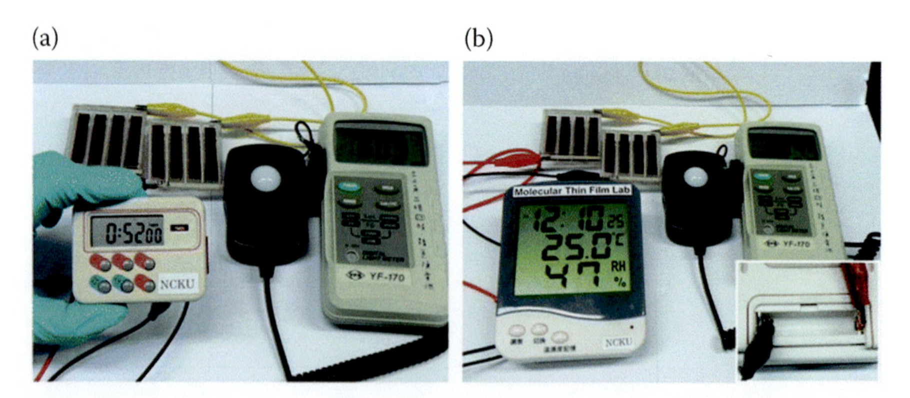

FIGURE 14.10 Demonstrations of small electronics powered by two dye-sensitized indoor photovoltaic module devices in the series connection under around 600-lux fluorescent lighting: (a) digital timer; (b) time/temperature/humidity display meter, with an inset showing connection on the back.

14.4.4 Printable Electrolytes

One of the main disadvantages of the PGEs is their poor penetration into the TiO_2. Therefore, the injection of electrolytes was mostly performed at elevated temperatures, making the PCEs of gel-state DSSCs lower than those of liquid-state cells. By using these hot electrolyte filling methods, gel-state DSSCs achieved efficiencies and stabilities higher than those of corresponding liquid-state cells. However, high performance was obtained only for small-area cells or laboratory cells utilizing the injection method for fabrication. In fact, this method is not suitable for large-area DSSCs. This problem was solved by using PELs that were prepared simply by using altering the composition of polymers into the LEs [31,36]. PELs with optimal viscosity were easily printed on the surface of the PEs either using a screen printing or doctor blade method. This technique can be useful for the strong penetration of highly viscous PGEs and the roll-to-roll coating process used in the preparation of large-area module cells.

The polyethylene oxide (PEO)/polyvinylidene fluoride (PVDF) iodide PELs were prepared by introducing various compositions of the polymers into iodide LEs containing 0.05 M iodine to prepare the optimal PELs [31]. The QS-DSSCs using these PELs exhibited similar efficiencies (~14.40%) under 200 lux illuminances. To overcome these problems, the performance of the cells was measured using various concentrations of iodide in the electrolytes. The J_{sc}, V_{oc}, and PCE decreased when the iodine concentration was increased from 0.01 to 0.07 M. The high J_{sc} and V_{oc} at a concentration of 0.01 M iodine was the reason for the high efficiency (15.23%) of the related cells (Figure 14.11a). This result was contrary to the results obtained under one-sun conditions. The J_{sc} and PCE increased with increases in the concentration of iodine in the PELs. However, high efficiency of 8.48% was obtained when the concentration of iodine was 0.05 M in the electrolytes (Figure 14.11a). These results clearly imply the importance of iodine concentration on the performance of QS-DSSCs under different light conditions. Generally, the ion diffusivity decreased and the ionic conductivity increased with an increase in the concentration of redox couples in the electrolytes. However, the high J_{sc} was

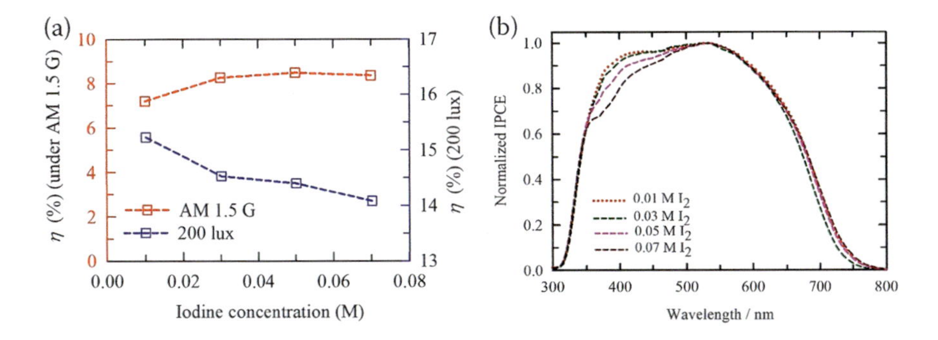

FIGURE 14.11 (a) Effect of iodine concentrations on the efficiencies of the QS-DSSCs under both one-sun and room light conditions, (b) IPCE spectra of the QS-DSSCs using various concentration of iodine.

obtained for the cells using electrolytes containing low concentrations of redox couples. This was due to the cells using low concentrations of redox couples having higher IPCE values than the cells using higher concentration of redox couples (Figure 14.11b). The poor absorption and high transmittance of light in the electrolytes containing low concentrations of redox couples was the main reason that improved the IPCE of the corresponding cells, even though conductivity of the electrolytes containing low concentrations of iodine was lower than that of the electrolyte containing high concentrations of redox couples. The chemical capacitance for the cells using 0.01M iodine was higher than that for the cells using high concentrations of iodine. This indicates that the Fermi level of the PEs downshifted, which is consistent with the high J_{sc} obtained for the related DSSC. However, the cells also had a high V_{oc}, which is contrary to what was expected from the downshift of the Fermi level. This contrast was attributed to high R_{ct} in the cells using 0.01 M iodine. Therefore, the V_{oc} of the cell did not decrease obviously due to the downshift of the Fermi level.

Apart from the iodine concentration, the performance of the DSSCs was increased by regulating the amount of chemical additives, in particular, 1, 3 dimethylimidazolium iodide (DMII) and *t*Bp in the iodide electrolytes [36]. (Table 14.2). The cells using 0.2 M DMII and 0.8 M *t*Bp achieved high PCEs of 16.06% and 16.32%, respectively. The high V_{oc} reached for the DSSCs utilizing these optimal additives concentrations contributed to the high efficiencies of the related cells. The R_{ct} generally improved in the presence of high concentration of *t*Bp. This was because a huge amount of *t*Bp adsorbed onto the surface of the PE, inhibiting the recombination reaction between the I_3^- and the electrons on the PE. The higher concentration of *t*Bp in the electrolyte increased the charge density of the TiO_2 CB, which is consistent with the high V_{oc} value obtained for the related DSSCs. The LE contains the optimal concentrations of redox couple and additives were utilized for the preparation of PEO PELs. The PCE of the cells using this PEO PEL was similar to that of the PEO/PMMA PELs (16.40%) and LEs (16.32%) [36].

To further increase the PCE, cobalt PEO PELs were prepared by mixing various amounts of PEO (3–12 wt.%) into an MPN-based cobalt LE at 120°C [36]. The 3 wt.% PEO and LE formed a PGE instead of a paste, which was more suitable for an injection process than for a printing process. The 6, 9, and 12 wt.% PEO formed moderate, optimal, and highly viscous pastes with the LEs. However, the electrolyte paste- formed using 9 wt.% PEO was the most convenient for the printing method. This is because of its adequate viscosity as compared to the 9, and 12 wt.% PEO-based electrolytes. A high PCE of 21.06% was obtained for the cells using cobalt PEO PELs. This efficiency was remarkably higher than the efficiencies of the DSSC utilizing cobalt PEO/PMMA PELs (18.67%) and iodide PEO PELs (16.40%) [36]. The cell using a 9 wt.% PEO-based PEL had high R_{ct} at the PE /electrolyte interface and a high charge transfer at the CE/electrolyte interface that contributed to the high V_{oc} and high J_{sc} of the related cell, respectively. Therefore, the cell using this PEL achieved an efficiency similar to that of its liquid counterpart. This difference is ascribed to the different compositions of Li^+ and Co^{3+} around the PE (Figure 14.12). The presence of Li^+ around the interface will repel Co^{3+} away from the interface,

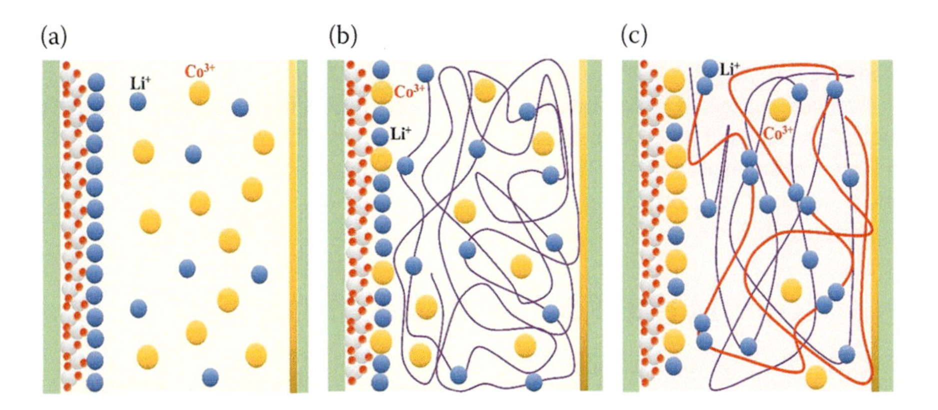

FIGURE 14.12 A proposed model of Li+ and Co^{3+} distribution for DSSCs using liquid electrolyte (a), PEO PE (b) and PEO/PMMA PE (c).

decreasing the recombination of excited electrons to Co^{3+}. According to the molecular structure, PMMA has more lone pair electrons to coordinate with Li^{+} ions, which will decrease the concentration of free Li^{+} more significantly than does by PEO (Figure 14.12c).

Therefore, the presence of PMMA will decrease and increase, respectively, the Li^{+} and Co^{3+} concentrations at the PE/electrolyte interface, resulting in more significant recombination of electrons to the -Co^{3+}. Consequently, the PCE of the PEO/PMMA cell is lower than that of the PEO cell. This effect doesn't occur in I^{-}/I_3^{-} system because the concentration variation of negatively charged ions did not affect significantly the recombination of the electrons at the interface. By using this cobalt PEL, a bifacial QS-DSSC achieved PCEs of 17.22% and 14.25%, respectively, under front-side and back-side illumination by 200 lux T5 light. A sub-module QS-DSSC using the cobalt PEL attained a PCE of 12.56%. In another study, 9 wt.% PEO/PVDF PELs with and without ZnO nanofillers were respectively coated onto the PE and CE (Figure 14.13a) [30]. This double-layered structure solved the problems of poor penetration related to the electrolyte layer with a

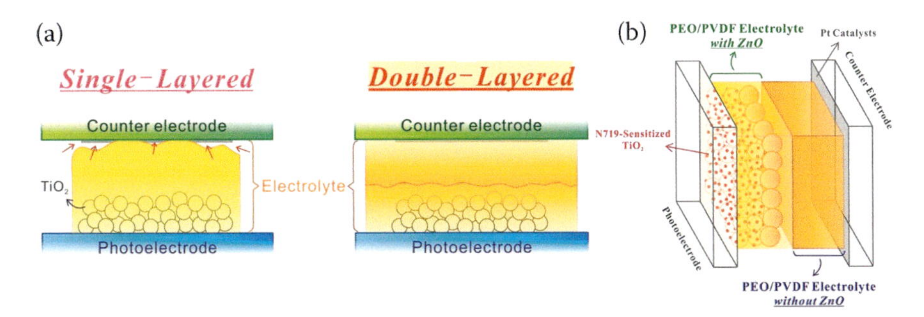

FIGURE 14.13 Scheme 1(a) Schematic illustration of single- and double-layered electrolyte architectures employed in DSSCs. (b) Schematic illustration of a DSSC featuring a "modified double-layered" electrolyte architecture.

CE (Figure 14.13a), which was observed in the cells using screen-printed or doctor-blade electrolytes.

The cells prepared using a double-layered structure (Figure 14.13b) maintained high V_{oc} value, and its J_{sc} value was relatively increased. Therefore, the related cells achieved efficiencies of 15.17% and 15.97% at room light illumination of 200 lux and 1,000 lux, respectively, and these values were superior to those of a related liquid version.

14.5 COUNTER ELECTRODE MATERIALS FOR DSSCS

14.5.1 Platinum-Based CEs

Apart from PEs and electrolytes, CEs play a predominant role in regulating the overall performance of DSSCs because they help reduce the oxidized redox couples by donating electrons to them (Table 14.3) [46]. Thus far, platinum (Pt)-based CEs have continued to be utilized for the DSSCs applications. This is due to their high catalytic activity, and high mechanical and chemical stability. The iodide cells using Pt-based electrodes achieved a high PCE of 13% under one-sun conditions [47]. In these cells, mostly thick Pt layers were utilized to obtain high catalytic activity towards the redox couples. These thick layers were not suitable for the room light DSSCs to achieve high cell efficiencies. By regulating the Pt-layer thicknesses, the photovoltaic performance of the cells under room light conditions was improved [35,36,38]. In a recent study, several CEs were prepared using ultrathin Pt layers with thicknesses of 0.55, 1.61, 5.19, 8.75, & 12.1 nm by regulating the deposition time in the sputtering methods [35]. The performance of the ACN and MPN-based cobalt cells using these CEs were studied. The J_{sc}, V_{oc}, and PCE increased when the Pt layer thickness was increased from 0.55 to 1.61 nm. The MPN-based cell with the 1.61 nm layer achieved PCEs of 22.66%, 23.48%, and 24.52%, respectively, under T5 light illumination of 200, 607.8, and 999.6 lux. These efficiencies were higher than those of the other cells. This result is different from the results obtained under one sun condition. The ACN-DSSCs with the 1.61 nm layer showed the highest efficiency of 6.83% compared with the other cells. The lower R_{Pt} value obtained for the 1.61 nm layer contributed to the PCEs of the related DSSCs (Figure 14.14a). The cells using a 1.61 nm layer were highly stable during a long-term performance test at both 35 and 50°C. The CE materials also play a crucial role in the performance of bifacial cells. Since the efficiency of the bifacial DSSCs is related to the high transparency of the CEs, the effect of the Pt layer thickness on the transmittance of the CEs and the corresponding performance of the DSSCs were generally evaluated.

For the UV-vis transmittance measurement, 0.55–12.1 nm Pt layers were prepared on the FTO glasses [35] (Figure 14.14b). The transmittance of the 0.55 nm layer was 76.02% at 580 nm. The transmittances of the 1.66 nm layer and bare FTO glass were 71.64% and 76.97%, respectively. The data revealed that the transparency of the 0.55 nm layer was close to that of the bare FTO glass and far better than that of the 1.66 nm layer. For the preparation of bifacial DSSCs with a

TABLE 14.3
Room Light DSSCs Using Various Counter Electrodes

DSSC Components	Illuminance [lux]	J_{sc} (μAcm^{-2})	V_{oc} (V)	FF	P_{in} (μW cm^{-2})	Pmax (μW cm^{-2})	η [%]
Pt (5s) (0.55 nm)[(a)]	201.8	23.19	0.765	0.726	63.84	12.89	20.19
Pt (15s) (1.61nm)[(a)]	201.8	25.78	0.785	0.714	63.84	14.47	22.66
Pt (45s) (5.19 nm)[(a)]	201.8	24.72	0.773	0.735	63.84	14.07	21.99
Pt (75s) (8.75 nm)[(a)]	201.8	24.46	0.765	0.733	63.84	13.73	21.50
Pt (105s) (12.10 nm)[(a)]	201.8	23.65	0.768	0.726	63.84	13.21	20.69
Pt (15s) (1.61 nm)[(a)]	607.8	74.06	0.827	0.729	190.26	44.69	23.48
Pt (15s) (1.61 nm)[(a)]	999.6	124.38	0.846	0.725	311.47	76.39	24.52
Pt (5s) (0.55 nm) (Front)[(a)]	201.8	19.94	0.718	0.789	63.84	11.31	17.42
Pt (5s) (0.55 nm) (Rear)[(a)]	201.8	17.48	0.735	0.782	63.84	10.05	15.48
Pt (5s) (0.55 nm)(Front)[(b)]	200	25.74	0.785	0.722	69.10	14.6	21.11
Pt (5s) (0.55 nm)(Rear)[(b)]	200	22.2	0.782	0.751	69.10	13.0	18.90
Pt (5s) (0.55 nm)(Front)[(b)]	1,000	139.39	0.831	0.730	337.57	85.13	25.04
Pt (5s) (0.55 nm)(Rear)[(b)]	1,000	127.33	0.833	0.760	337.57	80.59	23.70
PEDOT (Conventional)[(c)]	200	27.12	0.710	0.778	68.2		22.0
PEDOT (Conventional)[(c)]	1,000	137.6	0.770	0.786	329		25.3
PEDOT (Front)[(c)]	1,000	127.7	0.786	0.765	329		23.3
PEDOT (Rear)[(c)]	1,000	115.4	0.785	0.780	329		21.5
PEDOT (100 nm+Au)[(d)]	200	29.6	0.808	0.722	63.6	17.5	27.5

(Continued)

TABLE 14.3 (Continued)
Room Light DSSCs Using Various Counter Electrodes

DSSC Components	Illuminance [lux]	J_{sc} (μAcm^{-2})	V_{oc} (V)	FF	P_{in} (μW cm^{-2})	Pmax (μW cm^{-2})	η [%]
PEDOT[d]	500	74.3	0.850	0.750	159.1	47.1	30.8
PEDOT[d]	1,000	149.3	0.878	0.773	318.2	283	31.8
PEDOT[e]	1,000	131.2	0.860	0.780	303.1	88.5	29.2
PEDOT[f]	200	29	0.840	0.780	60.6	19	31.4
PEDOT[f]	1,000	147	0.910	0.77	303.1	103.1	34.0
PEDOT[g]	1,000	137	0.860	0.77	303.1		30.0

Notes
[a] 6 μm (30NR-D) + 4 μm (PST-400C) TiO_2 layer, Y123 dye, 0.11 M Co (II)/0.025 M Co (III), 1.2 M *t*BP, MPN.
[b] 4.16 μm (30NR-D)/3.63 μm (PST-400C) /4.16 μm (30NR-D) TiO_2 layer, Y123 dye, 0.11 M Co (II)/0.025 M Co (III), 1.2 M *t*BP, MPN.
[c] 6 μm (30NR-D) + 4 μm (PST-400C) TiO_2 layer, D35/XY1b (7/3) dye, PVDF-HFP (9/1), 0.11 M Co (II)/0.025 M Co (III), 1.2 M *t*BP, ACN.
[d] 4 μm (18 nm) +3 μm TiO_2 layer, XY1+ Y123, 0.04 M Cu(II)/0.20 M Cu(II), 0.1 M LiTFSi, and 0.6M MBI in ACN.
[e] 4 μm (30 NRD)/4 μm TiO_2 layer, D35/XY1B, 0.2 M Cu(I)/0.04M Cu(II), 0.1 M LiTFSI,0.6 M *t*BP in ACN or PPN.
[f] 4 μm (30 NRD)/4 μm TiO_2 layer, XY1+L1, 0.2 M Cu(I)/0.04M Cu(II), 0.1 M LiTFSI,0.6 M *t*BP in ACN or PPN.
[g] 4 μm (30 NRD)/4 μm TiO_2 layer, 0.2 M Cu(I)/0.04M Cu(II), 0.1 M LiTFSI,0.6 M *t*BP in ACN or PPN.

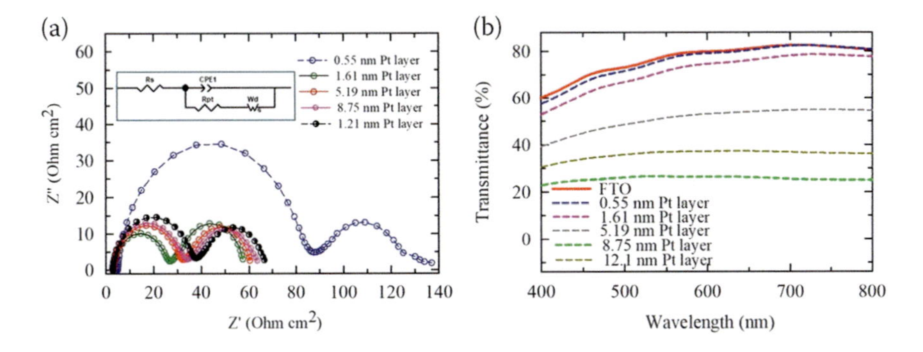

FIGURE 14.14 (a) EIS curves measured for the dummy cells using 0.55–1.21 nm Pt layer thickness, (b) Transmittance measured for the 0.55–1,21 nm Pt layers on the FTO surface.

semitransparent property, PEs without the TiO_2 SL (8μm), as well as an ultrathin Pt film (0.55 nm) were utilized. The PEs without an SL decreased the light scattering from the backside which was beneficial to the light absorption of rear light illumination. The DSSC using a 0.55 nm Pt layer had a higher J_{sc} than those using 1.61–12.1 nm Pt layers under front illumination. As a result, the bifacial cell gained a high PCE of 17.42%. This PCE was 2.18% higher than the PCE of the cells using the 12.1 nm Pt layer (15.24%). The J_{sc} increased with decreases in the Pt layer thickness, which had a marked effect on the efficiency of the DSSCs under rear illumination. The increase in the J_{sc} was attributed to the increase in the light transmittance with a decrease in the Pt layer thickness. In the presence of the 0.55 nm layer, the J_{sc} of the cell was higher than that of the cells using other Pt layer thicknesses, which led to a PCE higher than those of the other DSSCs.

14.5.2 Poly(3,4-ethylenedioxythiophene)-Based CEs

Since Pt is highly expensive, Poly(3,4-ethylenedioxythiophene) (PEDOT) is utilized as an alternative CE material for room light DSSCs. PEDOT CEs were prepared via electropolymerization of 3,4, ethylenedioxythiophene (EDOT) from a 0.01 mM aqueous solution with 0.1 M sodium dodecyl sulfate [37,39–42]. A smaller R_{ct} value related to the PEDOT CE compared to the Pt CE was measured, confirming the superior catalytic activity. Therefore, the liquid cells using the PEDOT CE have high PCEs under room light conditions. The cobalt PVDF-HFP-based cells using PEDOT CE have obtained a PCE of 25.3% under room light illumination of 1000 lux. The QS-DSSC bifacial cells using the PEDOT -CE and PE without the SL obtained the PCEs of 23.3% and 21.5% under front and rear illumination, respectively, which had an efficiency ratio of 92% [37]. The QS-DSSCs using the PEDOT CE showed high stability at room temperature and 50°C for a 2000 h test period. In another study, spacer-free DSSCs were proposed in which PEDOT CEs and dye-sensitized PEs were combined together without using any spacer in between them [40]. In this cell, the dye-sensitized TiO_2 layer and the PEDOT layer were served as the n-type inorganic semiconductor and p-type polymer semiconductor, respectively (Figure 14.15). This cell achieved a PCE of 32% under indoor light illuminance of 1000 lux. This PCE of

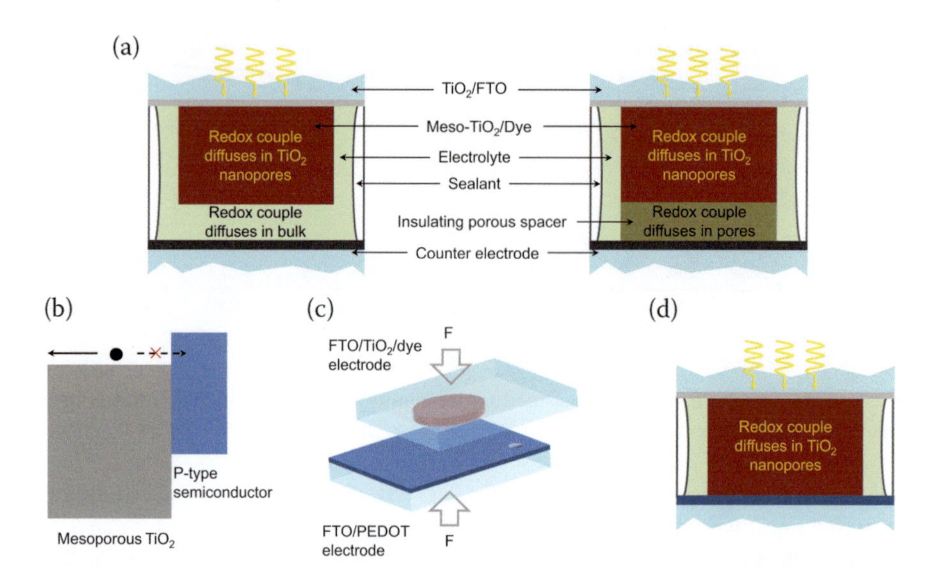

FIGURE 14.15 (a) Classical DSSC use either a thermoplastic or porous insulating spacer to avoid short circuit between the mesoporous TiO_2 and the CE. (b) Type II junction alignment of the band edges for the mesoporous TiO_2 film and a p-type semiconductor layer. The p-type semiconductor serves as an electron-blocking hole-selective charge collection layer. (c) The sensitized TiO_2 electrode and the PEDOT semi-conductor-based CE make direct contact via mechanical pressing and make a new DSSC embodiment. (d) In the DSSC with the contacted electrodes, the redox couple diffuses merely through the mesoscopic TiO_2 film.

the cell was further improved to 34.0%, 32.7%, and 31.4%, respectively under 1,000, 500, and 200 lux of fluorescent light, respectively. The spacer-free DSSCs had low diffusion resistance (Rdiff) which contributed to the high performance of the related cells. These room light cells were further utilized as the power supply cells for both IoT devices and a base station [42]. Following these studies, spacer-free so-called solid-state Zombie DSSCs [45] were fabricated by Michaels et al. In these cells, the copper electrolyte was dried out by evaporating its organic solvents through drilled holes on an FTO electrode, leaving the Cu redox couple and additives to serve as a solid-hole transport material. These cells had a high PCE of 30% under 1000 lux illumination conditions [42]. For the cells working under one-sun conditions, the use of the PEDOT CEs had the stable R_{ct} in a stability test. However, the R_{diff} of the DSSCs increased which affected the J_{sc} of the DSSCs. This was due to suspected electrolyte degradation in cells using PEDOT CE. To solve this problem, carbonaceous materials can be utilized as CEs for Co and Cu redox couple-based room light DSSCs.

14.6 SUMMARY AND CONCLUSION

Room light DSSCs are highly efficient in providing the minimum power required for small self-powering devices such as smart meters, wearable devices, and wireless sensor nodes interconnected with IoT. DSSCs are easily fabricated with simple procedures using inexpensive materials. Therefore, this technology is suitable for the preparation of high-performance indoor solar panels with simple, compact, and colorful designs for various practical applications. This chapter briefly describes the most important electrodes and electrolytes ever prepared to fabricate room light DSSCs. The optimal parameters for cells to perform under room light illuminations are different from those under sunlight. Therefore, the components of DSSCs are regulated for obtaining maximum PCEs under room light conditions. The TiO_2 PEs with the compact blocking layer prepared by the spray pyrolysis method works efficiently under room light conditions. The thicknesses of (4+4) 8 μm and (6+4) 10 μm TiO_2 layers, respectively were considered optimal for the iodide, copper, and cobalt LEs--based DSSCs. DSSCs based on the copper -LE, XY1/L1 dye and PEDOT CE have a high PCE of 34% under room light illumination of 1000 lux. This PCE is higher than the PCEs of the cells based on classical redox couples, dyes, and CEs. By using D35 and XY1b as co-sensitizers of a sandwich PE and Pt CE, the front-side and rear-side PCEs of the bifacial cells achieve 25.04 and 23.7%, respectively, under 1000-lux fluorescent lighting, which has a rear-to-front efficiency ratio of 95%. DSSCs using LE have various issues, including leakage and evaporation of the organic solvents, which affects the long-term stability of the cells. To overcome these disadvantages, LEs have been replaced with PGEs and PELs because of their promising physical and electrochemical properties. The QS-DSSCs based on cobalt PVDF-HFP PGEs (D35/XYIb) and PEO PELs (Y123 dye) have PCEs of 22.0% and 21.06%, respectively under T5 light illumination of 200 lux. These efficiencies are much higher than those of the cells using iodide polymer electrolytes and N719 dye. The cells using these QS-electrolytes shows high stability at high temperature. In the literature, copper and polymer-solid-state electrolytes have not yet been developed for room light DSSC applications. Therefore, the production of high-performance room light QS-DSSCs requires rapid research and development activities from educational institutions and industry. Through these activities room light DSSC technology can be provided to people around the world at reasonable prices.

ACKNOWLEDGMENT

The financial support by the Ministry of Science and Technology of Taiwan through grand MOST 108–2221-E-006 -158 -MY3 is acknowledged.

REFERENCES

1. B. Smit, M. W. Skinner, Adaptation options in agriculture to climate change: A typology. *Mitig. Adapt. Strateg. Glob. Chang.* 7 (2002) (1) 85–114. doi:10.1023/A:1015862228270
2. N. L. Panwar, S. C. Kaushik, S. Kothari, Role of Renewable Energy Sources in Environmental Protection: A Review. *Renew. Sustain. Energy Rev.* 15 (2011) (3) 1513–1524.
3. L. El, Chaar, L. A. Lamont, N. El Zein, Review of Photovoltaic Technologies. *Renew. Sustain. Energy Rev.* 15 (2011) (5) 2165–2175.

4. P. Asghari, A. M. Rahmani, H. H. S. Javadi, Internet of things applications: A systematic review. *Comput. Networks* 148 (2019) 241–261.
5. A. Raj, D. Steingart, Review-power sources for the internet of things. *J. Electrochem. Soc.* 165 (2018) B3130–B3136.
6. A. Blakers, N. Zin, K. R. Mclntosh, K. Fong, High efficiency silicon solar cells. *Energy Procedia* 10 (2013) 1–10.
7. K. Yoshikawa, H. Kawasaki, W. Yoshida, T. Irie, K. Konishi, K. Nakano, T. Uto, D. Adachi, M. Kanematsu, H. Uzu, K. Yamamoto, Silicon heterojunction solar cell with interdigitated back contacts for a photoconversion efficiency over 26%. *Nat. Energy* 2 (2017) 17032.
8. J. Yan, B. R. Sauders, Third generation solar cells: a review and comparison of polymer:fullerene, hybrid polymer and perovskite solar cells, *RSC Adv.* 4 (2014) 43286–43314.
9. A. Hagfeldt, M. Grätzel, Molecular photovoltaics, *Acc. Chem. Res.* 33 (2000) 269–277.
10. B. O'Regan , M. Grätzel, A low-cost, high-efficiency solar cell based on dye-sensitized colloidal TiO_2 films, *Nature.* 353 (1991) 737–740.
11. J. Y. Kim, J.-W. Lee, H.-S. Jung, H. Shin, N.-G. Park, High-efficiency pervoskite solar cells, *Chem. Rev.* 2020 (15) 7867–7918.
12. Y. Ren, D. Zhang, J. Suo, Y. Y. Cao, F. T. Eickemeyer, N. Vlachopoulos, S. M. Zakeeruddin, A. Hagfeldt, M. Gratzel, Hydroxamic acid pre-adsorption raises the efficiency of cosensitized solar cells. *Nature.* 613 (2022) 60–65.
13. R. Cheng, C.-C. Chung, H., Zhang, F. Liu, W.-T. Wang, Z. Zhou, S. Wang, A. D. Djurisic, S.-P. Feng, Tailoring triple-anion perovskite material for indoor light harvesting with restrained halide segregation and record high efficiency beyond 36%, *Adv. Energy Mater.* 9 (2019) 1901980.
14. M. Kokkonen, P. Talebi, J. Zhou, S. Asgari, S. A. Soomro, F. Elsehrawy, J. Halme, S. Ahmed, A. Hagfeldt, S. G. Hasmi, Advanced research trends in dye-sensitized solar cells, *J. Mater. Chem. A.* 9 (2021)10527.
15. C. T. Li, Y. L. Kuo, C. P. Kumar, P. T. Huang, J. T. Lin, Tetraphenylethylene tethered phenothiazine-based double-anchored sensitizers for high performance dye-sensitized solar cells, *J. Mater. Chem. A.* 7 (2019) 23225–23233.
16. C. Hora, F. Santos, M. G. F. Sales, D. Ivanou, A. Mendes, Dye-sensitized solar cells for efficient solar and artificial light conversion, *ACS Sustainable Chem. Eng.* 7 (2019) 13464–13470.
17. C. H. Huang, Y. W. Chen, C. M. Chen, Chromatic titanium photoanode for dye-sensitized solar cells under rear illumination, *ACS Appl. Mater. Interfaces.* 10 (2018) 2658–2666.
18. H. T. Chen, Y. J. Huang, C. T. Li, C. P. Lee, J. T. Lin, K. C. Ho, Boron nitride/sulfonated polythiophene composite electrocatalyst as the TCO and Pt-free counter electrode for dye-sensitized solar cells: 21% at dim light, *ACS Sustain. Chem. Eng.* 8 (2020) 5251–5259.
19. M. Chandra Sil, L. S. Chen, C. W. Lai, Y. H. Lee, C. C. Chang, C. M. Chen, Enhancement of power conversion efficiency of dye-sensitized solar cells for indoor applications by using a highly responsive organic dye and tailoring the thickness of photoactive layer, *J. Power Sources* 479 (2020) 229095.
20. V. S. Nguyễn, T. K. Chang, K. Kannankutty, J. L. Liao, Y. Chi, T. C. Wei, Novel ruthenium sensitizers designing for efficient light harvesting under both sunlight and ambient dim light, *Sol. RRL.* 4 (2020) 2000046.
21. K. S. K. Reddy, Y. C. Chen, C. C. Wu, C. W. Hsu, Y. C. Chang, C. M. Chen, C. Y. Yeh, Cosensitization of structurally simple porphyrin and anthracene-based dye for dye-sensitized solar cells, *ACS Appl. Mater. Interfaces.* 10 (2018) 2391–2399.

22. M. C. Tsai, C. L. Wang, C. W. Chang, C. W. Hsu, Y. H. Hsiao, C. L. Liu, C. C. Wang, S. Y. Lin, C. Y. Lin, A large, ultra-black, efficient and cost-effective dye-sensitized solar module approaching 12% overall efficiency under 1000 lux indoor light, *J. Mater. Chem. A.* 6 (2018) 1995–2003.
23. Y. S. Tingare, N. S. n. Vinh, H. H. Chou, Y. C. Liu, Y. S. Long, T. C. Wu, T. C. Wei, C. Y. Yeh, New acetylene-bridged 9,10-conjugated anthracene sensitizers: Application in outdoor and indoor dye-sensitized solar cells, *Adv. Energy Mater.* 7 (2017) 1700032
24. P. Zhai, H. Lee, Y. T. Huang, T. C. Wei, S. P. Feng, Study on the blocking effect of a quantum-dot TiO_2 compact layer in dye-sensitized solar cells with ionic liquid electrolyte under low-intensity illumination, *J Power Sources* 329 (2016) 502–509.
25. Y. C. Liu, H. H. Chou, F. Y. Ho, H. J. Wei, T. C. Wei, C. Y. Yeh, A feasible scalable porphyrin dye for dye-sensitized solar cells under one sun and dim light environments, *J. Mater. Chem. A.* 4 (2016) 11878–11887.
26. C. L. Wang, P. T. Lin, Y. F. Wang, C. W. Chang, B. Z. Lin, H. H. Kuo, C. W. Hsu, S. H. Tu,C. Y. Lin, Cost-effective anthryl dyes for dye-sensitized cells under one sun and dim light, *J. Phys. Chem. C.* 119 (2015) 24282–24289.
27. F. De Rossi, T. Pontecorvo, T. M. Brown, Characterization of photovoltaic devices for indoor light harvesting and customization of flexible dye solar cells to deliver superior efficiency under artificial lighting, *Appl. Energy.* 156 (2015) 413–422.
28. M. B. Desta, N. S. Vinh, C. Pavan Kumar, S. Chaurasia, W. T. Wu, J. T. Lin, T. C. Wei, E. Wei-Guang Diau, Pyrazine-incorporating panchromatic sensitizers for dye sensitized solar cells under one sun and dim light, *J. Mater. Chem. A.* 6 (2018) 13778–13789.
29. H. H. Chou, Y. C. Liu, G. Fang, Q. K. Cao, T. C. Wei, C. Y. Yeh, Structurally simple and easily accessible perylenes for dye-sensitized solar cells applicable to both 1 sun and dim-light environments, *ACS Appl. Mater. Interfaces.* 9 (2017) 37786–37796.
30. I. P. Liu, Y. Y. Chen, Y. S. Cho, L. W. Wang, C. Y. Chien, Y. L. Lee, Double-layered printable electrolytes for highly efficient dye-sensitized solar cells, *J. Power Sources* 482 (2021) 228962.
31. S. Venkatesan, I. P. Liu, C. M. W.-N. Hung, H.- Teng, Y.-L. Lee, Highly efficient quasi-solid-state dye-sensitized solar cells prepared by printable electrolytes for room light applications, *Chem. Eng. J.* 367 (2019) 17–24.
32. S. M. Feldt, E. A. Gibson, E. Gabrielsson, L. Sun, G. Boschloo, A. Hagfeldt, Design of organic dyes and cobalt polypyridine redox mediators for high-efficiency dye-sensitized solar cells, *J. Am. Chem. Soc.* 132 (2010) 16714–16724.
33. I. P. Liu, W. H. Lin, C. M. Tseng-Shan, Y. L. Lee, Importance of compact blocking layers to the performance of dye-sensitized solar cells under ambient light conditions, *ACS Appl. Mater. Interfaces* 10 (2018) 38900–38905.
34. S. Venkatesan, I. P. Liu, C. W. Li, C. M. Tseng-Shan, Y. L. Lee, Quasi-solid-state dye-sensitized solar cells for efficient and stable power generation under room light conditions, *ACS Sustain. Chem. Eng.* 7 (2019) 7403–7411.
35. S. Venkatesan, W.-H. Lin, H. Teng, Y.-L. Lee, High-efficiency bifacial dye-sensitized solar cells for application under room light conditions, *ACS Appl. Mater. Interfaces* 11 (2019) 42780–42789.
36. S. Venkatesan, I. P. Liu, C.-M. Tseng Shan, H.- Teng, Highly efficient indoor quasi-solid-state dye-sensitized solar cells using cobalt polyethylene oxide-based printable electrolytes, *Chem. Eng. J.* 394 (2020) 124954.
37. I. P. Liu, Y. S. Cho, H. Teng, Y. L. Lee, Quasi-solid-state dye-sensitized indoor photovoltaics with efficiencies exceeding 25%, *J. Mater. Chem. A.* 8 (2020), 22423.

38. S. Venkatesan, Y.-S. Cho, I.-Ping Liu, H.-Teng, Y.-L. Lee, Novel architecture of indoor bifacial dye-sensitized solar cells with efficiencies surpassing 25% and efficiency ratios exceeding 95%, *Adv. Opt. Mater.* (2021).
39. M. Freitag, J. Teuscher, Y. Saygili, X. Zhang, F. Giordano, P. Liska, J. Hua, S. M. Zakeeruddin, J. E. Moser, M. Grätzel, A. Hagfeldt, Dye-sensitized solar cells for efficient power generation under ambient lighting, *Nat. Photonics* 2011 (2017) 372–378.
40. Y. Cao, S. M. Zakeeruddin, A. Hagfeldt, M. Grätzel, Direct Contact of selective charge extraction layers enables high-efficiency molecular photovoltaics, *Joule* 2 (2018) 1108–1117.
41. E. Tanaka, H. Michaels, M. Freitag, N. Robertson, Synergy of co-sensitizers in a copper bipyridyl redox system for efficient and cost-effective dye-sensitized solar cells in solar and ambient light, *J. Mater. Chem. A* 8 (2020) 1279–1287.
42. H. Michaels, M. Rinderle, R. Freitag, I. Benesperi, T. Edvinsson, R. Socher, A. Gagliardi, M. Freitag, Dye-sensitized solar cells under ambient light powering machine learning: Towards autonomous smart sensors for the internet of things, *Chem. Sci.* 11 (2020) 2895–2906.
43. J. Wu, Z. Lan, J. Lin, M. Huang, Y. Huang, L. Fan, G. Luo, Electrolytes in dye-sensitized solar cells, *Chem. Rev.* 115 (2015) 2136–2173.
44. S. Venkatesan, Y. L. Lee, Nanofillers in the electrolytes of dye-sensitized solar cells – A short review, *Coord. Chem. Rev.* 353 (2017) 58–112.
45. Y. -Cao, Y. Saygili, A. Ummadisingu, J. Teuscher, J. Luo, N. Pellet, F. Giordano, S. M. Zakeeruddin, J. -E. Moser, M.- Freitag, A. Hangfeldt, M. Gratzel, 11% efficiency solid-state dye-sensitized solar cells with copper (II/I) hole transport materials, *Nat Commun.* (2017) (8) 15390.
46. J. Wu, Z. Lan, J. Lin, M. Huang, Y. Huang, L. Fan, G. Luo, Y. Lin, Y. Xie, Y. Wei, Counter electrodes in dye-sensitized solar cells, *Chem. Soc. Rev.* 46 (2017) 5975–6023.
47. S. Mathew, A. Yella, P. Gao, R. Humphry-Baker, B. F. E. Curchod, N. Ashari-Astani, I. Tavernelli, U. Rothlisberger, M. K. Nazeeruddin, M. Grätzel, Dye-sensitized solar cells with 13% efficiency achieved through the molecular engineering of porphyrin sensitizers, *Nat. Chem.* 6 (2014) 242–247.

15 Progress and Prospects of Intermediate-Temperature Solid Oxide Fuel Cells

Shu-Yi Tsai and Kuan-Zong Fung
Hierarchical Green-Energy Materials (Hi-GEM) Research Center, National Cheng Kung University, Tainan City, Taiwan

CONTENTS

15.1 INTRODUCTION

A fuel cell operates like a battery but does not need to be recharged, and continuously produces power when supplied with fuel and oxidant. Fuel cells' efficiencies are not limited by the Carnot cycle of a heat engine, and the magnitudes of pollutant output from fuel cells are lower than from conventional technologies. Recently, fuel cells have high efficiencies, low noise and pollutant output, modular construction to suit load, and excellent load-following capability, promising to improve the power generation industry with a shift from central power stations and long transmission lines to disperse power generation at user sites. Fuel cells can be classified into several different types according to the type of electrolyte used: (1) proton exchange membrane

DOI: 10.1201/9781003367215-15

fuel cell (PEMFC) [1], (2) phosphoric acid fuel cell (PAFC) [2], (3) alkaline fuel cell (AFC), (4) molten carbonate fuel cell (MCFC) [3,4] and (5) solid oxide fuel cell (SOFC). Table 15.1 lists the common fuel cell types and their respective operating temperatures, efficiencies, and electrolytes.

Among several types of fuel cells, solid oxide fuel cell has shown its advantages such as high conversion efficiency, and no need for noble metal catalysts drastically reduced overall costs. The operation at high temperatures allows SOFCs to oxidize several types of fuel, not relying solely on pure H_2. The SOFC can utilize fuel sources such as natural gas (methane, carbon monoxide), liquid hydrocarbon fuels (biodiesel, ethanol, etc.), and also sour fuels containing sulfides. Despite these advantages, SOFCs have failed to find commercialize, mainly due to high materials costs for the high-temperature operation. In order to increase their commercial viability, an intense area of research focus on optimizing the materials to reduce the overall cost of the SOFC device.

A typical single SOFC is composed of a dense oxygen-ion conducting solid electrolyte separating two porous electrode catalysts – a cathode on the oxidant (oxygen) side and an anode on the fuel side as shown in Figure 15.1. At the cathode, also called the oxygen electrode, the oxygen molecule is reduced and oxide ions, O^{2-}, are produced. Oxygen molecules are fed to the cathode side and combined with electrons and reduced into ions and then injected into the electrolyte which conducts the oxygen ions, but blocks electrons. These oxide ions are transported through the solid electrolyte to the anode, thus necessitating the need for electrolyte materials to have excellent ionic conductivity. The oxygen ions emerge on the anode side where they react with H_2 to form H_2O and electrons. Electrons produced during the reaction are released to the external circuit. Thus, the circuit is fulfilled and the electrical power can be used in the external circuit. The reactions at both electrodes (*an anode and a cathode*), which can be described as follows:

$$\text{Anode: } \frac{1}{2}O_2 + 2e^- \rightarrow O^{2-} \cdots \cdots \cdots \tag{15.1}$$

$$\text{Cathode: } H_2 + O^{2-} \rightarrow H_2O + 2e^- \cdots \cdots \cdots \tag{15.2}$$

$$\text{Total reaction: } H_2 + \frac{1}{2}O_2 \rightarrow H_2O \cdots \cdots \cdots \tag{15.3}$$

SOFCs can work in both fuel cell mode and electrolysis mode. Excess energy can be stored in chemicals (e.g., H_2, CO, CH_4) through Solid Oxide Electrolyzer Cell(SOEC) [5], and when needed, can be converted to electricity again. Such grid-scale storage systems will be needed to match fluctuations in supply and demand when renewables grow to comprise a large proportion of total electrical grid generation capacity.

In recent years, reducing the operating temperature of SOFC to the intermediate temperature range of 550°C to 800°C is a very important direction in many research topics [6–9]. However, due to the high operating temperature of SOFC, the overall material selection is limited and the startup is slow. Working at lower temperatures

TABLE 15.1
The Types of Fuel Cell Devices and Their Respective Parameters

	Anode	Electrolyte	Cathode	Operating Temperature (°C)	Efficiency (%)	Application
Alkaline fuel cell (AFC)	Pt/Pd	Liquid -KOH	Pt/Au	60–90	50–60	Space and transport
Polymer electrolyte fuel cell (PEMFC)	Polytetrafluoroethylene-Pt-C	Polymer	Polytetrafluoroethylene-Pt-C	50–80	50–60	Space, transport and portable power station
Phosphoric acid fuel Cell (PAFC)	Polytetrafluoroethylene-Pt-C	H_3PO_4	Polytetrafluoroethylene-Pt-C	160–220	55	Distributed generation
Molten carbonate fuel cell (MCFC)	Li-NiO	Li_2CO_3	Ni	620–600	60–65	Electric utility and distributed generation
Solid oxide fuel cell (SOFC)	Ni-YSZ	YSZ ($Y2O_3$-ZrO_2) SDC ($Sm_xCe_{1-x}O_2$) GDC ($Ga_xCe_{1-x}O_2$)	LSM ($LaSrMnO_3$)	600–1,000	55–65	Auxiliary power, electric utility and distributed generation

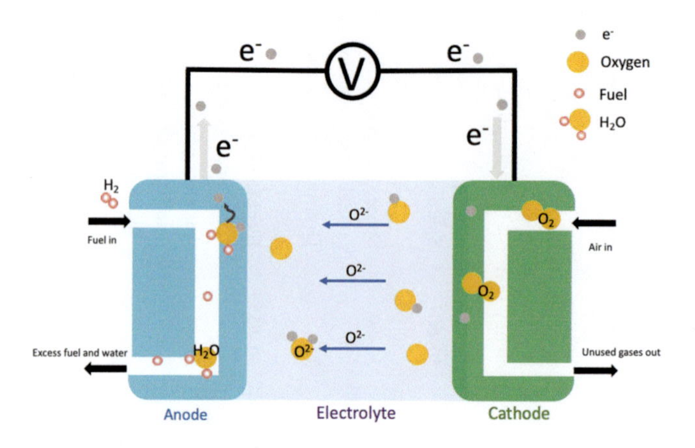

FIGURE 15.1 Principle of operation of a SOFC.

broadens the choices of cost-efficient metallic-based interconnect materials and thus will help market penetration. Another important advantage of lower temperature operation offers more rapid start-up and shut-down procedures and significantly reduced corrosion rates. But a big problem arises, when the operating temperature is high, the polarization loss is negligible, but once the temperature is lowered to the middle range, the polarization loss becomes significant, thereby reducing the kinetics related to ORR and the energy at the cathode of the battery. Charge transport negatively affects cell efficiency. A few review articles focusing on intermediate temperature SOFC(IT-SOFC) have been found to develop a technology that still maintains high power density at low temperatures.

15.2 ANODE MATERIAL

Fuel flexibility is a major advantage of solid oxide fuel cells. The most common fuel is hydrogen, other fuel sources come from any hydrocarbon that can decompose hydrogens, such as natural gas, alcohol, and methane. The anode materials should have high electrical conductivity, and high catalytic efficiency. Due to the fuel gas directly contacting the anode materials that should be also stable in the reduction environment of the fuel cell. The fuel gas should through the anode side easily so the anode must have a porous structure. However, when fueled with hydrocarbons, SOFCs encountered great difficulty in both performance and stability, which should be attributed to the sluggish hydrocarbon oxidizing reactions, the severe carbon deposition reactions, and the possible sulfur poisoning reactions in the anode. As well, the anode must satisfy the same conditions, i.e., chemically compatible with the electrolyte, thermally stable, sufficiently high ionic and electronic conductivities at the operating temperature, and must have a porous structure.

15.2.1 Ni–YSZ Cermet Anode Materials

The YSZ provides a thermal expansion match with the YSZ electrolyte and also ionic conductivity to extend the reaction zone in the anode, in addition

FIGURE 15.2 SEM images of Ni/YSZ after exposure to methane for 4 h at 800°C.

to functioning as a structural support for the anode that prevents Ni sintering [10]. Ni provides reactivity for fuel oxidation and electronic conductivity. However, the Ni metal phase in conventional Ni–YSZ anodes causes several issues for SOFCs under practical conditions, such as carbon deposition in hydrocarbon fuels and poisoning by trace amounts of impurities from the fuel such as sulfur-containing species as shown in Figure 15.2. Therefore, it is still facing a great challenge in the development of new anode materials in SOFCs.

15.2.2 Ni–Ceria Cermet Anode Materials

CeO_2 shares a similar fluorite-type structure with YSZ and also has good ionic conductivity [11]. Under reducing atmospheres, it shows mixed ionic and electronic conductivity (MIEC), about 1 S/cm at 900°C at PO_2 of 10^{-18} atm. The ionic conductivity of ceria is further improved by doping with heterovalent metal ions, such as rare earth metals Gd and Sm. Hibino et al. [12,13] studied a thin ceria-based electrolyte film SOFC with a Ru/Ni–GDC ($Ce_{0.9}Gd_{0.1}O_{1.95}$) anode and directly operated on organic chemical compounds consisting of hydrogen and carbon. The results revealed the Ru catalyst promoted the reforming reactions in the anode reaction. The maximum power density of 0.769 Wcm^{-2} in wet (2.9 vol.% H_2O) hydrogen atmosphere.

15.2.3 Perovskite Structure

Perovskites offer unique advantages over traditional metal-cermet type anodes and, in the past decade, have been an area of extensive research in the SOFC field. The properties of perovskite-based materials with a general formula, ABO_3, may be improved based on the theory of defect chemistry. The combination of an inactive A-site ion (such as La^{3+} or Sr^{2+}) with an active B-site ion (such as Co^{3+}, Fe^{3+}, Mn^{3+}, Ti^{4+}) in the perovskite structure provides a priori choice of the active site for charge compensation by the addition of dopants. In 2003, Tao et al. [14] reported an oxygen-deficient perovskite $La_{0.75}Sr_{0.25}Cr_{0.5}Mn_{0.5}O_3$, with comparable electrochemical performance to that of Ni/YSZ cermet and with good catalytic activity for the electro-oxidation of CH_4 at high temperatures. Unfortunately, it has low electronic conductivity in the reducing anodic atmosphere and is not stable to sulfur impurities in the fuel [15].

Liu et al [16,17], reported that a strontium-doped lanthanum vanadate with a nominal composition of $La_{1-x}Sr_xVO_3$ may be used as an anode material for SOFC operating on fuels containing high concentration of H_2S. Strontium doped lanthanum vanadate, with a formula of $La_{1-x}Sr_xVO_3$, for instance, shows interesting conducting behavior from one end member of semiconducting orthorhombic $LaVO_3$ to another end member of $SrVO_3$ metallic conductor. The conductivity reached 193 S/cm^{-1} at 800°C

Goodenough et al. [18] reported some promising results of double perovskite structure $Sr_2Mg_{1-x}Mn_xMoO_{6-\delta}$ as anode for SOFC. The major attraction of such oxides is the possibility of an enhanced reaction zone over the three-phase boundary due to the high electronic conductivity as compared to conventional YSZ. The mixed valence Mo(VI)/Mo(V) sub-array provides electronic conductivity with a large enough work function to accept electrons from a hydrocarbon.

Cenk Gumeci et al. [19] studied the oxygen reduction reaction (ORR) kinetics of $PrBa_{0.5}Sr_{0.5}Co_{1.5}Fe_{0.5}O_{5+}\delta$ (PBSCF). The resulting peak power density reached 2539 mW/cm^2 with a very low polarization resistance of ~0.025 Ω-cm, which was comparable to LSCF cathode performance of 1974 mW/cm^2 at 750°C. It was also demonstrated that a three-dimensional nanofiber network has large interfacial contact areas, leading to an improved oxygen reduction reaction.

15.2.4 Pyrochlores

Pyrochlore-type oxides, $A_2B_2O_7$, belongs to the family of cubic structure with Fd 3m space group and lattice parameter of a≈10.29Å. It has been studied in order to clarify the relationship between electrical conductivity and cation and/or oxide-ion disordering. The structure can be transformed into other structures such as fluorite-type or rare-earth C-type structures depending on the cation radius ratio r(A^{3+})/r (B^{4+}). Bismuth ruthenate is more of interest used in solid oxide fuel cells because of its better stability and less toxicity than Lead(II) ruthenate at a temperature range of 500–700°C. Low-temperature bismuth ruthenate $Bi_2Ru_2O_{7.3}$ with cubic structure transforms irreversibly to high-temperature pyrochlore $Bi_2Ru_2O_7$ beyond 950°C [20]. Takeshi Hagiwara et al [21], investigated three different pyrochlore systems,

$Ln_2Zr_2O_7$ (Ln = La, Nd, Eu), that increase in the 8b site occupancy means the increase of incompleteness of the pyrochlore-type structure. It is indicated that the pyrochlore solid solution $La_2Zr_2O_7$ may be a possible candidate as an anode material in a SOFC system.

15.3 SOLID ELECTROLYTES MATERIAL

The SOFC electrolyte plays a key role in conducting ionic species between the electrodes, completing the electrical circuit. The electrolyte is a dense ceramic and must have sufficiently high ionic and an electronic insulator to avoid a short circuit across the cell. Additionally, the material must be stable in both reduction and oxidation atmospheres at the operating temperature(500–1,000°C). The ohmic losses in the electrolyte increase dramatically as the operating temperature at less than 700°C. There are voltage losses while oxygen ions transport from the cathode to the electrolyte by ionic resistivity and transition of electrons through the cathode to the anode by electronic resistivities. The electrolyte causes more ohmic losses, especially IT-SOFCs with thick electrolytes. The ionic resistivity of the electrolyte is greater than the electronic resistivity of the cathode and the anode. The solution to this problem is reducing the thickness of electrolytes in IT-SOFC and choosing electrolyte material that has high ionic conductivity.

15.3.1 Stabilized Zirconium Oxide

Most conventional fast oxide-ion conducting materials have crystal structures of fluorite type, AO_2, where A is a tetravalent cation. For example, acceptor-doped ZrO_2 is a well-known fluorite-type oxygen ion conductor. Pure zirconium oxide doesn't exhibit good ion conductivity, and only reveals the fluorite structure above 2300°C. Acceptor dopants or stabilizers are introduced into the crystal lattice of zirconium oxide to stabilize the fluorite structure at room temperatures and to increase oxygen vacancies concentration. With proper addition of larger cation of lower valence, such as Y^{+3} and Ca^{+2}, not only R_{cation}/R_{anion} is greater than 0.73, the positive oxygen vacancies are also created. Thus, 8 mol% Y_2O_3 stabilized ZrO_2 (YSZ), is the most widely used electrolyte in SOFCs. It exhibits adequate oxygen-ion conductivity (~0.02 S/cm at 800°C) as well as required stability in both oxidizing and reducing atmospheres. However, YSZ is limited to high-temperature operation due to its poor ionic conductivity at temperatures below 800°C. A target temperature for SOFC operation is 500°C, reducing the requirement for high-temperature materials and lowering the cost.

15.3.2 Bismuth Oxide

There are four polymorphs of Bi_2O_3 that have been reported in the literature, monoclinic α-phase stable above 730°C, tetragonal β-phase, rhombohedral γ-phase, and cubic δ-phase. Two stable polymorphs are monoclinic α-phase and cubic δ-phase, while tetragonal β-phase and rhombohedral γ-phase included in two metastable polymorphs. For it to crystallize in the CaF_2-type(fluorite) structure

(δ-Bi_2O_3), it is imperative that 25% of the oxygen sites are vacant. The presence of a large concentration of anion vacancy has contributed to the high ionic conductivity of δ-Bi_2O_3.

Besides the presence of high oxygen vacancy in δ- Bi_2O_3, the weak Bi-O bond and high polarizability of Bi^{3+} with its lone pair of 6s electrons have contributed to the high anion mobility that will affect the increase of ionic conductivity. The conductivity of δ- Bi_2O_3 is greater than stabilized zirconia up to two orders of magnitude.

Hence, obtaining pure cubic bismuth oxide is limited to the narrow temperature range (730°C–824°C). Cubic bismuth oxide melts above 824°C, while pure cubic bismuth oxide will transform to the monoclinic phase when cooling below 730°C. The cubic to the monoclinic phase transformation not only reduces in significant volume change but also drastically decreases the conductivity due to the more compact atomic arrangement. Thus, the phase transformation of bismuth oxide needs to be avoided to prevent the volume and conductivity changes. Therefore, many studies have focused on the stabilization of cubic Bi_2O_3 by doping with other elements to obtain high oxygen ion conductivity.

Most solid solutions of Bi_2O_3–M_2O_3 (where M denotes rare-earth oxide) are effective oxygen ion conductors based on their either cubic or rhombohedral structure. Yb_2O_3, Er_2O_3, Y_2O_3, Dy_2O_3, Gd_2O_3, and have all been extensively investigated. In a single-doped bismuth oxide system, if the dopant concentration decreased to enhance the conductivity, the lattice constant of the system will be increased. Then it will contribute to the phase instability of the single-doped Bi_2O_3 system. To overcome this problem, using a multi-doped system could be a promising solution. Fung and his coworkers then developed double-doped bismuth oxide to achieve stabilized bismuth oxide system with higher conductivity but also have proper phase stabilization by decreasing the dopant concentration.

15.3.3 Doped Cerium Oxide

Another potential electrolyte material with a fluorite-type structure is cerium oxide. Unlike pure ZrO_2, pure CeO_2 shows a stable fluorite structure for a wide temperature range. In undoped CeO_2, very limited intrinsic oxygen defects suppress its oxygen ion conductivity in the range of 10^{-3}~10^{-4} S/cm at 800°C. The ionic conductivity of ceria is further improved by doping with heterovalent metal ions, such as rare earth metals Gd and Sm. Although doped ceria shows much higher conductivities than YSZ, they tend to exhibit partially electronic conduction at low oxygen partial pressure or at temperatures greater than 650°C. It is believed that the n-type semi-conducting behavior of ceria is due to some cerium (IV) being reduced to cerium (III). Additionally, there is also a mechanical issue due to lattice volume expansion that occurs due to the reduction in charge density.

15.4 CATHODE MATERIAL

The high operating temperature of solid oxide fuel cells from 800°C to 1000°C which promotes physical and chemical corrosion for material compatibility. A reduction in SOFC operating temperatures allows for the use of inexpensive stainless steel to operate and have greater durability. However, decreasing the operating temperature below 800°C causes lower cell performance due to less active electrodes and poor conductive electrolytes. The key issue to improve the electrochemical performance is to reduce cathodic polarization resistance and to minimize ohmic resistance resulting from electrolytes. The polarization losses at the cathode of SOFCs are much higher than at the anode because a greater energy barrier is required for the reduction of oxygen. Thus, the cathode materials should have highly the site for the oxygen reduction reaction (ORR). The improvement in the cathode kinetics can directly increase the operating efficiency of SOFCs. Furthermore, the electrons are necessary for the reduction of oxygen molecules, and superior electronic conductivity of the cathode material is the primary requirement. The porous microstructure of the cathode material further enhances the ORR as it ensures the access of oxygen molecules to the cathode surface, and thus extends the active ORR region beyond the narrow triple phase boundary.

15.4.1 Perovskite Structure

The most common Perovskite type of the SOFC cathode was single phase electronic conductor such as noble metal or electronically conducting ceramic material, $La_{1-x}Sr_xMnO_{3\pm\delta}$ (LSM; For the classic composition x = 0.2 which has high chemical stability and sufficient conductivity, δ ranges from +0.01 to +0.06 at 1000°C and 700°C). However, the insufficient ion conductivity of LSM restricted its oxygen reduction reaction (ORR) to the cathode/electrolyte interface which consists of an electronic conductor, ionic conductor, and gas phase (Triple phase boundary, TPB). To increase the ORR rate, a composite cathode material was considered an ideal ionic conductor because of its excellent catalytic ability and high O^{2-} ion conductivity. Ostergard et al. [22,23] by using the composites of LSM+ YSZ reduced area specific resistance (ASR) from 2.7 Ωcm^2 obtained by using pure LSM to 0.5 Ωcm^2 for an LSM+YSZ composite operating at 1000°C. However, because of long-term thermal and mechanical degradation problems with LSM cathode materials and its low inherent oxide ion conductivity, a search for better materials for cathode materials. There has thus recently been significant research activity in developing MIEC materials for SOFC cathodes based on perovskites, double perovskites, Ruddlesden-Popper phases, and other layered oxide materials. Several researchers had reported a method to significantly decrease the cathode polarization resistance (R_P) by introducing composite cathode, such as Fung et al. by using the yttria-stabilized bismuth oxides (YSB) mixed with Sr-doped $LaMnO_3$ (LSM) which exhibit high ionic conductivity and oxygen electrocatalysis into SOFCs cathode. The major finding of this work was that the composite cathode is able to extend a triple-phase boundary and provide numerous reaction sites for electro-chemical reactions to occur as shown in Figure 15.3.

A new class of materials, derived from the perovskite structure are the oxide materials known as double perovskites with the general formula $AA'B_2O_{5+\delta}$ where A is a rare earth cation, A' is an alkaline earth metal cation and B is a transition metal cation. This leads to a doubling of the c in an oxygen-deficient

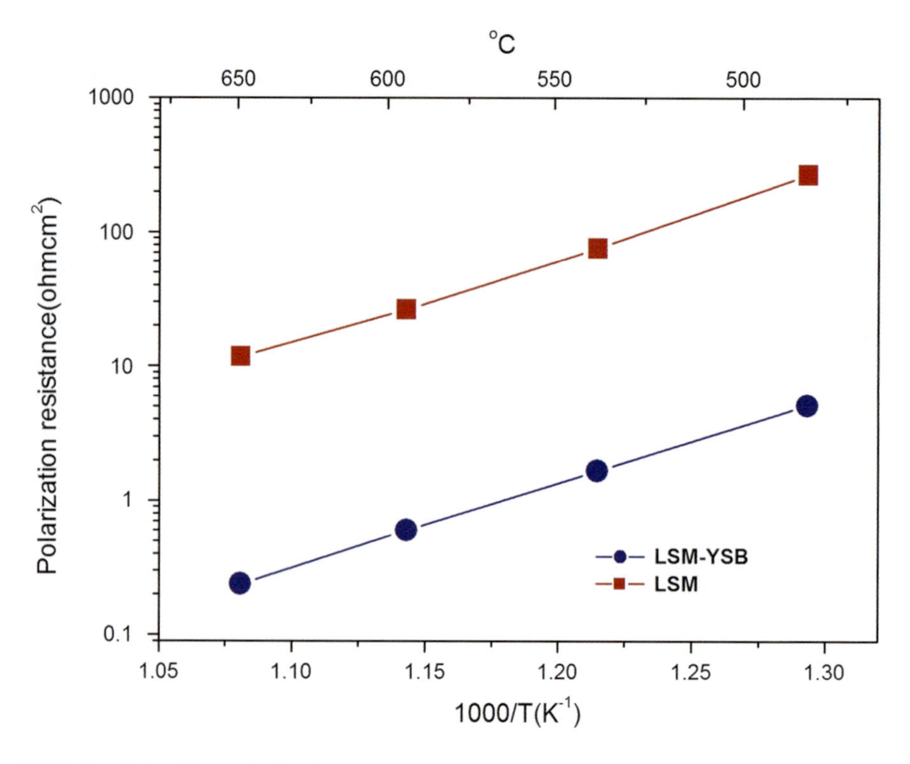

FIGURE 15.3 The polarization resistance of LSM-YSB composite cathode on YSB electrolyte.

system with oxygen vacancies mainly located in the rare-earth layer. Double perovskites can accommodate large amounts of oxygen non-stoichiometry and have been studied extensively for their magnetic properties. Recent research focuses on lanthanide (Ln)-containing oxide materials doped with alkaline elements (Ba, Sr, Ca, etc.) and transition metals (Cr, Mn, Fe, etc.). Rongzheng Ren et al [24], by using the $PrBaFe_2O_{5+\delta}$ (PBF) modified by Zn^{2+} doping. Owing to the influence of charge compensation, the low-valence Zn2+ ions can stimulate the formation of both holes and oxygen vacancies, thus regulating the mixed oxygen-ionic and electronic properties. Unfortunately, the electrochemical activity of Fe-based layered perovskites is inferior to that of Co-based analogs. In addition, these materials still exhibit slow oxygen transportation kinetics at an operator temperature of 500–650°C.

15.4.2 Mixed Ionic-Electronic Conductor (MIEC)

The particularity of MIEC is that it conducts both ion and electron carrier materials and mixed conductivity in electrodes enhances the performance of SOFCs. The mixed conductivity of MIEC cathode materials allows the system to extend the ORR region beyond the narrow TPB zone and thus increase the efficiency of the cell. The main advantage of MIECs lies in their ORR activity which is not restricted to the TPB. $La_{1-x}Sr_xCo_{1-y}Fe_yO_{3-\delta}$ (LSCF) have been identified as possible cathode materials to replace the conventional LSM cathode [25]. The ionic conductivity of 10^{-2} S cm^{-1} at 800°C has been reported and the material also has TEC in the range of 15×10^{-6} K^{-1}, which is in the compatible range with other cell components [26]. However, because of long-term thermal and mechanical degradation problems with LSCF cathode materials and its l degradation issues and strontium segregation, a search for better materials for IT- SOFCs continues.

15.5 SUMMARY

Climate changes observed during the last decades have originated a concern in the search for environmentally friendly resources for the generation of energy. Fuel cells are one of the indispensable empowered technologies for next-generation hydrogen energy production. Among several types of fuel cells, solid oxide fuel cell has shown advantages such as high conversion efficiency, no need for noble metal catalysts, use of hydrocarbon fuel, and no liquid. Although SOFC technology has many advantages and prospects, its commercialization is still not expected. A key factor for the successful commercialization of SOFC is that they must be economically competitive. The current trends development of electrolytes, cathode, and anode materials are reviewed in detail. Nowadays some new materials and new characterization techniques seem to be developed to overcome known shortcomings, and improve cell performance and economic viability. As a final note, due to growing awareness of global warming and air pollution and the environmental impact of the Paris Agreement, governments and companies around the world are actively turning to other possible solutions. Thus, SOFCs are experiencing tremendous growth recently as they are the best solution to provide energy efficiency and low greenhouse effect than conventional power plants.

REFERENCES

1. G. Das, J.H. Choi, P.K.T. Nguyen, D.J. Kim, Y.S. Yoon, Anion exchange membranes for fuel cell application: A review, *Polymers-Basel*, 14 (2022) 1197.
2. S. Wilailak, J.H. Yang, C.G. Heo, K.S. Kim, S.K. Bang, I.H. Seo, U. Zahid, C.J. Lee, Thermo-economic analysis of phosphoric acid fuel-cell (PAFC) integrated with Organic Ranking Cycle (ORC), *Energy*, 220 (2021) 119744.
3. R.R. Contreras, J. Almarza, L. Rincon, Molten carbonate fuel cells: A technological perspective and review, *Energ Source Part A*, (2021).
4. L. Chen, C.Y. Yuh, Hardware materials in molten carbonate fuel cell: A review, *Acta Metall Sin-Engl*, 30 (2017) 289–295.
5. M.K. Mahapatra, K. Lu, Glass-based seals for solid oxide fuel and electrolyzer cells – A review, *Mat Sci Eng R*, 67 (2010) 65–85.
6. S.C. Shirbhate, K. Singh, S.A. Acharya, A.K. Yadav, Review on local structural properties of ceria-based electrolytes for IT-SOFC, *Ionics*, 23 (2017) 1049–1057.
7. A. Arabaci, Ceria-based solid electrolytes for IT-SOFC applications, *Acta Phys Pol A*, 137 (2020) 530–534.
8. S.H. Woo, K.E. Song, S.W. Baek, H. Kang, W. Choi, T.H. Shin J.Y. Park, J.H. Kim, Pr- and Sm-Substituted layered perovskite oxide systems for IT-SOFC cathodes, *Energies*, 14 (2021) 6739.

9. D. Garces, A.L. Soldati, H. Troiani, A. Montenegro-Hernandez, A. Caneiro, L.V. Mogni, La/Ba-based cobaltites as IT-SOFC cathodes: A discussion about the effect of crystal structure and microstructure on the O-2-reduction reaction, *Electrochim Acta*, 215 (2016) 637–646.
10. A. Atkinson, S. Barnett, R.J. Gorte, J.T.S. Irvine, A.J. Mcevoy, M. Mogensen, S.C. Singhal, J. Vohs, Advanced anodes for high-temperature fuel cells, *Nat Mater*, 3 (2004) 17–27.
11. D.G. Lamas, M. Bellora, C.H. Iriart, L. Toscani, R. Bacani, S. Larrondo, M. Fantini, In-situ DXAS study of NiO/CeO_2-Sm_2O_3 nanocomposites for IT-SOFC anodes, *Acta Crystallogr A-Found Adv*, 73 (2017) C284–C284.
12. A. Hashimoto, D. Hirabayashi, T. Hibino, M. Sano, Intermediate-temperature SOFCs with ru-catalyzed Ni-cermet anodes, *Electroceram Jpn VII*, 269 (2004) 151–154.
13. T. Hibino, A. Hashimoto, M. Yano, M. Suzuki, M. Sano, Ru-catalyzed anode materials for direct hydrocarbon SOFCs, *Electrochim Acta*, 48 (2003) 2531–2537.
14. S. Tao, J.T. Irvine, A redox-stable efficient anode for solid-oxide fuel cells, *Nat Mater*, 2 (2003) 320–323.
15. S. Zha, P. Tsang, Z. Cheng, M. Liu, Electrical properties and sulfur tolerance of $La_{0.75}Sr_{0.25}Cr_{1-x}Mn_xO_3$ under anodic conditions, *J Solid State Chem*, 178 (2005) 1844–1850.
16. C.Y. Liu, S.Y. Tsai, C.T. Ni, K.Z. Fung, C.Y. Cho, Enhancement on densification and crystallization of conducting $La_{0.7}Sr_{0.3}VO_3$ perovskite anode derived from hydrothermal process, *Jpn J Appl Phys*, 58 (2019) SDDG03.
17. C.Y. Liu, S.Y. Tsai, C.T. Ni, K.Z. Fung, Interfacial reaction between YSZ electrolyte and $La_{0.7}Sr_{0.3}VO_3$ perovskite anode for application, *J Aust Ceram Soc*, 55 (2019) 97–102.
18. Y.H. Huang, R.I. Dass, Z.L. Xing, J.B. Goodenough, Double perovskites as anode materials for solid-oxide fuel cells, *Science*, 312 (2006) 254–257.
19. C. Gumeci, J. Parrondo, A.M. Hussain, D. Thompson, N. Dale, Praseodymium based double-perovskite cathode nanofibers for intermediate temperature solid oxide fuel cells (IT-SOFC), *Int J Hydrogen Energ*, 46 (2021) 31798–31806.
20. Z.M. Zhong, Bismuth ruthenate-based pyrochlores for IT-SOFC applications, *Electrochem Solid St*, 9 (2006) A215–A219.
21. T. Hagiwara, H. Yamamura, H. Nishino, Relationship between oxide-ion conductivity and ordering of oxygen vacancy in the $Ln_2Zr_2O_7$ (Ln = La, Nd, Eu) system using high temperature XRD, *J Fuel Cell Sci Tech*, 8 (2011) 051020–051025.
22. M.J.L. Ostergard, C. Clausen, C. Bagger, M. Mogensen, Manganite-Zirconia composite cathodes for Sofc – Influence of structure and composition, *Electrochim Acta*, 40 (1995) 1971–1981.
23. M.J.L. Ostergard, M. Mogensen, Ac-Impedance study of the oxygen reduction-mechanism on La1-Xsrxmno3 in solid oxide fuel-cells, *Electrochim Acta*, 38 (1993) 2015–2020.
24. R.Z. Ren, Z.H. Wang, X.G. Meng, C.M. Xu, J.S. Quo, W. Sun, K.N. Sun, Boosting the electrochemical performance of Fe-based layered double perovskite cathodes by Zn2+ doping for solid oxide fuel cells, *Acs Appl Mater Inter*, 12 (2020) 23959–23967.
25. C. Lei, M.F. Simpson, A.V. Virkar, Investigation of ion and electron conduction in the mixed ionic-electronic conductor- La-Sr-Co-Fe-Oxide (LSCF) using alternating current (AC) and direct current (DC) techniques, *J Electrochem Soc*, 169 (2022).
26. Y.H. Liu, F. Zhou, X.Y. Chen, C. Wang, S.H. Zhong, Enhanced electrochemical activity and stability of LSCF cathodes by Mo doping for intermediate temperature solid oxide fuel cells, *J Appl Electrochem*, 51 (2021) 425–433.

16 Concluding Remarks

Ngoc Thanh Thuy Tran
Hierarchical Green-Energy Materials (Hi-GEM) Research Center, National Cheng Kung University, Tainan City, Taiwan

Chin-Lung Kuo
Department of Materials Science and Engineering, National Taiwan University, Taipei, Taiwan

Ming-Fa Lin
Department of Physics and Hierarchical Green-Energy Material (Hi-GEM) Research Center, National Cheng Kung University, Tainan City, Taiwan

Wen-Dung Hsu
Department of Materials Science and Engineering and Hierarchical Green-Energy Materials (Hi-GEM) Research Center, National Cheng Kung University, Tainan City, Taiwan

Jeng-Shiung Jan
Department of Chemical Engineering, National Cheng Kung University, Tainan City, Taiwan

Hong-Ping Lin
Department of Chemistry, National Cheng Kung University, Tainan City, Taiwan

Chia-Yun Chen
Department of Materials Science and Engineering, National Cheng Kung University, Tainan City, Taiwan

Yuh-Lang Lee
Department of Chemical Engineering, and Hierarchical Green-Energy Material (Hi-GEM) Research Center, National Cheng Kung University, Tainan City, Taiwan

Shu-Yi Tsai
Hierarchical Green-Energy Materials (Hi-GEM) Research Center, National Cheng Kung University, Tainan City, Taiwan

DOI: 10.1201/9781003367215-16

Jow-Lay Huang
Department of Materials Science and Engineering, Center for Micro/Nano Science and Technology and Hierarchical Green-Energy Materials (Hi-GEM) Research Center, National Cheng Kung University, Tainan City, Taiwan

CONTENT

In general, this book covers the development of theoretical frameworks, experimental fabrication techniques, high-resolution measurements, and excellent performance of energy storage and conversion materials, including lithium-ion-based batteries, supercapacitors, solar cells, and fuel cells. It can significantly promote the full understanding of basic sciences, application engineering, and commercial products. To fully understand the optimal growth processes, the theoretical simulations need to be done the full cooperation between the first-principles, molecular dynamics, and machine learning methods. The up-to-date experimental observations are only focused on chemical syntheses and X-ray patterns, while most of the examinations are absent because of intrinsic limits under the quasi-Moire superlattices (1) and quantum confinements (2). How to overcome them and achieve the simultaneous progresses between experimental and theoretical researches will be discussed in detail.

In Chapter 2, we used our ReaxFF parameter and reproduced the two-stage lithiation in amorphous silicon. The faster lithiation rate for the amorphous silicon has been found to cause the 2-stage lithiation for amorphous silicon. Besides, a concentration gradient is found in the Li–Si alloy lithiated from amorphous silicon. The silicon lithiation serves as a reaction-diffusion system, and the Si–Si bond-breaking reaction rate is found to be the controlling factor for the overall process. Finally, the formation of c-$Li_{15}Si_4$ is found to lower the toughness of Li–Si alloy, which should be the reason for the cracking of the silicon anode. The lower Li-concentrated Li–Si alloy generated from amorphous silicon results in its higher critical size than crystalline silicon. Besides, the longer diffusion path in Li–Si alloy for larger silicon nanoparticles makes the formation of c-$Li_{15}Si_4$ easier and leads to the critical size issue in nanostructured silicon anode.

Graphite magnesium compounds, which belong to 3D donor-type metals, present diverse quasiparticle phenomena. For example, there exist certain important differences among group-II [Be/Mg/Ca/Sr/Ba] and group-I Li/Na/K/Rb/Cs] by modulating the distinct guest atoms (3), the intercalant distributions and concentrations, as well as, their normal/ irregular stacking orderings (4). In Chapter 3, the delicate calculations are consistent with one another through the obvious effects due to the well-behaved carbon- and magnesium layer, the zone folding (5,6), and the interlayer charge transfers (7). The 3D layered structures clearly indicate the highly anisotropic and active environments of chemistry, physics, and material engineering. The various crystals can be generated by different methods, being very helpful in establishing an enlarged quasiparticle framework. The featured quasiparticle states are clearly shown in 3D band structures along the high-symmetry points of the first Brillouin zone, e.g., the great enhancement of asymmetric hole and electron spectra about the Fermi level after the Mg-adatom interactions (the significant C- and Mg-layer interactions (8)), E_F far away the K or G valleys (the more charge transfer from the guest atoms; the higher electron affinity of host atoms), the observable variations of subband number, energy dispersions, their initial energies and stable valleys, π- and σ- electronic states. Obviously, each carbon honeycomb lattice displays the perpendicular π and σ bodings, in which the former and the latter have very high and middle carrier densities. It keeps a planar structure after the prominent intercalations, clearly illustrating the almost unchanged σ bondings and the π-induced interlayer interactions. The side-view charge densities of host and guest layers present important variations. The above-mentioned features are further reflected in the atom and orbital-decomposed density of states, such as the blue shift of the Fermi level under the n-type doping, and the merged van Hove singularities due to the orbital hybridizations of active orbitals.

The successful commercialization of Li-ion batteries (LIBs) causes their widespread and mass production. Meanwhile, the problem of lithium availability leads researchers to consider the next-generation ion batteries without lithium. For this purpose, sodium-ion (Na-ion) batteries as potential candidates are studied. The early studies demonstrate poor energy densities as well as the lack of suitable electrolyte and anode materials. This situation has been kept for a long period. Recent developments have overcome the shortage in energy densities, i.e., Na-ion batteries with energy densities comparable to LIBs have been reported. This is a crucial signal that Na-ion batteries once again become potential candidates for next-generation ion batteries. In Chapter 4, the physical and electronic properties of Na-intercalation compounds are given by first-principles calculations, such as geometric structure, band structure, density of states, and spatial charge distribution. These fundamental understandings are beneficial for researchers to develop better materials for designing Na-ion batteries with high energy density, well rate performance, long cycle life, and stable quality.

In Chapter 5, the 3D quaternary $LiLaTiO_4$ compound, which is a candidate for cathode materials of lithium-ion batteries, is predicted to exhibit rich and unique lattice symmetries, energy band structures, charge density, and atom-/orbital-projected density of states. Besides, $LiLaTiO_4$ has rich chemical bonding due to the rare-earth- and transition-metal atoms with seven and five kinds of f- and d- orbital in a periodic table. The accurate analyses, which are made from the exact first-principles calculations are available in identifying the significant single-/multi-orbital hybridizations and spin configurations in La-O, Ti-O, La-Ti, and Li-O. The multiplicated band structure with various atom dominations, the sizable indirect gap of 2.19 eV. The charge density distribution and a lot of van Hove singularities originated from extreme point dispersions.

The Si-Li_2S hetero-structure, a candidate for anode-electrolyte materials of lithium-ion batteries, exhibits rich and unique lattice symmetries, quasi-energy band structures, spatial charge distributions, and atom-/orbital-projected density of states. The delicate analyses established from the first-principles calculations in Chapter 4 are available in showing the significant orbital hybridizations in the chemical bonds. Particularly, the Si-Li_2S compound presents the huge Moire superlattice with an extremely complicated geometric structure, three kinds of chemical bonds as Si-Si, Li-S and S-Si survive in the large unit cell, the complicated electronic band structure with various atom dominations, the metallic behavior with n-type free carriers, the charge distributions, and a lot of merges van Hove singularities in density of states (DOS) due to the significant multi-orbital hybridizations. As a consequence, the critical mechanisms, the important multi-orbitals in Si-Si, Li-S, and Si-S bonds, could be well characterized. The calculated results clearly indicate that the Si-Li_2S heterostructure could serve as a candidate electrode-electrolyte component in Lithium-based batteries. Our prediction provides meaningful information about the critical physical/chemical pictures in LIBs. The current theoretical framework is appropriate for fully comprehending the diversified properties in anode/cathode/electrolyte and other emerging materials.

In Chapter 3, based on Vienna Ab initio Simulation Package (VASP) analysis results, we can confirm that the $LiMnO_2$ compound has fascinating electronic and magnetic properties. Specifically, the $LiMnO_2$ configuration has multi-orbital hybridization in chemical bonds, with monoclinic and orthorhombic phases, respectively. Both the Li-O and Mn-O bonds have active multi-orbital hybridization of 2s-[2s, $2p_x$, $2p_y$, $2p_z$], [4s, $3d_{x^2-y^2}$, $3d_{z^2}$, $3d_{xy}$, $3d_{yz}$, $3d_{xz}$] - [2s, $2p_x$, $2p_y$, $2p_z$], respectively. The different orbitals and atoms dominate at distinct regions; however, both configurations have indirect bandgaps and exhibit semiconductor behavior. In addition, spin polarization in the band structure and state density graph; positive magnetic moment values indicate that the $LiMnO_2$ structure exhibits ferromagnetic behavior.

Olivine-structured cathode materials, such as lithium nickel phosphate ($LiNiPO_4$) (LNP) and lithium cobalt phosphate ($LiCoPO_4$), etc, suffer from low conductivity and ionic conductivity, so it is necessary to modify the cathode material to improve the battery performance. Unfortunately, the surface properties and structure of LNP cathode are not well understood at present. In Chapter 8, the surface properties of LNP are investigated by first-principles calculation, by static calculation, the surface model, surface energy calculation, and Wulff shape are established to understand the surface structure and morphology of LNP at thermodynamic equilibrium condition, later by the density of state analysis, the electronic properties difference between bulk and surfaces models can be observed. The result obtained from first-principles calculation method can be a reference for later studies about interfacial properties of the olivine-type cathode, which can explain the experimental phenomenon.

In recent years, machine learning (ML) has already made a significant contribution to the discovery and study of energy materials. However, there remains considerable room for improving both the predictive performance and interpretation ability of models. With the development of better theoretical methods, data collection, databases, and other techniques, these challenges can be overcome and ML will become an integral complement to existing experimental and computational techniques for materials science. An important long-term objective would be the combination of DFT, ML, and experiments to arrive at a structural solution that would provide a comprehensive understanding of batteries' atomistic and nanometer structures as discussed in Chapter 9. ML will play a key role in the formulation of future scientific theories. There is no doubt that ML potentials are likely to become an integral part of battery modelers' toolkits in the coming years, and it will be exciting to see how this will affect the next decade of energy material development.

In Chapter 10, SnOx-RGO nanocomposite was prepared by a chemical reduction procedure at room temperature. SnOx quantum dots were grown uniformly over the RGO sheets and the composite delivered significant capacity (938 $mAhg^{-1}$) in the first cycle and that retain nearly 88 % after 50 cycles. The high rate capability indicates very good stability of the integrated structure and the synergistic effect of RGO led to a large enhancement of the electrochemical properties. We have also used the identical procedure to dope SnO2 with Mo to prepare Mo-doped SnO2 QDs/rGO nanocomposite. The as-prepared composite showed an excellent rate capability of (~587 mAh g−1@1.5 Ag^{-1}). The significantly enhanced battery characteristics are due to the narrow size of Mo-doped SnO2, uniform distribution of the quantum dots over the RGO, and enhanced electrical conductivity attributed to RGO layers.

In Chapter 11, the series of polymer electrolytes were successfully synthesized through the UV *in-situ* polymerization of PETMP, dicationic imidazolium-based crosslinker (VIm-PEG400-Vim), PEO crosslinker PEGMEMA and LiTFSI in plasticizer (PEGDME) additive. The electrolytes show high ionic conductivity, and advantageous electrochemical stabilities. In addition, the electrolytes have good Coulombic efficiency during long-term cycling tests at 0.2°C under 60°C due to strong adhesion between electrolyte and cathode. These properties above demonstrate that the method of in-situ polymerization is a potential candidate for lithium-ion battery technologies.

The biomass materials used for synthesis are relatively environmentally friendly, and using biomass as raw material comes from the fact that the net CO_2 emissions from the utilization process are considered to be zero. These advantages are beneficial for future economical large-scale production. In addition to using agricultural wastes, some food wastes (e.g., eggshells or shells) consisting of $CaCO_3$ and proteins can be used as both activating agents and nitrogen dopants to prepare the nitrogen-containing multiporous carbons of large surface areas and pore volumes as shown in Chapter 12. The method substantially reduces the cost of the manufacturing process and is consistent with the goals of the circular economy. The synthesized multiporous carbons have not only a high specific surface area but also both microporous and mesoporous structures; thus, these materials can be used in supercapacitors to provide appropriate structural permeability and enhance the retention rate of supercapacitors during rapid charging and discharging. Future research could apply multiporous carbons for CDI with the goal of adsorbing heavy metal ions in aqueous solutions to reduce water pollution and improve water recyclability; CDI is thus a promising technology for next-generation water treatments.

For the past few decades, solar power has been taken as an unprecedented development to help humanity live in a technologically advanced and sustainable world. In Chapter 13, the involvement of low-dimensional structures in solar cells stands for one of the promising strategies because they allow efficiency improvement while sustaining the minimum material usage and fabrication simplification, paving the critical rule to impact the next-generation photovoltaic applications.

Chapter 14 presents the idea of producing next-generation dye-sensitized solar cells because the novel and optimal classical DSSC materials utilized to improve the performance of DSSCs under ambient lighting have been reviewed here. Using these materials, it is also possible to prepare advanced room light energy harvest electrolytes based on printable and solid-state polymer electrolytes for room light DSSC applications. DSSCs using solid-state electrolytes can act as efficient and stable power supply cells for portable electronics and IoT devices. Therefore, rapid research and development activities can be expected from many research laboratories and industries, which can accelerate the mass production of indoor light DSSCs for practical applications.

Fuel cells are one of the most potential technologies for next-generation hydrogen energy production and become the alternative energy source other than oil and gas with lower pollution emitted. Although SOFC is a technology with a lot of advantages and prospects, the commercialization of SOFC is still in development. The report finds that SOFC is currently enjoying unprecedented political and business momentum, with the number of policies and projects around the world expanding rapidly. As indicated in Chapter 15, a key factor for the successful commercialization of SOFC is that they must be economically competitive. It concludes that now is the time to scale up technologies and work together in the development and commercialization of fuel cells to achieve a better environment for the next generation.

REFERENCES

1. Liu Y, Zeng C, Yu J, Zhong J, Li B, Zhang Z, et al. Moiré superlattices and related moiré excitons in twisted van der Waals heterostructures. *Chemical Society Reviews*. 2021;50(11):6401–6422.
2. Takagahara T, Takeda K. Theory of the quantum confinement effect on excitons in quantum dots of indirect-gap materials. *Physical Review B*. 1992;46(23):15578.
3. Daley AJ. Quantum computing and quantum simulation with group-II atoms. *Quantum Information Processing*. 2011;10(6):865–884.
4. Yuan L, Chung T-F, Kuc A, Wan Y, Xu Y, Chen YP, et al. Photocarrier generation from interlayer charge-transfer transitions in WS2-graphene heterostructures. *Science Advances*. 2018;4(2):e1700324.
5. Boykin TB, Klimeck G. Practical application of zone-folding concepts in tight-binding calculations. *Physical Review B*. 2005;71(11):115215.
6. Sato K, Saito R, Cong C, Yu T, Dresselhaus MS. Zone folding effect in Raman G-band intensity of twisted bilayer graphene. *Physical Review B*. 2012;86(12): 125414.
7. Khusayfan NM, Khanfar HK. Impact of Mg layer thickness on the performance of the Mg/Bi2O3 plasmonic interfaces. *Thin Solid Films*. 2018;651:71–76.
8. Ban X, Sun K, Sun Y, Huang B, Jiang W. Enhanced electron affinity and exciton confinement in exciplex-type host: power efficient solution-processed blue phosphorescent OLEDs with low turn-on voltage. *ACS Applied Materials & Interfaces*. 2016; 8(3):2010–2016.

17 Energy Resources and Challenges

Thi Dieu Hien Nguyen, Wei-Bang Li, Vo Khuong Dien, and Nguyen Thi Han
Department of Physics, National Cheng Kung University, Tainan City, Taiwan

Hsien-Ching Chung
R D Dept. Super Double Power Technology Co. Ltd., Changhua City, Changhua County, Taiwan

Shih-Yang Lin
Department of Physics, National Chung Cheng University, Chiayi, Taiwan

Le Vo Phuong Thuan
Department of Physics, National Cheng Kung University, Tainan City, Taiwan

Ngoc Thanh Thuy Tran
Hierarchical Green-Energy Materials (Hi-GEM) Research Center, National Cheng Kung University, Tainan City, Taiwan

Jheng-Hong Shih
Department of Physics, National Cheng Kung University, Tainan City, Taiwan

Ming-Fa Lin
Department of Physics and Hierarchical Green-Energy Material (Hi-GEM) Research Center, National Cheng Kung University, Tainan City, Taiwan

CONTENTS

DOI: 10.1201/9781003367215-17

This book has successfully presented a lot of merits and drawbacks in the basic and applied science research closely related to green energies. Certain significant issues under the current and near-future studies are expressed as follows.

17.1 CHALLENGES OF BASIC AND APPLIED SCIENCES

The basic science researches, which have been developed in the published books for ion-based batteries, semiconductor compounds, and emergent materials under a unified quasiparticle framework, clearly show a lot of merits and drawbacks in terms of positive progresses. The composite quasiparticles represent the enlarged science viewpoints of physics, chemistry, and material engineering. They are clearly revealed as the unusual behaviors during the charging and discharging processes with the chemical reactions in the cathode [1,2], electrolyte [3–7], and anode [8–11]. How to directly link the intrinsic interactions closely related to one another is the necessary search strategy in finding and determining the critical ion transport mechanisms [12–14]. That is to say, their physical pictures should be in great contrast to those of electron Fermions. The significant differences need to be systematically investigated under the simultaneous progresses of the first-principles simulations [15–17]. phenomenological models [18,19] and experimental examinations [3,20–22]. Obviously, this strategy might induce a breakthrough in boson transport theories and battery sciences [23,24].

In order to thoroughly comprehend lithium-related batteries, the composite quasiparticles in ternary [2,8,25–30] or quaternary [31–33] lithium oxides have been successfully studied by using the VASP simulations, e.g, the geometric [34], electronic [16,35], optical [36], and quantum phonon properties [37] of cathodes [1], electrolytes [3,6,38], and anodes [5], respectively, corresponding to LiFeO/LiFePO/LiCoO/LiNiO [Figures 17.1(a)-17.1(b); Figures 17.3(a)-17.3(b)], LiSiO/LiGeO/LiSnO] [Figures 17.2(a) and 17.4(a)], and LiScO/LiTiO [Figures 17.2(b) and 17.4(b)]. Specifically, the transition metal host atoms can create the spin-configuration ferromagnetisms, with the net magnetic moments [39,40], in which the spatial spin density distributions mainly come from the atom- and 3d-orbital-projected contributions [41,42], The charge- and spin-dominated intrinsic interactions, which survive in the Hamiltonian simultaneously, will play the critical roles for creating the diverse quasiparticle phenomena. Their single-particle kinetic energies and on-site Coulomb

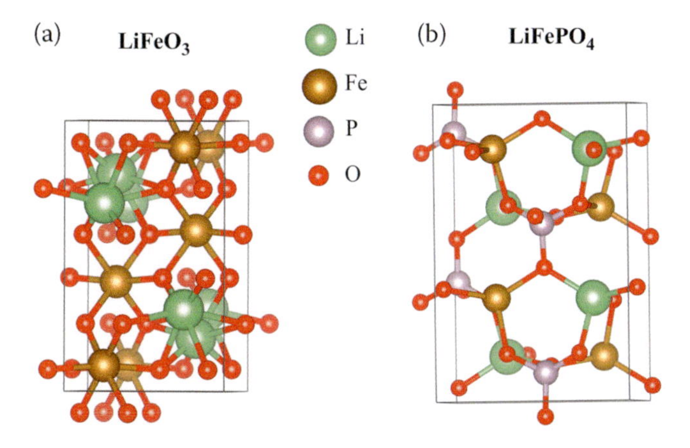

FIGURE 17.1 The optimal stability of cathode materials with high non-uniform environment: (a) ternary LiFeO compound and (b) quaternary LiFePO material.

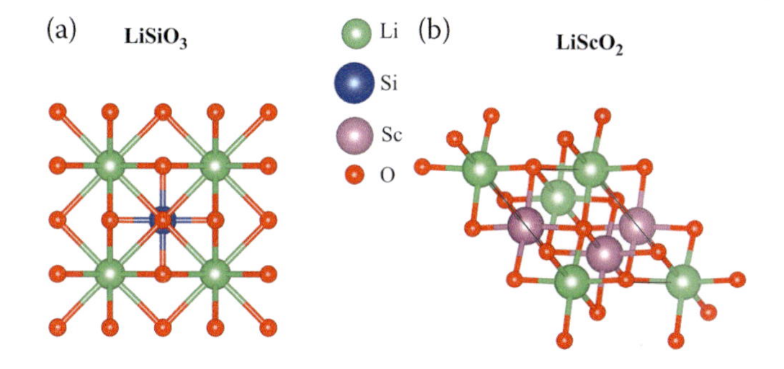

FIGURE 17.2 The optimal stability structures for different electrolyte and anode materials for ternary compounds of (a) LiSiO and (b) LiScO, in which these two compounds presents high non-uniform environment.

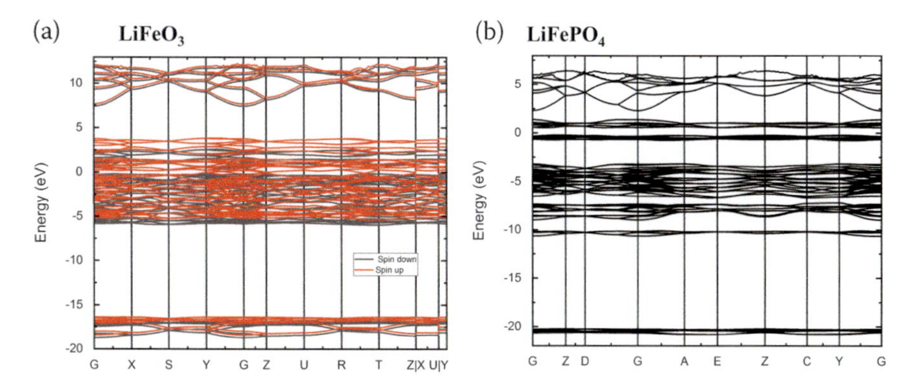

FIGURE 17.3 The wave-vector dependent band structure with diverse sub-bands in occupied and unoccupied states for (a) ternary LiFeO and (b) quaternary LiFePO materials.

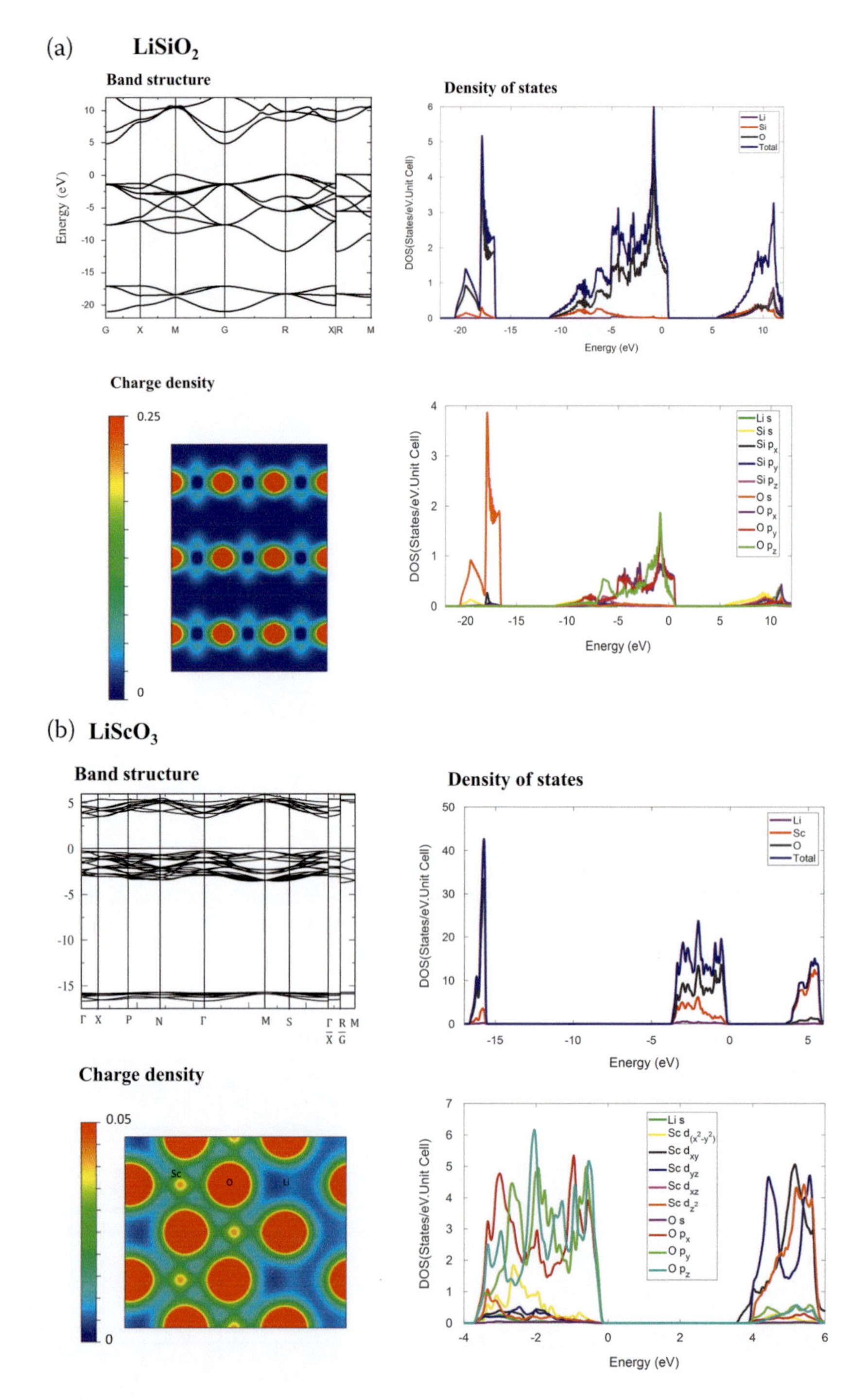

FIGURE 17.4 Electronic structures: Band characteristics, atom-,orbital- density of states, charge densities for difference electrolyte and anode materials.

potentials [43,44], as well as, the many-body exchange, correlation [45], and the spin-dependent on-site electron-electron interactions [the Hubbard-like ones; [44]]. Most important, the unusual heterojunctions between cathode/electrolyte and electrolyte/anode should be very interesting search topics because of three kinds of chemical bondings/spin distributions within two subsystems and a spatially modulated boundary. A giant Moire superlattice, which has a commensurate lattice of two distinct subsystems, comes to exist with totally different chemical bonds/spin configurations. The rich and unique composite quasiparticles will clearly illustrate the diverse quasiparticle phenomena under the near-future theoretical and experimental studies.

The dynamic chemical reactions will dominate the intermediate crystal configurations, and the time-dependent crystal structures appear in cathode, electrolyte, and anode materials during the charging and discharging processes; that is, intercalations and de-intercalations are capable of leading to the dramatic transformations of crystal symmetries. How to fully comprehend such unusual phenomena from theoretical predictions [17] and high-resolution X-ray observations [46] is urgently requested under the current explorations. The direct combinations among the first-principles methods [15–17], molecular dynamics [47,48], and chemical reaction equations can achieve the temporary crystal configuration with many atoms in a primitive unit cell [14,21,22]. The relative atom concentration only presents a very slight variation after the stationary follow through the ion intercalations/de-intercalations [49–52]. Of course, the composite quasiparticles are totally different from one another during the electrical operations of ion-based batteries. Their time-dependent behaviors are never explored in detail in previous scientific works. On the experimental side, the time-resolved X-ray diffraction peaks are required using the up-to-date femtosecond technique [53], in which a lot of quasi-stable crystal structures could be examined under the delicate measurements and analyses of the minimum time scale.

Each crystal structure consists of position-dependent chemical bonds, in which a Moire superlattice presents a highly non-uniform and anisotropic environment. In general, the spatial charge density distribution is very unique and complicated during the battery operations [9,31,54,55], e.g., the coexistences of quaternary LiFePO with Li-O, Fe-O, and P-O bonds, respectively, through 2s-[2s, $2p_x$, $2p_y$, $2p_z$], [$3d_{xz}$, $3d_{zy}$, $3d_{yx}$, $3d_{x^2-y^2}$, $3d_{z^2}$]-[2s, $2p_x$, $2p_y$, $2p_z$], and [3s, $3p_x$, $3p_y$, $3p_z$]-[2s, $2p_x$, $2p_y$, $2p_z$]. Their sensitive variation should accompany by very rapid chemical reactions. The static Coulomb fields, which correspond to the various intermediate crystal configurations, might be sufficient in driving the heavy and stationary ion transports. While the macroscopic phenomena are utilized to evaluate the ion flow, the perturbation and intrinsic forces should be in the equivalent status. A finite voltage between cathode and anode materials will provide the necessary energy sources in creating the intercalation/de-intercalation processes. How to express the suitable formulas under reliable chemical reactions and physical perturbations is the most important studying focus. Very interesting, both ion and electron transports are totally different from each other. The up-to-date theoretical studies show that the Kubo formula, with the Fermion quantum statistics [56], is available for the latter of any condensed-matter systems within the linear or non-linear regimes. The elastic and inelastic scatterings, which respectively, arise from charged impurities/defects/

boundaries [12,57–60] and quantum phonons [37,61,62], are able to generate the residual and temperature-dependent resistances [63]. The microscopic quasiparticle frameworks play critical roles in identifying concise physical/chemical/material pictures. Its strong partnerships with the macroscopic ion behavior, e.g., the prominent charge screening effects of electrons and ions, will greatly promote the development of composite quasiparticles.

A research approach can either be basic or applied. Basic research aims at expanding knowledge by creating new theories and modifying existing ones, while applied research focuses on providing practical solutions to specific problems by combining empirical data. Researchers need to understand the similarities and differences between applied and basic research methods. The main difference between applied and basic research is the purpose of the research and research outcomes. Basic research is useful for generating novel theories or improving existing theories that explain a concept or phenomenon. In other words, applied research provides solutions to specific problems or creates new technology.

Basic research is also known as fundamental research to achieve a better and more detailed understanding of a research subject or phenomenon, not to solve a specific problem. Applied research focuses on providing practical solutions to a specific problem. It is a form of investigation generally employing empirical methodologies. On the other hand, basic research is theory-oriented, and applied research is practical-oriented. Applied research is the following design for basic research. It further creates innovative solutions to specific problems, while basic research is an approach to expanding knowledge in a field of study. It means that applied research is solution-driven while basic research is knowledge-specific. Basic research results in new concepts, and it also expands existing knowledge, while applied research does not lead to new theories, and only focuses on how to apply these theories to solve existing problems. Hence, basic research is universal while applied research is limited in nature. Applied research solves specific problems by subjecting empirical evidence to standardize procedures and making a more valid research method. In contrast, basic research focuses simply on expanding theories without providing solutions to existing problems. However, basic research proffers solutions to a specific problem. For example, in material sciences, the outcome of basic research in material sciences lies in new dimensions, new combinations to existing structures that improve performance behaviors, and predicts other phenomena. In contrast, applied research uses theories to develop techniques that alter material structures. Basic and applied research adopt similar data collection processes to collect relevant data and achieve the most objective research outcomes. The research outcomes of basic research often serve as the bedrock of applied research.

17.2 ION-BASED BATTERIES

The wide application of lithium-ion (Li-ion) batteries in various fields, from 3C products to grid-scale energy storage systems (ESSs), has revolutionized our way of life [25,28]. In 1985 at Asahi Kasei Corporation, Akira Yoshino [64,65] assembled a full rechargeable battery including the petroleum coke anode with Goodenough's

$LiCoO_2$ cathode [60,66–71] based on Whittingham's framework [72,73], preventing the risk of dendrite-formation-induced thermal runaway and ensuring the stability for the commercial market. This battery was later commercialized by Sony in 1991 with a gravimetric energy capacity of 80 Wh/kg and volumetric energy capacity of 200 Wh/L [74]. The 2019 Nobel Prize in Chemistry has been awarded to John B. Goodenough, M. Stanley Whittingham, and Akira Yoshino for their contributions to developing Li-ion batteries. Currently, Li-ion batteries have exhibited many branches based on their active materials, performing various applications [2]. They are commonly named by the cathode (positive) materials, such as lithium cobalt oxide ($LiCoO_2$ or LCO) [60,61,74], lithium manganese oxide ($LiMnO_2$ or LMO) [75–79], lithium iron phosphate ($LiFePO_4$ or LFP) [62,63,80,81], lithium nickel cobalt manganese oxide ($Li(Ni_xCo_yMn_z)O_2$ or NCM) [82–84], and lithium nickel cobalt aluminum oxide ($LiNi_{0.8}Co_{0.15}Al_{0.05}O_2$ or NCA) [85,86]. Some are named by their anode (negative) materials, such as lithium titanium oxide ($Li_4Ti_5O_{12}$ or LTO) [87–102]. There are still many challenges in studying the anode and cathode materials, electrolytes and separators, making better Li-ion batteries. Based on the development of low-dimensional materials, gapless graphene [3,17–19,103–106], insulating hexagonal boron nitride (h-BN) [107], and semiconducting transition metal dichalcogenides (TMDs) [108] and group III-V compounds [16], these materials are expected to improve the performance of Li-ion batteries. From an industrial viewpoint, the successful commercialization and popularization of electric vehicles (EVs) worldwide [109] cause the widespread and mass production of Li-ion batteries. The retired power batteries have largely increased, causing waste of resources and environmental protection threats. Hence, recycling and utilization of such retired batteries have been promoted [110–112]. Some retired power batteries still possess about 80% initial capacity [20,26,27,30,113–116]. In other words, they can be repurposed and utilized once again, e.g., serving as the battery modules in the stationary energy storage system [117–120]. Governments in various countries have acknowledged this emergent issue and prepared to launch their policies to deal with the recovery and reuse of repurposing batteries, such as coding principles, traceability management system, manufacturing factory guidelines, dismantling process guidelines, residual energy measurement, federal and state tax credits, rebates, and other financial support [29,121,122]. Back to the fundamental researches, the successful and wide application of Li-ion batteries also stimulate scientists to consider other next-generation ion batteries, such as Na-ion batteries.

17.3 OPTICAL ELECTRONICS

The efficient use of solar cell technology on a global scale demands efficient breakthroughs in materials and devices to lower fabrication costs and enhance power conversion efficiency [123]. To date, Perovskite solar cell devices have become one of the dominant systems in fundamental materials research and engineering applications, owing to their ability to supply eco-friendly energy, efficiency, and unique routes to convert photon energy into electricity [87]. The absorber layer is sandwiched between the electron-transporting and hole-transporting layers in the basic design of a perovskite solar cell (Figure 17.5)

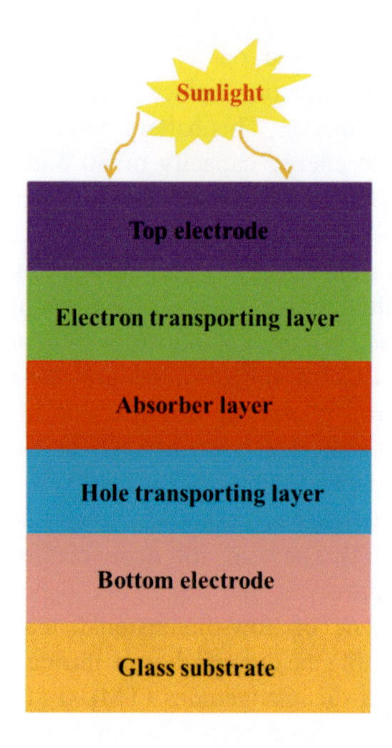

FIGURE 17.5 The scheme of the standard structure of a solar cell with some components, includes the absorber layer sandwich between the electron- and the hole-transporting layers, glass substrate.

[88,89]. The first component is critical because it absorbs the electromagnetic wave (EM) or photon and produces electron-hole pairs. The geometric, electrical, and optical excitation mechanisms of these components appear to be very important shared traits. These factors should be critical conditions in determining the efficiency of light-generated current and searching for an optimal match of three kinds of core components. The crucial mechanism of solar cells based on photovoltaic effects involves two key processes [90,91] (Figure 17.6). In the first process, the electron-hole pairs (excitons) are generated in the absorber layer upon the absorption of a suitable photon. In the second process, charge carriers (electrons and holes) are produced upon the action of the electric field existing at the interface between the absorber and the charge-transporting layers. When an electron is detached from a hole and injected into an electron-transporting layer, it migrates to the anode. The hole is simultaneously injected into the hole-transporting layer and migrates to the cathode. The electrons and holes are collected by working and counter electrodes respectively and transported to the external circuit to produce current. Generally, the electron-transporting and the hole-transporting layers of solar cell panel are produced by the different semiconductors (or same semiconductor with different types of conduction), while semiconductor materials used for the absorber layer must be dissimilar to the charge-transporting layers in order to generate the built-in electric field and to carry the electric current [92,93].

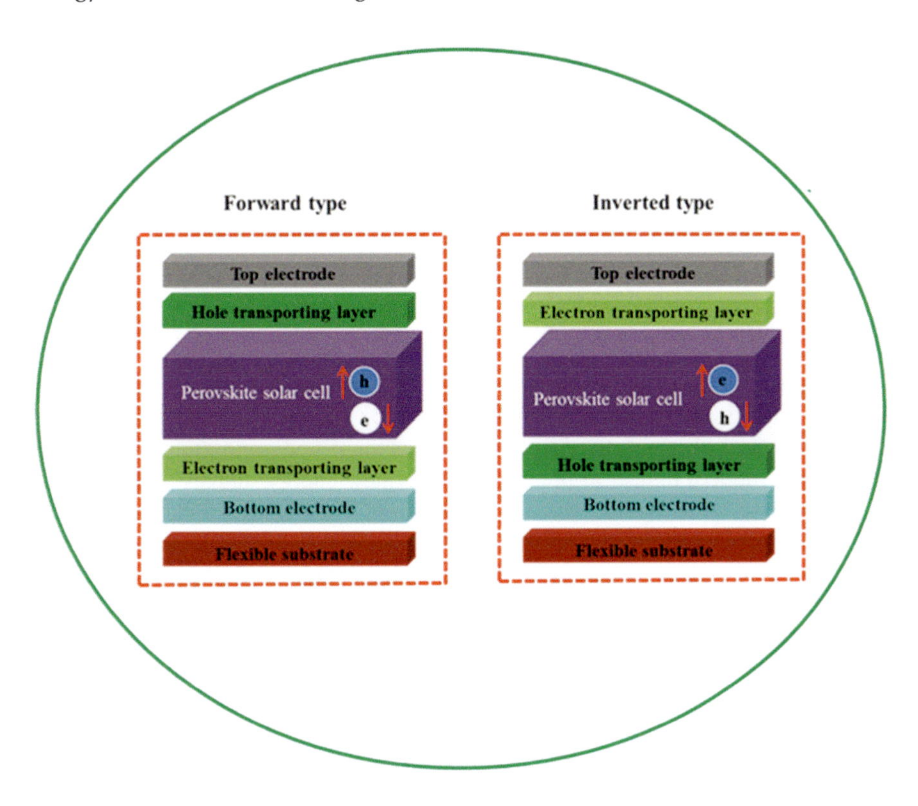

FIGURE 17.6 The well-characterized architecture of the perovskite solar cells: forward and inverted types.

On the theoretical side, investigations indicated that the first-principles calculations (DFT method and GW one) [94–96] are very efficient and reliable in determining the rich essential properties of the 3D compounds, such as the geometric, electronic, and optical properties of the perovskite materials [97–99]. Fortunately, the previous results calculated by DFT based on VASP package are a full understanding of the diversified physical and chemical phenomena. Indeed, systematic studies have been successfully conducted on layered 2D graphene/silicene with or without chemical modifications (absorptions and substitutions) [100,101], 1D graphene nanoribbons [102] and the 3D ternary cathode/anode/electrolyte compounds [25,124]. Through delicate analyses, the critical mechanisms and a concise picture of the geometric, electronic, magnetic, and optical properties could be firmly obtained. As a result, such state-of-the-art calculations might be suitable for exploring the diverse fundamental properties of a lot of complicated perovskites compound for solar cell applications. The main characteristics are clearly revealed in the optical gap [threshold excitation frequency], prominent absorption structures, absorption reflection coefficient, and energy loss function with charge screenings. The position-dependent chemical environment can create the strongly anisotropic excitation spectra, being sensitive to the directions of electric polarizations. Moreover, the many-body effects, the excitonic ones, might play an important role in determining the featured optical properties, e.g., the red-shift of initial absorption frequencies due to the stable excitons. How to use the first-principles results for examining the critical orbital hybridizations is the main studying focus. The solar cell materials are chosen as the typical systems to test the developed theoretical framework. Obviously, the optical measurements, such as absorption, transmission, and reflectance spectroscopies combine with Kramer–Kronig relations can give information on the optical excitations [125].

17.4 WIND FORCE FIELDS

Wind energy is one of the earliest sources of energy discovered and used by humans, at least 3,000 years ago. Initially, this type of energy was used mainly to provide mechanical power to operate the water pumps and grind grain. In the early years of the modern industrial revolution, many new energy sources took the place of wind energy, typically now thermal energy from fossil fuels and the electric grid, because it is considered a stable source of energy. However, by the early 1970s, a dramatic increase in oil prices led to a spectacular return to wind power [126]. At the time, instead of providing mechanical energy, wind energy is mainly utilized to produce electricity. The first wind turbine, which generated electricity, was built in the early 20th century. With the continuous improvement of technology, by the early 1980s, the cost of wind energy production decreased by 1/6 compared to before, and the amount of electricity supplied tended to increase [127]. Consequently, wind energy has become one of the most important and sustainable energy sources. Today wind energy is mainly used in electricity generation. By 2020, the wind will provide nearly 1,600 TWh of electricity, accounting for 5% of worldwide electricity production and 2% of the energy consumed [128–131].

Principally, wind power is the conversion of wind energy into electrical or mechanical energy through the movement of wind turbines [132–134]. Wind energy generates by the wind moving the blades and creating torque on the rotor [135–141]. In general, the amount of electricity obtained depends on the turbine blade size and wind speed. To improve the output of electricity from wind energy, many researchers focus on developing wind turbines. The classification of different types of wind turbines is based on the wind energy conversion system within it [142–147]. These systems can divide into aerodynamic drag-dependent systems and aerodynamic lift-dependent systems. Modern wind turbines rely heavily on an aerodynamic lift [148].

Wind energy production holds great potential but also faces significant challenges [149]. Indeed, the prospect of wind energy in many developing and emerging countries is enormous. In many places, generating electricity from wind energy is an economically viable solution to replace fossil energy sources. Especially in developing countries, the wind turbine is considered the optimal solution for power plants. It cost savings, environmental friendliness, and reduces dependence on imports of fossil fuels. However, wind energy production still has many challenges to be solved. First, wind power does not have consistent output as fossil fuel plants. Small-scale wind generators need battery storage to enable low or no wind seasons. Second, the generator must be designed within a specific wind speed range. This means that the technology can only be used in areas with sufficient wind. Finally, the initial investment cost for the construction of wind generators is considerable [150].

17.5 HYDROGEN ENERGIES

The most abundant substance in the universe, hydrogen, frequently comes to exist in the diverse forms of single-atom/molecular gases, liquids, and solid-state molecule crystals [151]. As a result of its rather active half-occupied 1s orbital, hydrogen atom plays one of the critical roles in the basic and applied sciences, such as the outstanding candidates for chemical modifications [152–154], energy storages [103,104,155], and theoretical models under the single-and many-particle pictures [105,106,153,154,156]. The active chemical environments are very convenient for observing the rich and unique physical/chemical/material phenomena. Increasing greenhouse effects accompanying the energy demand have encouraged researchers to develop a new energy resource. Compare with other energy generator systems such as solar cells, wind energy, and others, hydrogen is pre-eminent since it has no pollution during system operation, and thus, it is clean and provides huge thermal energy [104]. Using hydrogen as an alternative fuel source for petrol gas and other fossil fuel is one of the critical components of the renewable and clean energy plans of many countries, including the USA, European countries, and some Asian countries like Japan and Taiwan. In the past decade, enormous efforts have been devoted to hydrogen energy applications, including hydrogen production, transport, storage, and utilization. Highly efficient fuel cells have been developed and the technology of converting conventional fuels into hydrogen is also matured for application. Whereas hydrogen storage remains a big challenge and the ultimate solution is not clear until now [157,158].

Layered/nanostructured materials exhibit a lot of outstanding features. For example, layered group-IV elements, which exist in four significant [s, p_x, p_y, p_z]-orbitals, clearly indicate the rather strong covalent bonding. The single- and multi-orbital hybridizations are frequently revealed in the main-stream 2D materials, the well-separated p_z-π & (s, p_x, p_y)-σ bondings for the graphene-related systems [159], and the non-negligible mixing of π & σ bonding due to the buckled silicene/germanene/tinene/plumbene crystals [160,161]. As for group-IV nanotubes, the p_z and (s, p_x, p_y) orbitals, respectively, induce π and σ energy bands, the curvature effects, the misorientation of $p_z\pi$ orbitals and mixing of $p_z\pi$ and $sp^2\sigma$ orbitals have rather strong effects on the low-frequency electronic excitations of carbon-nanoribbons [162]. These unusual quasiparticle properties, as done in a series of books [160–162], are basic science interests and are useful for many technical applications. Very interestingly, the rather active π bond makes these systems become strong candidates for hydrogen storage materials.

The first idea of hydrogen storage using layered/nanostructured materials has been considered since 2005 [163]. The spark came in 2009 when hydrogen-passivated graphene, graphane, was first demonstrated [164]. Graphane is graphene with atomic hydrogen chemically bonded to each graphene lattice atom. It is predicted to be an insulator with a calculated band gap between 3.5 eV. The insulating behavior results from forming C–H bonds, localizing all the delocalized π-electrons in the graphene lattice. Continuing this success, Balog et al. can modulate the adatom adsorption by exposing graphene to hydrogen for different lengths of time, clearly illustrating that hydrogen concentration and energy gap grow with increasing exposure time [165]. Apparently, the electronic properties are drastically changed by the density of adsorbed hydrogen atoms. Graphene has taken center stage in the field of hydrogen storage due to its high surface area and vast possibilities for chemical functionalization. Beyond graphene, a plethora of experimental studies on the adsorption of an H atom on other group-IV surfaces, such as silicene and germanene, have been well proceeded [166,167].

On the theoretical side, the optimal crystal symmetries and featured energy spectra have been studied for all the hydrogenated group-IV materials [168,169]. A new hydrocarbon material, graphene with 100% percentage hydrogenation, is predicted by Sofo et al. in 2007 [170] through the first-principles density functional theory. Only a few works [153,154,171] are conducted on the double- and single-side hydrogenations under various concentrations and arrangements. The top-side H-adatoms can create or enhance the buckled structure of a honeycomb lattice because of the strong four-orbital s-sp^3 bondings, e.g., a dramatic transformation from the planar into the deformed crystal for the hydrogen-adsorbed graphene systems at higher concentrations. Very interesting, certain adsorption cases even present the ferromagnetic spin configurations [171]. The highly non-uniform chemical environments, atom-dominated band structures, spatial charge densities before and after hydrogenation, atom- and orbital-decomposed magnetic moments, as well as, atom-, orbital- and spin-projected density of states, are utilized to identify the critical charge and spin mechanisms of the quasiparticle framework [171].

The absorption/adsorption of hydrogen atoms/molecules on surfaces of nanostructure materials, being closely related to hydrogen storage applications, can express a lot of outstanding electronic and electronic properties. Even many hydrogenated systems are explored through various theoretical methods and experimental measurements. However, there are only a few works conducted on the main intrinsic features, e.g., electronic and optical properties of the host materials under different concentrations and arrangements of hydrogen atoms, the single-/multi-orbital hybridizations of hydrogen atoms with the absorbents, the close connections of the geometric, electronic and the optical excitations of these systems. Generally, how to achieve the physical and chemical pictures, especially for the specific orbital hybridizations in the chemical bonds with regard to the main quasi-particle features of hydrogen storage materials are the main challenge of current studies.

17.6 WATER RESOURCES

Water is another natural resource that can be used to generate renewable energy, along with solar power, wind power, and thermal power. With its sustainability and low operation costs, hydropower is considered the best power source among water resources [1–4]. Hydropower works by utilizing the power of flowing water, without consuming or vesting it. Because hydropower plants produce clean energy, they can reduce air pollution and slow down global warming. No air pollutants or toxic waste are produced under hydropower, as compared to other renewable energy types such as batteries and solar cells. In hydropower plants, people use water stored in dams and flowing in rivers in order to create electricity. The dams have served not only as a source of electricity but also as irrigation, water supply systems, etc. for humans' welfare. Dams need to be repaired and maintained to stay productive renewable sources of clean energy.

17.7 NATURE GASES

Natural gases are naturally occurring hydrocarbon gas, and it consists of methane and various amounts of other higher alkanes, carbon dioxide, nitrogen, hydrogen sulfide, or helium. It used to be stored in gas reservoirs, which are the rocks that trap and hold gas molecules. The size of the rock pore system could affect the transport and recovery of the gas. Some of the reservoirs belong to unconventional gas reservoirs because of their complex nanopore system, while the conventional gas reservoirs naturally produce by pressure depletion. A suitable estimation of the gas recovery of unconventional reservoirs is needed to enhance the economic feasibility of the recovery process. Density functional theory is the most suitable method to approach for molecular-level study of the adsorption and diffusion of the gas. The experimental and theoretical approaches were utilized to understand the adsorption of CH_4 and CO_2 on various types and models of surfaces, but they the cylindrical pores were not included. The cylindrical pores then thus needed to be investigated to reveal a complete picture.

Previous DFT study has used to exhibit the adsorption phenomena of CO_2, N_2, H_2O, and CH_4 on calcite surfaces. However, various adsorption surfaces can result in different adsorption properties, which needs to be further investigated. Moreover, a previous study used DFT to explore and compare the CH_4 and CO_2 adsorption on different surfaces, and the effect of the adsorption surface morphology is revealed. On the other side, molecular dynamics calculations can be used to study gas transport and adsorption on different types of surfaces. A more comprehensive study that considers differently shaped nanopores and rock surfaces is also needed to provide more information on the gas interactions.

17.8 NUCLEAR POWERS

In past years, electricity was mainly produced by the combustion of coal, oil, and natural gas, it helped improve human life and the development of industry. However, fossil fuel causes drastic climate problems with no doubt. Therein many renewable energy resources such as wind, solar, and hydropower are to be vastly expanded to replace the current use of fossil fuels. Recently, the nuclear power resource has been seen as one of the solutions to the climate crisis [172–174]. It is fortunate that the probabilities of accidents are minor even if the accident will cause very serious damage. However, the nuclear power is not a sustainable resource because it is still a finite resource; also, there exist other problems, e.g., the risk of nuclear radiation due to incidents, and the nuclear waste. Many researchers are attracted to investigating the improvement of nuclear power resources in theoretical calculations and experiments [175–177] and are based on primary energy factors

(PEF), which describes how much energy from the primary resources is being used per unit of energy delivered. According to the PEF [178], researchers could evaluate the efficiency of energy transformation and usage in detail. But there still exist some problems needed to be solved in the future. The nuclear waste must be one of the main problems, which includes the spent fuel and radioactive waste. The latter is a form of energy, and that means the radioactive waste covers a lot of primary energy. How to reprocess the radioactive waste and reduce the spent fuel is an important issue to be solved. The reuse of reprocessed used fuel is an environmentally friendlier option for nuclear power resources which provides a safer nuclear waste management route [179].

17.9 COAL MINERALS

The thermal power, which includes natural gas, nuclear, and coal mineral, accounted for 75% of the global net electricity generation in 2017 [180]. For instance, coal power generation is dominant in developing Asia. Many power plants must use water for cooling the equipment when working. However, the electrical outputs of these plants are increasingly constrained due to low water availability or high intake water temperature [181–183]. Moreover, the changes in streamflow and water temperature due to climate change exacerbate the abovementioned constraints [182,184–187]; also, some previous studies show that the results will decrease the power generation by 1.6%–16% in the U.S., 6.3%–19% in Europe, and 8%–16% on average globally by the middle of the 21st century. In addition, the coal chemical industry, as one of the most important fields for coal use, is also a primary contributor to CO_2 emissions. How to realize coal conversion with high efficiency and low CO_2 emissions is one of the keys to the further development of the coal chemical industry. And this is followed by another issue related to the reuse of CO_2. CO_2 is not just a greenhouse gas, but also an important source of carbon for making organic chemicals, materials, and carbohydrates. Hence, how to capture CO_2 from power plants efficiently and even from the atmosphere to be recycled to produce added chemicals must be the problem to be resolved in the future.

17.10 UNDERGROUND THERMALS

It is known that energy is supplied to a storage system for removal and use at a later time [188]. A complete process involves three steps: charging, storing, and discharging. In practical systems, some of the steps may occur simultaneously, and each step can happen more than once in each storage cycle [189]. The underground thermal energy storage (UTES) with both boreholes (BTES) and aquifers (ATES) is the most developed storage concept and the mostly used for seasonal storage. The heat storage in ATES consists in extracting groundwater from a well, heating this water with an available heat source, and then reinjecting it back into the aquifer in another well. The estimated heat storage capacity of 105 m^3 of the aquifer is 3 MJ for each 10 K temperature range [64]. A lot of factors are taken into consideration in deciding on the type and the design of any thermal storage system, and one of the key issues is the thermal capacity. However, selection of an appropriate system depends on many factors, such as cost-benefit considerations, technical criteria, and environmental criteria. Also, to realize a greater level of geothermal deployment, long-term policies for decarbonization need to be established. These long-term policies would enable geothermal to compete more effectively for heating applications with natural gas and other hydrocarbon resources, which experience fluctuations and uncertainties in costs and availability. With numerous barriers limiting the rapid deployment of advanced nuclear energy systems in the near future, more attention should be given to developing geothermal technologies to lower their technical risks and to demonstrate geothermal direct use for a range of geologic conditions.

17.11 OTHER CHEMICAL MATERIALS

Chemical energy comes into existence in chemical systems due to the significantly broken bonds. The critical mechanism of chemical bonding can be explained under the single or multi-hybridization theories. For example, ethylene consists of the σ and π bonds, which are formed from s-sp^2, sp^2-sp^2, and $2p_z$-$2p_z$, respectively. Principally, the chemical interactions of diverse materials will release or absorb the energy under the thermal form. At the same time, the initial condition requires the heated precursor and can proceed with the chemical storage. It is well-known that chemical energy can appear in any material that can be utilized as a fuel, which presents in Figure 17.7. Some kinds of energies have been discussed, which show the potential for noteworthy practical applications. Remarkably, the certain

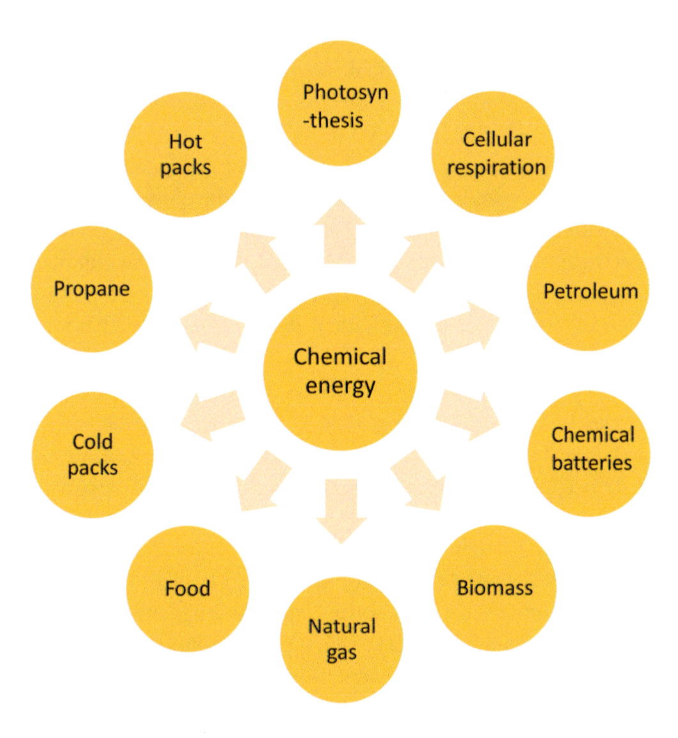

FIGURE 17.7 Chemical energy.

conversion of chemical energy into other forms of energy or versa might be prominent in the general mechanism. For example, the digestion process will convert the chemical energy in food into mechanical and thermal energy of the cells, while another chemical energy in wood will convert to optical and thermal energy under the combustion reaction. The useful form of energy mainly is produced from the reaction of oxygen and other organic materials in nature play a vital role in biofuel production. The semiquantitative analysis of Klaus [65] showed that combustions can yield 418 KJ per mole of oxygen. This calculation is useful for estimating the heat of any complex systems from the elemental components of combustions without understanding the molecular structure. Therefore, biofuel production also can be explained by the certain above analysis. Biofuels are known as renewable energy, which mainly originate from biomass. The close combination between biofuels and their petroleum components will determine the critical roles of the industrial requirements. Popular examples of biofuels can be taken into consideration, e.g., ethanol, biobutanol, biodiesel, naphtha, aviation fuel, and olefins. Some of them might produce hazards, or damage to environments, particularly hydrotreated biofuels. On the other hand, several biofuels provide a good chance to dismiss global warming and improve energy quality [66]. The issues of how to use biofuels in nanotechnology applications might be promising research at this time.

17.12 SIGNIFICANT APPLICATIONS: SEMICONDUCTOR CHIPS

It is well known that semiconductor compounds are outstanding candidates in the basic and applied sciences, as well as, the high-potential industry products of specific chips. Furthermore, in general, they play critical roles in dominating the economic developments of all states in this world, e.g., the first- [e.g., silicon and germanium [67]], second- [GaAs; [68–70]], third- [GaN/SiC; [71–76]], and fourth-generation ones. There exists competition among the super high-level companies in creating the artworks about the nanoscale integrated circuits, such as the close partnerships of TSMC, INTEL, and SAMSUNG. The strong cooperation of universities, institutes, and companies can greatly promote the composite quasiparticles under an enlarged framework. Through many theoretical [77,78] and experimental efforts, the main-stream semiconductor materials cover the single-element crystals [e.g., the diamond-like silicon and germanium], binary compounds [III-VI/III-V/IV-IV/II-VI/II-V; [76,79–83]] and ternary ones [III-VI-VI/III-III-VI/III-V-V/III-III-V/IV-IV-IV/II-VI-VI/II-II-VI]. Very interestingly, these condensed-matter systems are outstanding candidates for the systematic science investigations [Refs] and thus determine the diverse quasiparticle behaviors [84–86]. The various crystal phases, featured hole and electron states, unique charge density distributions, and rich van Hove singularities, which are consistent with one another, can delicately identify the concise pictures of the significant multi-orbital hybridizations about the chemical bonds [107,108]. The significant mechanisms are responsible for the other essential properties, such as magnetic properties/rich quantization [161], quasiparticle decay rates [86,109,110], optical transitions [111–113], Coulomb-field electronic excitations/magneto-plasmon modes [114–116,162], transport properties/quantum Hall effects [117–119], thermal capacities [61,120], mechanical strains [109,121,122,190], and n- or p-type doping effects [191–193].

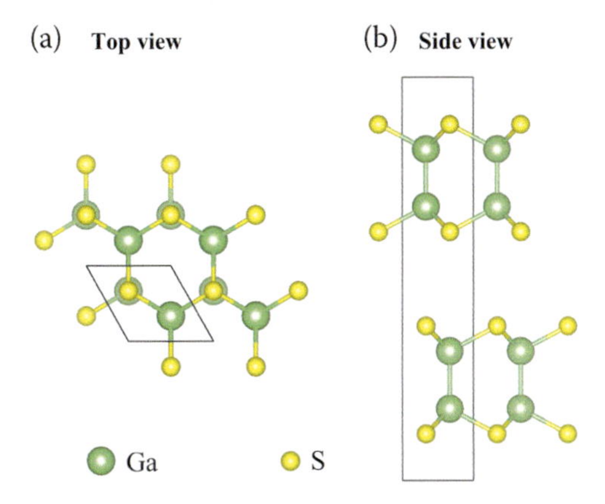

FIGURE 17.8 The layer structure in 3D gallium sulfur compound in different sides: (a) Top view and (b) side view. This structure presents unusual electronic and optical properties based on its geometric characteristics.

According to the up-to-date successful syntheses, the popular semiconductor compounds could be classified into certain crystal symmetries: (1) 3D diamond, (2) hexagonal structure [e.g., GaS/GaSe/GaTe/InSe/InTe; Figure 17.8], (3) others [trigonal or monoclinic crystal of gallium oxide; Figure 17.9], (4) 2D few-layer crystals with the distinct configurations and buckled degrees [GaS/GaSe/GaTe/InSe/InTe; Figure 17.10], (5) 1D cylindrical nanotubes/quantum-confinement nanoribbons, and (6) 0D quantum dots/fullerenes/rings/lines. The various crystal symmetries could be summarized together to present the highly non-uniform and anisotropic environments of chemistry, physics, and material engineering. The critical roles mainly come from dimensionalities [120], stacking configurations [194,195], layer numbers [196], buckled structures [197], quantum confinements [198], and chiral/achiral open edges [199–203]. Their prominent partnerships among them are the studying focuses. The delicate VASP

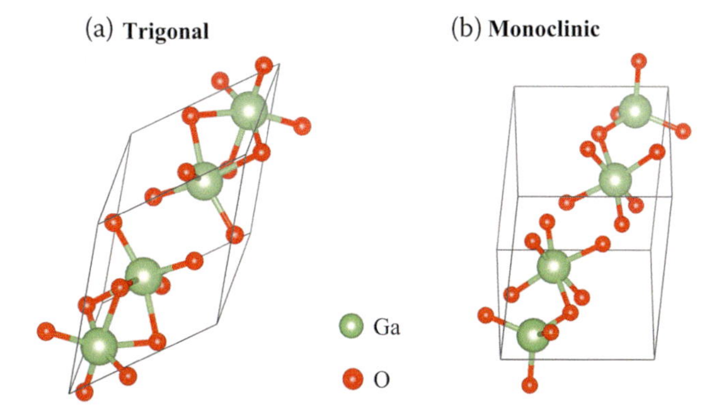

FIGURE 17.9 (a) Trigonal or monoclinic crystal structure, corresponding to the distinct growth environments of binary gallium oxides.

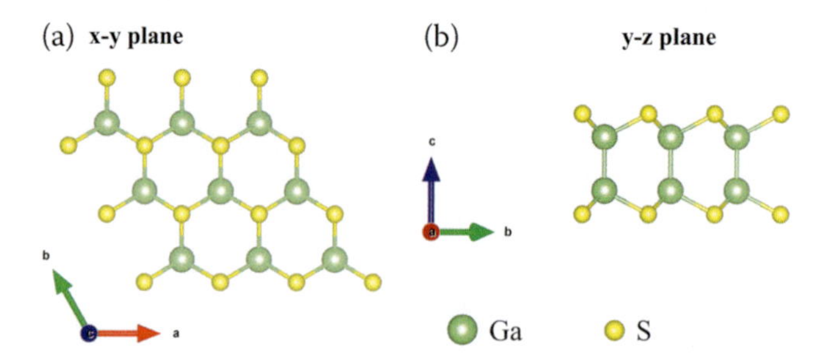

FIGURE 17.10 Monolayer gallium sulfide projected on the (a) (x, y)- and (b) (y, z)-planes.

simulations could be further utilized to predict the other crystal structures, being very useful in developing the modulation strategy under a greatly enlarged quasiparticle framework. The dimension crossover will become very interesting research topics through the near-future investigations of a series of book proposals.

The critical mechanisms, the atom- and orbital-dependent hybridizations [107,108], have been thoroughly examined from all calculated results [details in books of [152]]. For example, binary gallium sulfur compounds in 3D [Figures 17.8(a)-17.8(b)] and 2D [Figs 17.10(a)-17.10(b)] crystal structures clearly illustrate the concise pictures of [4s, $4p_x$, $4p_y$, $4p_z$]-[4s, $4p_x$, $4p_y$, $4p_z$] and [4s, $4p_x$, $4p_y$, $4p_z$]-[3s, $3p_x$, $3p_y$, $3p_z$] in Ga-Ga and Ga-S bonds, respectively, The similar multi-orbital mixings are expected to come to exist in other binary/ternary/quaternary semiconductor compounds [79,80,83]. The various chemical pictures are fully assisted by the distinct chemical/physical/material environments [discussed earlier], thus, leading to the diversified phenomena of composite quasiparticles, for example, the binary-ternary oxygen-substituted gallium sulfur compounds can exhibit strong chemical modifications by the gradual increase of the guest-atom concentrations [GaO_xS_{1-x}, such as the enlarged unit cells, the dramatic transform about the crystal symmetries from the graphite-like layered structure [x = 0 in Figures 17.8(a)-17.8(b)] to the trigonal or monoclinic system [x = 1 in Figures 17.9(a)-17.9(b)]. The chemical substitutions and absorptions are frequently utilized to create diverse quasiparticle behaviors. Of course, the distinct physical conditions are outstanding modulation factors, such as the drastic changes of spatial charge density distributions closely related to the cylindrical surfaces, open chiral or achiral edges [199], finite-width confinements, sliding translations, bilayer low-symmetry rotations, intrinsic helical structures, folded crystals, and bended chains. The VASP simulations are available in the further developments of diversified composite quasiparticles.

The multi-orbital hybridizations are very sensitive to the significant variations in the crystal symmetries, leading to giant difficulties in determining the suitable parameters in the phenomenological models [details in books of [152]]. The semiconductor compounds, which only possess the active outmost- [s, p_x, p_y, p_z] orbitals from group-II to group-VI elements [204], can create pure or impure sp^3-sp^3 mixings. The atom- and orbital-dependent hopping integrals and on-site ionization energies are strongly dependent on the various modulations of geometric symmetries [discussed earlier], e.g., the totally different among the 3D graphite-like, 2D few-layer, 1D nanotube/nanoribbon, and 0D toroidal structures. How to create the intrinsic interactions has generated numerical barriers, as clearly indicated by the almost absence of such works in the previously published scientific research. The prominent relations with the first-principles methods [15–17,205] and experimental examinations [discussed later] should be built as possible as, e.g., the full consistence of the low-lying valence and conduction subbands [206]. A reliable research strategy will be urgently finished for the semiconductor compounds with relatively concise crystal symmetries, e.g., band structures of monolayer GaS/GaSe/GaTe under the tight-binding model [207,208]. And then, this is generalized to any condensed-matter systems. Of course, certain important quasiparticle phenomena are worthy of systematic investigations, covering electron-electron interactions without/with the electron-phonon scatterings [209], and the diverse [momentum frequency]-excitation phases in the presence/absence of doping effects [42,191], the unusual magnetic quantization/Hall transports [161,210–213], the stable/quasi-stable excitons [25,214,215], and the geometry-/chemical-modification-/physics-condition-enriched quantum phonon modes [61].

In general, the experimental sides cover the various syntheses by the chemical/physical/material engineering methods, the high-resolution examinations [e.g., STM/TEM/RLEED/ARPES/STS/optical spectroscopies; [216–227]] and the industrial products the diversified chips in controlling computations, drawings, communications, electrical resistances, optical modulators, optical sensors, laser resources, biophysical detectors, and biomedicine nanoscale devices. Most of the low-dimension systems are absent up to now, while the opposite is true for carbon-sp^2-related mainstream materials [228–230]. Very interestingly, such composite quasiparticles, corresponding to the basic and applied sciences, should be summarized together to present the strong partnerships under a grand quasiparticle framework. According to recent scientific research, there exist many drawbacks in fully finding the experimental verifications, since certain merits mainly come from the part of sample syntheses and X-ray diffraction patterns. However, the high-precision measurements and analyses about the featured band structures and van Hove singularities are almost mission impossible because of the intrinsic physics limits. How to overcome the first-principles simulations is under the current investigations. Strong cooperation between the theoretical and experimental subgroups is necessary for near-future studies.

The information about the phonon properties is very useful for the basic science as well as the high-tech applications, e.g., superconducting applications. How to get reliable phonon energy spectra, its wave functions and related phonon properties using first-principles calculations are the main studying focus.

The vibrations of the ion cores around their equivalence center are caused by the temperature and external perturbations. For most excitations, the atom's displacements are very small compared to the lattice constant and thus, the potential U could be expressed as the second harmonic oscillator, in which, their motions simultaneously satisfy Newton's second law and Hooke's law. The phonon dispersions for each polarization could be achieved by diagonalizing the Dynamically matrix. Apparently, to get the appropriate phonon spectrum, we need to consider the force constant of at least several neighboring atoms, e.g., the interatomic interactions of up to four neighboring atoms can get the suitable phonon spectrum of graphene.

The harmonic approximation can provide accurate phonon results for a variety of materials, ranging from metal to semiconductors to complex oxides. However, for some systems with the destruction of the symmetry (defect, impurity, vacancy) and/or high temperatures, anharmonic terms can modify the phonon structure significantly. Moreover, under harmonic approximation, the ion core's displacement is almost the same at any temperature and thus, cannot explain the thermal expansion phenomena. The current theoretical framework also cannot explain the three-phonon scattering process and thus, phonon quasi-particles in the systems have infinite lifetimes, phonon mean-free-paths, and thus cannot explain the heat transport properties. How to build a unified theory to deal with such problems of anharmonic oscillations could be considered an open issue.

At very high temperatures, all of the vibration modes are fully excited. The heat capacity could be easy to deal with the Dulong-Petit theory at the classical limit. However, at a very low temperature below the Debye temperature, there are only certain phonon modes are active, e.g., the three acoustic modes at low frequency. The characteristics of such phonon mode will strongly affect the thermal properties of materials at low temperatures. For example, the higher specific heat capacity of graphite compares with the diamond at low temperatures due to the contribution of prominence Van Hove singularities caused by the weak van der wall interactions. The phonon properties from the first-principles calculations could be examined/compared to the phenomena model and experimental, such as the tight-binding model, neutron inelastic scattering, electron inelastic scattering, and Raman spectroscopy.

REFERENCES

1. Mukherjee, S., et al., Novel mesoporous microspheres of Al and Ni doped LMO spinels and their performance as cathodes in secondary lithium ion batteries. *International Journal of Green Energy*, 2017. **14**(7): p. 656–664.
2. Godshall, N., I. Raistrick, and R. Huggins, Thermodynamic investigations of ternary lithium-transition metal-oxygen cathode materials. *Materials Research Bulletin*, 1980. **15**(5): p. 561–570.
3. Marchini, F., et al., The hidden side of nanoporous β-Li3PS4 solid electrolyte. *Advanced Energy Materials*, 2021. **11**(34): p. 2101111.
4. Saroja, A.P.V.K., S.S. Samantaray, and R. Sundara, A room temperature multivalent rechargeable iron ion battery with an ether based electrolyte: A new type of post-lithium ion battery. *Chemical Communications*, 2019. **55**(70): p. 10416–10419.
5. Wang, A., et al., Review on modeling of the anode solid electrolyte interphase (SEI) for lithium-ion batteries. *npj Computational Materials*, 2018. **4**(1): p. 1–26.
6. Liang, J., et al., In Situ Li3PS4 Solid-State Electrolyte Protection Layers for Superior Long-Life and High-Rate Lithium-Metal Anodes. *Advanced Materials*, 2018. **30**(45): p. 1804684.
7. de Klerk, N.J., E. van der Maas, and M. Wagemaker, Analysis of diffusion in solid-state electrolytes through MD simulations, improvement of the Li-ion conductivity in β-Li3PS4 as an example. *ACS Applied Energy Materials*, 2018. **1**(7): p. 3230–3242.

8. Musa, N., et al., Optimization of Li 2 SnO 3 synthesis for anode material application in Li-ion batteries. *Materials Today: Proceedings*, 2017. **4**(4): p. 5169–5177.
9. Wenzel, S., et al., Interphase formation and degradation of charge transfer kinetics between a lithium metal anode and highly crystalline Li7P3S11 solid electrolyte. *Solid State Ionics*, 2016. **286**: p. 24–33.
10. Li, Q.-F., et al., Theoretical prediction of anode materials in Li-ion batteries on layered black and blue phosphorus. *The Journal of Physical Chemistry C*, 2015. **119**(16): p. 8662–8670.
11. Snyder, M.Q., et al., Synthesis and characterization of atomic layer deposited titanium nitride thin films on lithium titanate spinel powder as a lithium-ion battery anode. *Journal of Power Sources*, 2007. **165**(1): p. 379–385.
12. Kuganathan, N., L.H. Tsoukalas, and A. Chroneos, Defects, dopants and Li-ion diffusion in Li2SiO3. *Solid State Ionics*, 2019. **335**: p. 61–66.
13. Dathar, G.K.P., et al., Li-ion site disorder driven superionic conductivity in solid electrolytes: A first-principles investigation of β-Li 3 PS 4. *Journal of Materials Chemistry A*, 2017. **5**(3): p. 1153–1159.
14. Yang, J. and S.T. John, First-principles molecular simulations of Li diffusion in solid electrolytes Li3PS4. *Computational Materials Science*, 2015. **107**: p. 134–138.
15. Hafner, J., Ab-initio simulations of materials using VASP: Density-functional theory and beyond. *Journal of Computational Chemistry*, 2008. **29**(13): p. 2044–2078.
16. Hafner, J., Materials simulations using VASP—A quantum perspective to materials science. *Computer Physics Communications*, 2007. **177**(1–2): p. 6–13.
17. Hafner, J. and G. Kresse, *The vienna ab-initio simulation program VASP: An efficient and versatile tool for studying the structural, dynamic, and electronic properties of materials*, in *Properties of complex inorganic solids*. 1997, Springer. p. 69–82.
18. Schulz, S. and G. Czycholl, Tight-binding model for semiconductor nanostructures. *Physical Review B*, 2005. **72**(16): p. 165317.
19. Goringe, C.M., D.R. Bowler, and E. Hernández, Tight-binding modelling of materials. *Reports on Progress in Physics*, 1997. **60**(12): p. 1447–1512.
20. Phuc, N.H.H., et al., Chemical synthesis of Li3PS4 precursor suspension by liquid-phase shaking. *Solid State Ionics*, 2016. **285**: p. 2–5.
21. Liu, Z., et al., Anomalous high ionic conductivity of nanoporous β-Li3PS4. *Journal of the American Chemical Society*, 2013. **135**(3): p. 975–978.
22. Homma, K., et al., Crystal structure and phase transitions of the lithium ionic conductor Li3PS4. *Solid State Ionics*, 2011. **182**(1): p. 53–58.
23. Ma, L., et al., Nanomaterials: Science and applications in the lithium–sulfur battery. *Nano Today*, 2015. **10**(3): p. 315–338.
24. Zeng, X., J. Li, and N. Singh, Recycling of spent lithium-ion battery: A critical review. *Critical Reviews in Environmental Science and Technology*, 2014. **44**(10): p. 1129–1165.
25. Han, N.T., V.K. Dien, and M.-F. Lin, Excitonic effects in the optical spectra of Li2SiO3 compound. *Scientific Reports*, 2021. **11**(1): p. 1–10.
26. Dien, V.K., et al., Orbital-hybridization-created optical excitations in Li 2 GeO 3. *Scientific Reports*, 2021. **11**(1): p. 1–10.
27. Scott, I.D., et al., Ultrathin Coatings on Nano-LiCoO2for Li-Ion Vehicular Applications. *Nano Letters*, 2011. **11**(2): p. 414–418.
28. Yi, T.-F., et al., Recent development and application of Li4Ti5O12 as anode material of lithium ion battery. *Journal of Physics and Chemistry of Solids*, 2010. **71**(9): p. 1236–1242.
29. Molenda, J., Structural, electrical and electrochemical properties of LiNiO2. *Solid State Ionics*, 2002. **146**(1–2): p. 73–79.

30. Kanno, R., et al., Synthesis, structure, and electrochemical properties of a new lithium iron oxide, LiFeO2, with a corrugated layer structure. *Journal of the Electrochemical Society*, 1996. **143**(8): p. 2435.
31. Nakayama, M., et al., Density functional studies of olivine-type LiFePO4 and NaFePO4 as positive electrode materials for rechargeable lithium and sodium ion batteries. *Solid State Ionics*, 2016. **286**: p. 40–44.
32. Fongy, C., et al., Ionic vs electronic power limitations and analysis of the fraction of wired grains in LiFePO4 composite electrodes. *Journal of the Electrochemical Society*, 2010. **157**(7): p. A885.
33. Ong, S.P., et al., Li– Fe– P– O2 phase diagram from first principles calculations. *Chemistry of Materials*, 2008. **20**(5): p. 1798–1807.
34. Wang, V., et al., VASPKIT: A user-friendly interface facilitating high-throughput computing and analysis using VASP code. *Computer Physics Communications*, 2021. **267**: p. 108033.
35. Hacene, M., et al., Accelerating VASP electronic structure calculations using graphic processing units. *Journal of computational chemistry*, 2012. **33**(32): p. 2581–2589.
36. Rani, P., G.S. Dubey, and V. Jindal, DFT study of optical properties of pure and doped graphene. *Physica E: Low-dimensional Systems and Nanostructures*, 2014. **62**: p. 28–35.
37. Kuznetsov, A.V. and C.J. Stanton, Theory of coherent phonon oscillations in semiconductors. *Physical Review Letters*, 1994. **73**(24): p. 3243–3246.
38. Tan, D.H., et al., Elucidating reversible electrochemical redox of Li6PS5Cl solid electrolyte. *ACS Energy Letters*, 2019. **4**(10): p. 2418–2427.
39. Červenka, J., M. Katsnelson, and C. Flipse, Room-temperature ferromagnetism in graphite driven by two-dimensional networks of point defects. *Nature Physics*, 2009. **5**(11): p. 840–844.
40. Pietzsch, O., et al., Real-space observation of dipolar antiferromagnetism in magnetic nanowires by spin-polarized scanning tunneling spectroscopy. *Physical review letters*, 2000. **84**(22): p. 5212.
41. Sahin, H. and F.M. Peeters, Adsorption of alkali, alkaline-earth, and 3dtransitionmetal atoms on silicene. *Physical Review B*, 2013. **87**(8): p. 085423.
42. Dierolf, V., I. Ferguson, and J.M. Zavada, *Rare earth and transition metal doping of semiconductor materials: Synthesis, magnetic properties and room temperature spintronics*. 2016: Woodhead Publishing.
43. Schüler, M., et al., Optimal Hubbard models for materials with nonlocal Coulombinteractions: Graphene, silicene, and benzene. *Physical Review Letters*, 2013. **111**(3): p. 036601.
44. Hubbard, J., Electron correlations in narrow energy bands. *Proceedings of the Royal Society of London. Series A. Mathematical and Physical Sciences*, 1963. **276**(1365): p. 238–257.
45. Kira, M. and S.W. Koch, Many-body correlations and excitonic effects in semiconductor spectroscopy. *Progress in Quantum Electronics*, 2006. **30**(5): p. 155–296.
46. Bunaciu, A.A., E.G. Udriştioiu, and H.Y. Aboul-Enein, X-Ray diffraction: Instrumentation and applications. *Critical Reviews in Analytical Chemistry*, 2015. **45**(4): p. 289–299.
47. Rapaport, D.C. and D.C.R. Rapaport, *The art of molecular dynamics simulation*. 2004: Cambridge University Press.
48. Alder, B.J. and T.E. Wainwright, Studies in molecular dynamics. I. General method. *The Journal of Chemical Physics*, 1959. **31**(2): p. 459–466.

49. Li, W.-B., et al., Essential geometric and electronic properties in stage-n graphite alkali-metal-intercalation compounds. *RSC Advances*, 2020. **10**(40): p. 23573–23581.
50. Li, W.-B., et al., Essential electronic properties on stage-1 Li/Li+-graphite-intercalation compounds for different concentrations. arXiv preprint arXiv:2006. 12055, 2020.
51. Lin, M. and K.W.-K. Shung, Self-energy of electrons in graphite intercalation compounds. *Physical Review B*, 1996. **53**(3): p. 1109.
52. Dresselhaus, M. and G. Dresselhaus, Intercalation compounds of graphite. *Advances in Physics*, 1981. **30**(2): p. 139–326.
53. Ding, Y., et al., Femtosecond x-ray pulse characterization in free-electron lasers using a cross-correlation technique. *Physical Review Letters*, 2012. **109**(25): p. 254802.
54. Zhang, L., et al., Revealing the electrochemical charging mechanism of nanosized Li2S by in situ and operando X-ray absorption spectroscopy. *Nano Letters*, 2017. **17**(8): p. 5084–5091.
55. Kang, B. and G. Ceder, Battery materials for ultrafast charging and discharging. *Nature*, 2009. **458**(7235): p. 190–193.
56. Dieks, D., Quantum statistics, identical particles and correlations. *Synthese*, 1990. **82**(1): p. 127–155.
57. Cheianov, V.V. and V.I. Fal'ko, Friedel oscillations, impurity scattering, and temperature dependence of resistivity in graphene. *Physical Review Letters*, 2006. **97**(22): p. 226801.
58. Li, T.S. and M.F. Lin, Impurity states in semiconducting carbon nanotubes. *Journal of the Physical Society of Japan*, 2005. **74**(1): p. 425–429.
59. Özçelik, V.O., H.H. Gurel, and S. Ciraci, Self-healing of vacancy defects in single-layer graphene and silicene. *Physical Review B*, 2013. **88**(4): p. 045440.
60. Ravadgar, P., R.H. Horng, and T.Y. Wang, Healing of surface states and point defects of single-crystal β-Ga2O3 epilayers. *ECS Journal of Solid State Science and Technology*, 2012. **1**(4): p. N58–N60.
61. Nika, D.L. and A.A. Balandin, Two-dimensional phonon transport in graphene. *Journal of Physics: Condensed Matter*, 2012. **24**(23): p. 233203.
62. Young, J.A. and J.U. Koppel, Phonon spectrum of graphite. *The Journal of Chemical Physics*, 1965. **42**(1): p. 357–364.
63. Kittel, C., P. McEuen, and P. McEuen, *Introduction to solid state physics*. Vol. 8. 1996: Wiley New York.
64. Hasnain, S., Review on sustainable thermal energy storage technologies, Part I: heat storage materials and techniques. *Energy Conversion and Management*, 1998. **39**(11): p. 1127–1138.
65. Schmidt-Rohr, K., Why combustions are always exothermic, yielding about 418 kJ per mole of O2. *Journal of Chemical Education*, 2015. **92**(12): p. 2094–2099.
66. Chum, H.L. and R.P. Overend, Biomass and renewable fuels. *Fuel Processing Technology*, 2001. **71**(1–3): p. 187–195.
67. Kern, W., Chemical etching of silicon, germanium, gallium arsenide, and gallium phosphide. *RCA Review*, 1978. **39**(2): p. 278–308.
68. Passlack, M., R. Droopad, and G. Brammertz, Suitability study of oxide/gallium arsenide interfaces for MOSFET applications. *IEEE Transactions on Electron Devices*, 2010. **57**(11): p. 2944–2956.
69. Colombo, C., et al., Gallium arsenide p-i-n radial structures for photovoltaic applications. *Applied Physics Letters*, 2009. **94**(17): p. 173108.

70. Blakemore, J.S., Semiconducting and other major properties of gallium arsenide. *Journal of Applied Physics*, 1982. **53**(10): p. R123–R181.
71. She, X., et al., Review of silicon carbide power devices and their applications. *IEEE Transactions on Industrial Electronics*, 2017. **64**(10): p. 8193–8205.
72. Baliga, B.J., Gallium nitride devices for power electronic applications. *Semiconductor Science and Technology*, 2013. **28**(7): p. 074011.
73. Denis, A., G. Goglio, and G. Demazeau, Gallium nitride bulk crystal growth processes: A review. *Materials Science and Engineering: R: Reports*, 2006. **50**(6): p. 167–194.
74. Izhevskyi, V., et al., silicon carbide. Structure, properties and processing. *Cerâmica*, 2000. **46**(297): p. 4–13.
75. Casady, J. and R.W. Johnson, Status of silicon carbide (SiC) as a wide-bandgap semiconductor for high-temperature applications: A review. *Solid-State Electronics*, 1996. **39**(10): p. 1409–1422.
76. Harris, G.L., *Properties of siliconcarbide*. 1995: IET.
77. Bechstedt, F., F. Fuchs, and G. Kresse, Ab-initio theory of semiconductor band structures: New developments and progress. *Physica Status Solidi (B)*, 2009. **246**(8): p. 1877–1892.
78. Habenicht, B.F., et al., Ab initio study of vibrational dephasing of electronic excitations in semiconducting carbon nanotubes. *Nano Letters*, 2007. **7**(11): p. 3260–3265.
79. Afzaal, M. and P. O'Brien, Recent developments in II–VI and III–VI semiconductors and their applications in solar cells. *Journal of Materials Chemistry*, 2006. **16**(17): p. 1597–1602.
80. Tamargo, M.C., *II-VI semiconductor materials and their applications*. Vol. 12. 2002: CRC Press.
81. Walden, R.H., A review of recent progress in InP-based optoelectronic integrated circuit receiver front-ends. *International Journal of High Speed Electronics and Systems*, 1998. **9**(02): p. 631–642.
82. Mohammad, S.N. and H. Morkoç, Progress and prospects of group-III nitride semiconductors. *Progress in Quantum Electronics*, 1996. **20**(5–6): p. 361–525.
83. Turner, W.J., A.S. Fischler, and W.E. Reese, Physical properties of several II-V semiconductors. *Physical Review*, 1961. **121**(3): p. 759–767.
84. Hüser, F., T. Olsen, and K.S. Thygesen, Quasiparticle GW calculations for solids, molecules, and two-dimensional materials. *Physical Review B*, 2013. **87**(23).
85. Grüneis, A., et al., Tight-binding description of the quasiparticle dispersion of graphite and few-layer graphene. *Physical Review B*, 2008. **78**(20).
86. Giuliani, G.F. and J.J. Quinn, Lifetime of a quasiparticle in a two-dimensional electron gas. *Physical Review B*, 1982. **26**(8): p. 4421.
87. Pathak, C. and S.K. Pandey, Design, performance, and defect density analysis of efficient eco-friendly perovskite solar cell. *IEEE Transactions on Electron Devices*, 2020. **67**(7): p. 2837–2843.
88. Xu, Q., et al., Perovskite solar absorbers: materials by design. *Small Methods*, 2018. **2**(5): p. 1700316.
89. Bendib, T., et al., Combined optical-electrical modeling of perovskite solar cell with an optimized design. *Optical Materials*, 2020. **109**: p. 110259.
90. Green, M.A., A. Ho-Baillie, and H.J. Snaith, The emergence of perovskite solar cells. *Nature Photonics*, 2014. **8**(7): p. 506–514.
91. Sahoo, S.K., B. Manoharan, and N. Sivakumar, *Introduction: Why perovskite and perovskite solar cells?*, in *Perovskite photovoltaics*. 2018, Elsevier. p. 1–24.
92. Fonash, S., *Solar cell device physics*. 2012: Elsevier.

93. Shah, A., et al., Material and solar cell research in microcrystalline silicon. *Solar Energy Materials and Solar Cells*, 2003. **78**(1–4): p. 469–491.
94. Gross, E.K. and R.M. Dreizler, *Density functional theory*. Vol. 337. 2013: Springer Science & Business Media.
95. Bruneval, F., et al., Many-body perturbation theory using the density-functional concept: Beyond the G W approximation. *Physical Review Letters*, 2005. **94**(18): p. 186402.
96. García-González, P. and R.W. Godby, Self-consistent calculation of total energies of the electron gas using many-body perturbation theory. *Physical Review B*, 2001. **63**(7): p. 075112.
97. Idrissi, S., et al., DFT and TDDFT studies of the new inorganic perovskite CsPbI3 for solar cell applications. *Chemical Physics Letters*, 2021. **766**: p. 138347.
98. Haq, B.U., R. Ahmed, and S. Goumri-Said, DFT characterization of cadmium doped zinc oxide for photovoltaic and solar cell applications. *Solar Energy Materials and Solar Cells*, 2014. **130**: p. 6–14.
99. Taouali, W., et al., Theoretical design of new small molecules with a low band-gap for organic solar cell applications: DFT and TD-DFT study. *Computational Materials Science*, 2018. **150**: p. 54–61.
100. Tran, N.T.T., et al., Coverage-dependent essential properties of halogenated graphene: A DFT study. *Scientific Reports*, 2017. **7**(1): p. 1–13.
101. Lin, S.-Y., et al., Feature-rich electronic properties in graphene ripples. *Carbon*, 2015. **86**: p. 207–216.
102. Chang, S.-L., et al., Geometric and electronic properties of edge-decorated graphene nanoribbons. *Scientific Reports*, 2014. **4**(1): p. 1–8.
103. Wang, L., et al., Graphene oxide as an ideal substrate for hydrogen storage. *ACS Nano*, 2009. **3**(10): p. 2995–3000.
104. Tozzini, V. and V. Pellegrini, Prospects for hydrogen storage in graphene. *Physical Chemistry Chemical Physics*, 2013. **15**(1): p. 80–89.
105. Chang, S.-L., et al., Curvature effects on electronic properties of armchair graphene nanoribbons without passivation. *Physical Chemistry Chemical Physics*, 2012. **14**(47): p. 16409–16414.
106. Lin, S.-Y., et al., *Structure-and adatom-enriched essential properties of graphene nanoribbons*. 2018: CRC Press.
107. Alabugin, I.V., S. Bresch, and G. dos Passos Gomes, Orbital hybridization: A key electronic factor in control of structure and reactivity. *Journal of Physical Organic Chemistry*, 2015. **28**(2): p. 147–162.
108. Pauling, L., The nature of the chemical bond. Application of results obtained from the quantum mechanics and from a theory of paramagnetic susceptibility to the structure of molecules. *Journal of the American Chemical Society*, 1931. **53**(4): p. 1367–1400.
109. Lu, P., et al., Quasiparticle and optical properties of strained stanene and stanane. *Scientific Reports*, 2017. **7**(1): p. 1–8.
110. Bostwick, A., et al., Quasiparticle dynamics in graphene. *Nature Physics*, 2007. **3**(1): p. 36–40.
111. Ho, Y.-H., et al., Optical transitions between Landau levels: AA-stacked bilayer graphene. *Applied Physics Letters*, 2010. **97**(10): p. 101905.
112. Chen, G.D., et al., Fundamental optical transitions in GaN. *Applied Physics Letters*, 1996. **68**(20): p. 2784–2786.
113. Dumke, W., Optical transitions involving impurities in semiconductors. *Physical Review*, 1963. **132**(5): p. 1998.
114. Ho, J.H., et al., Coulomb excitations in AA- and AB-stacked bilayer graphites. *Physical Review B*, 2006. **74**(8).

115. Yan, H., et al., Infrared spectroscopy of tunable Dirac terahertz magneto-plasmons in graphene. *Nano Letters*, 2012. **12**(7): p. 3766–3771.
116. Batke, E., D. Heitmann, and C. Tu, Plasmon and magnetoplasmon excitation in two-dimensional electron space-charge layers on GaAs. *Physical Review B*, 1986. **34**(10): p. 6951.
117. Hancock, Y., et al., Generalized tight-binding transport model for graphene nanoribbon-based systems. *Physical Review B*, 2010. **81**(24): p. 245402.
118. Ando, T., Theory of electronic states and transport in carbon nanotubes. *Journal of the Physical Society of Japan*, 2005. **74**(3): p. 777–817.
119. Guinea, F., M. Katsnelson, and A. Geim, Energy gaps and a zero-field quantum Hall effect in graphene by strain engineering. *Nature Physics*, 2010. **6**(1): p. 30–33.
120. Ghosh, S., et al., Dimensional crossover of thermal transport in few-layer graphene. *Nature Materials*, 2010. **9**(7): p. 555–558.
121. Yan, W., et al., Strain and curvature induced evolution of electronic band structures in twisted graphene bilayer. *Nature Communications*, 2013. **4**(1): p. 1–7.
122. Wong, J.-H., B.-R. Wu, and M.-F. Lin, Strain effect on the electronic properties of single layer and bilayer graphene. *The Journal of Physical Chemistry C*, 2012. **116**(14): p. 8271–8277.
123. Huang, X., et al., Enhancing solar cell efficiency: The search for luminescent materials as spectral converters. *Chemical Society Reviews*, 2013. **42**(1): p. 173–201.
124. Han, N.T., et al., First-principles studies of electronic properties in lithium metasilicate (Li 2 SiO 3). *RSC Advances*, 2020. **10**(41): p. 24721–24729.
125. Kittel, C. and P. McEuen, *Kittel's introduction to solid state physics*. 2018: John Wiley & Sons.
126. Allamehzadeh, H. Wind energy history, technology and control. in *2016 IEEE Conference on Technologies for Sustainability (SusTech)*. 2016. IEEE.
127. Kaldellis, J.K. and D. Zafirakis, The wind energy (r) evolution: A short review of a long history. *Renewable Energy*, 2011. **36**(7): p. 1887–1901.
128. Bashir, A. and S. Khan, *Renewable energy sources: A study focused on wind energy*, in *Mitigating Climate Change*. 2022, Springer. p. 99–118.
129. Abbey, C. and G. Joos, Supercapacitor energy storage for wind energy applications. *IEEE Transactions on Industry Applications*, 2007. **43**(3): p. 769–776.
130. Akdağ, S.A. and A. Dinler, A new method to estimate Weibull parameters for wind energy applications. *Energy Conversion and Management*, 2009. **50**(7): p. 1761–1766.
131. Cardenas, R., et al., Overview of control systems for the operation of DFIGs in wind energy applications. *IEEE Transactions on Industrial Electronics*, 2013. **60**(7): p. 2776–2798.
132. Adams, A. and D. Keith. Wind energy and climate: Modeling the atmospheric impacts of wind energy turbines. in *AGU Fall Meeting Abstracts*. 2007.
133. Hahn, B., M. Durstewitz, and K. Rohrig, *Reliability of wind turbines*, in *Wind energy*. 2007, Springer. p. 329–332.
134. Belabes, B., et al., Evaluation of wind energy potential and estimation of cost using wind energy turbines for electricity generation in north of Algeria. *Renewable and Sustainable Energy Reviews*, 2015. **51**: p. 1245–1255.
135. Anaya-Lara, O., et al., *Wind energy generation: Modelling and control*. 2011: John Wiley & Sons.
136. Canale, M., L. Fagiano, and M. Milanese, KiteGen: A revolution in wind energy generation. *Energy*, 2009. **34**(3): p. 355–361.
137. Abad, G., et al., *Doubly fed induction machine: Modeling and control for wind energy generation*. 2011: John Wiley & Sons.

138. Xu, L. and P. Cartwright, Direct active and reactive power control of DFIG for wind energy generation. *IEEE Transactions on Energy Conversion*, 2006. **21**(3): p. 750–758.
139. Canale, M., L. Fagiano, and M. Milanese, Power kites for wind energy generation [applications of control]. *IEEE Control Systems Magazine*, 2007. **27**(6): p. 25–38.
140. Canale, M., L. Fagiano, and M. Milanese, High altitude wind energy generation using controlled power kites. *IEEE Transactions on Control Systems Technology*, 2009. **18**(2): p. 279–293.
141. Mahela, O.P. and A.G. Shaik, Comprehensive overview of grid interfaced wind energy generation systems. *Renewable and Sustainable Energy Reviews*, 2016. **57**: p. 260–281.
142. Johnson, G.L., *Wind energy systems*. 1985: Citeseer.
143. Spera, D.A., *Wind turbine technology*. 1994.
144. Blaabjerg, F. and K. Ma, Wind energy systems. *Proceedings of the IEEE*, 2017. **105**(11): p. 2116–2131.
145. Al-Bahadly, I.H., *Wind turbines*. 2011: BoD–Books on Demand.
146. Harrison, R., Hau, E., Snel, H. *Large wind turbines: Design and economics*. 2000: Wiley.
147. Hirsh, R.F. and B.K. Sovacool, Wind turbines and invisible technology: Unarticulated reasons for local opposition to wind energy. *Technology and Culture*, 2013. **54**(4): p. 705–734.
148. Blaabjerg, F. and Z. Chen, Power electronics for modern wind turbines. *Synthesis Lectures on Power Electronics*, 2005. **1**(1): p. 1–68.
149. Maradin, D., Advantages and disadvantages of renewable energy sources utilization. *International Journal of Energy Economics and Policy*, 2021. **11**(3): p. 176.
150. Chandrasekar, A., D. Flynn, and E. Syron, Operational challenges for low and high temperature electrolyzers exploiting curtailed wind energy for hydrogen production. *International Journal of Hydrogen Energy*, 2021. **46**(57): p. 28900–28911.
151. Sun, T., et al., Influence of synthesis gas components on hydrogen storage properties of sodium aluminum hexahydride. *Hydrogen*, 2021. **2**(1): p. 147–159.
152. Tran, N.T.T., et al., *Geometric and electronic properties of graphene-related systems: Chemical bonding schemes*. 2017: CRC Press.
153. Huang, H.-C., et al., Configuration-and concentration-dependent electronic properties of hydrogenated graphene. *Carbon*, 2016. **103**: p. 84–93.
154. Lin, S.-Y., et al., H–Si bonding-induced unusual electronic properties of silicene: A method to identify hydrogen concentration. *Physical Chemistry Chemical Physics*, 2015. **17**(39): p. 26443–26450.
155. Niaz, S., T. Manzoor, and A.H. Pandith, Hydrogen storage: Materials, methods and perspectives. *Renewable and Sustainable Energy Reviews*, 2015. **50**: p. 457–469.
156. Lin, S.-Y., et al., *Hydrogen-adsorbed silicene materials. Silicene-based layered materials*. 2020: IOP Publishing Limited.
157. Moradi, R. and K.M. Groth, Hydrogen storage and delivery: Review of the state of the art technologies and risk and reliability analysis. *International Journal of Hydrogen Energy*, 2019. **44**(23): p. 12254–12269.
158. Mohan, M., et al., Hydrogen storage in carbon materials—A review. *Energy Storage*, 2019. **1**(2): p. e35.
159. Geim, A.K. and K.S. Novoselov, The rise of graphene. *Nature Materials*, 2007. **6**(3): p. 183–191.
160. Lin, S.-Y., et al., *Silicene-based layered materials*. 2020: IOP Publishing Limited.
161. Lin, C.-Y., et al., *Diverse quantization phenomena in layered materials*. 2019: CRC Press.

162. Lin, C.-Y., et al., *Coulomb excitations and decays in graphene-related systems*. 2019: CRC Press.
163. Patchkovskii, S., et al., *Graphene nanostructures as tunable storage media for molecular hydrogen. Proceedings of the National Academy of Sciences*, 2005. **102**(30): p. 10439–10444.
164. Elias, D.C., et al., Control of graphene's properties by reversible hydrogenation: Evidence for graphane. *Science*, 2009. **323**(5914): p. 610–613.
165. Balog, R., et al., Bandgap opening in graphene induced by patterned hydrogen adsorption. *Nature Materials*, 2010. **9**(4): p. 315–319.
166. Nakamura, D. and H. Nakano, Liquid-phase exfoliation of Germanane based on Hansen solubility parameters. *Chemistry of Materials*, 2018. **30**(15): p. 5333–5338.
167. Qiu, J., et al., From silicene to half-silicane by hydrogenation. *ACS Nano*, 2015. **9**(11): p. 11192–11199.
168. Cadelano, E., et al., Elastic properties of hydrogenated graphene. *Physical Review B*, 2010. **82**(23): p. 235414.
169. Gmitra, M., D. Kochan, and J. Fabian, Spin-Orbit coupling in hydrogenated graphene. *Physical Review Letters*, 2013. **110**(24): p. 246602.
170. Sofo, J.O., A.S. Chaudhari, and G.D. Barber, Graphane: A two-dimensional hydrocarbon. *Physical Review B*, 2007. **75**(15): p. 153401.
171. Tran, N. T. T., et al., *Geometric and electronic properties of graphene-related systems: Chemical bondings*. arXiv e-prints, 2017: p. arXiv: 1702.02031.
172. Sailor, W.C., et al., A nuclear solution to climate change? *Science*, 2000. **288**(5469): p. 1177–1178.
173. Assonov, S., *Summary and recommendations from the International Atomic Energy Agency technical meeting on the development of stable isotope reference products (21–25 November 2016)*. 2018, Wiley Online Library.
174. Change, I.P.O.C., *Ipcc*. Climate change, 2014.
175. Modahl, I.S., et al., How methodological issues affect the energy indicator results for different electricity generation technologies. *Energy Policy*, 2013. **63**: p. 283–299.
176. Wilby, M.R., A.B.R. González, and J.J.V. Díaz, Empirical and dynamic primary energy factors. *Energy*, 2014. **73**: p. 771–779.
177. Alvarenga, R.A., I.d.O. Lins, and J.A.d. Almeida Neto, Evaluation of abiotic resource LCIA methods. *Resources*, 2016. **5**(1): p. 13.
178. Gode, J., et al., *Miljöfaktaboken 2011-Estimated emission factors for fuels, electricity, heat and transport in Sweden*. Värmeforsk, Stockholm, 2011.
179. Ojovan, M.I., W.E. Lee, and S.N. Kalmykov, *An introduction to nuclear waste immobilisation*. 2019: Elsevier.
180. Dudley, B., BP statistical review of world energy. *BP Statistical Review*, **6**(2018): p. 00116.
181. Spanger-Siegfried, E., *If you can't take the heat: How summer 2012 strained US power plants*. 2013, Union of Concerned Scientists, Cambridge, MA, USA. See http://blog.ucsusa … .
182. Wang, Y., et al., Vulnerability of existing and planned coal-fired power plants in Developing Asia to changes in climate and water resources. *Energy & Environmental Science*, 2019. **12**(10): p. 3164–3181.
183. Förster, H. and J. Lilliestam, Modeling thermoelectric power generation in view of climate change. *Regional Environmental Change*, 2010. **10**(4): p. 327–338.
184. Van Vliet, M.T., et al., Power-generation system vulnerability and adaptation to changes in climate and water resources. *Nature Climate Change*, 2016. **6**(4): p. 375–380.

185. Liu, L., et al., Vulnerability of US thermoelectric power generation to climate change when incorporating state-level environmental regulations. *Nature Energy*, 2017. **2**(8): p. 1–5.
186. Bartos, M.D. and M.V. Chester, Impacts of climate change on electric power supply in the Western United States. *Nature Climate Change*, 2015. **5**(8): p. 748–752.
187. Behrens, P., et al., Climate change and the vulnerability of electricity generation to water stress in the European Union. *Nature Energy*, 2017. **2**(8): p. 1–7.
188. De Gracia, A., et al., A simple model to predict the thermal performance of a ventilated facade with phase change materials. *Energy and Buildings*, 2015. **93**: p. 137–142.
189. Gil, A., et al., State of the art on high temperature thermal energy storage for power generation. Part 1—Concepts, materials and modellization. *Renewable and Sustainable Energy Reviews*, 2010. **14**(1): p. 31–55.
190. Qin, R., et al., First-principles calculations of mechanical and electronic properties of silicene under strain. *AIP Advances*, 2012. **2**(2): p. 022159.
191. Pham, H.D., et al., Rich p-type-doping phenomena in boron-substituted silicene systems. *Royal Society Open Science*, 2020. **7**(12): p. 200723.
192. Lv, R., et al., Nitrogen-doped graphene: Beyond single substitution and enhanced molecular sensing. *Scientific Reports*, 2012. **2**(1).
193. Han, W., et al., Boron-doped carbon nanotubes prepared through a substitution reaction. *Chemical physics letters*, 1999. **299**(5): p. 368–373.
194. Bao, C., et al., Stacking-dependent electronic structure of trilayer graphene resolved by nanospot angle-resolved photoemission spectroscopy. *Nano Letters*, 2017. **17**(3): p. 1564–1568.
195. Fu, H., et al., Stacking-dependent electronic structure of bilayer silicene. *Applied Physics Letters*, 2014. **104**(13): p. 131904.
196. Hao, Y., et al., Probing layer number and stacking order of few-layer graphene by Raman spectroscopy. *Small*, 2010. **6**(2): p. 195–200.
197. Tabert, C.J. and E.J. Nicol, Magneto-optical conductivity of silicene and other buckled honeycomb lattices. *Physical Review B*, 2013. **88**(8): p. 085434.
198. Huang, Y., C. Chang, and M.-F. Lin, Magnetic and quantum confinement effects on electronic and optical properties of graphene ribbons. *Nanotechnology*, 2007. **18**(49): p. 495401.
199. Patlatiuk, T., et al., Evolution of the quantum Hall bulk spectrum into chiral edge states. *Nature Communications*, 2018. **9**(1).
200. Neyts, E.C., A.C. Van Duin, and A. Bogaerts, Changing chirality during single-walled carbon nanotube growth: A reactive molecular dynamics/Monte Carlo study. *Journal of the American Chemical Society*, 2011. **133**(43): p. 17225–17231.
201. Shyu, F.-L., et al., Magnetoelectronic properties of chiral carbon nanotubes and tori. *Journal of Physics: Condensed Matter*, 2006. **18**(35): p. 8313.
202. Ding, J., X. Yan, and J. Cao, Analytical relation of band gaps to both chirality and diameter of single-wall carbon nanotubes. *Physical Review B*, 2002. **66**(7): p. 073401.
203. Saito, R., et al., Electronic structure of chiral graphene tubules. *Applied Physics Letters*, 1992. **60**(18): p. 2204–2206.
204. Cai, H., et al., Synthesis and emerging properties of 2D layered III–VI metal chalcogenides. *Applied Physics Reviews*, 2019. **6**(4): p. 041312.
205. Friesner, R.A., Ab initio quantum chemistry: Methodology and applications. *Proceedings of the National Academy of Sciences*, 2005. **102**(19): p. 6648–6653.
206. Do, T.-N., et al., Valley-and spin-dependent quantum Hall states in bilayer silicene. *Physical Review B*, 2019. **100**(15): p. 155403.

207. Bassani, F. and G.P. Parravicini, Band structure and optical properties of graphite and of the layer compounds GaS and GaSe. *Il Nuovo Cimento B (1965–1970)*, 1967. **50**(1): p. 95–128.
208. Doni, E., et al., Electronic properties of the III–VI layer compounds GaS, GaSe and InSe. I: Band structure. *Il Nuovo Cimento B (1971–1996)*, 1979. **51**(1): p. 154–180.
209. Tsang, J., et al., Doping and phonon renormalization in carbon nanotubes. *Nature Nanotechnology*, 2007. **2**(11): p. 725–730.
210. Huang, Y.-K., et al., Feature-rich magnetic quantization in sliding bilayer graphenes. *Scientific Reports*, 2014. **4**(1): p. 1–10.
211. Ezawa, M., Quantum Hall effects in silicene. *Journal of the Physical Society of Japan*, 2012. **81**(6): p. 064705.
212. Gusynin, V. and S. Sharapov, Unconventional integer quantum Hall effect in graphene. *Physical Review Letters*, 2005. **95**(14): p. 146801.
213. Schakel, A.M.J., Relativistic quantum Hall effect. *Physical Review D*, 1991. **43**(4): p. 1428–1431.
214. Tan, P.H., et al., Photoluminescence spectroscopy of carbon nanotube bundles: Evidence for exciton energy transfer. *Physical Review Letters*, 2007. **99**(13).
215. Mahan, G.D., Excitons in degenerate semiconductors. *Physical Review*, 1967. **153**(3): p. 882–889.
216. Binnig, G., et al., Surface studies by scanning tunneling microscopy. *Physical Review Letters*, 1982. **49**(1): p. 57.
217. Muller, D.A., Structure and bonding at the atomic scale by scanning transmission electron microscopy. *Nature Materials*, 2009. **8**(4): p. 263–270.
218. Lagally, M.G. and J.A. Martin, Instrumentation for low-energy electron diffraction. *Review of Scientific Instruments*, 1983. **54**(10): p. 1273–1288.
219. Jona, F., J. Strozier Jr, and W. Yang, Low-energy electron diffraction for surface structure analysis. *Reports on Progress in Physics*, 1982. **45**(5): p. 527.
220. Feenstra, R.M., Scanning tunneling spectroscopy. *Surface Science*, 1994. **299–300**: p. 965–979.
221. Chen, C.J., Theory of scanning tunneling spectroscopy. *Journal of Vacuum Science & Technology A: Vacuum, Surfaces, and Films*, 1988. **6**(2): p. 319–322.
222. Sobota, J.A., Y. He, and Z.-X. Shen, Angle-resolved photoemission studies of quantum materials. *Reviews of Modern Physics*, 2021. **93**(2).
223. Cattelan, M. and N. Fox, A perspective on the application of spatially resolved ARPES for 2D materials. *Nanomaterials*, 2018. **8**(5): p. 284.
224. Damascelli, A., Probing the electronic structure of complex systems by ARPES. *Physica Scripta*, 2004. **T109**: p. 61.
225. Wooten, F., *Optical properties of solids*. 2013: Academic Press.
226. Weightman, P., et al., Reflection anisotropy spectroscopy. *Reports on Progress in Physics*, 2005. **68**(6): p. 1251.
227. Gangopadhyay, P., et al., Optical absorption and photoluminescence spectroscopy of the growth of silver nanoparticles. *Physical Review Letters*, 2005. **94**(4): p. 047403.
228. Castro Neto, A.H., et al., The electronic properties of graphene. *Reviews of Modern Physics*, 2009. **81**(1): p. 109–162.
229. Novoselov, K.S., Electric field effect in atomically thin carbon films. *Science*, 2004. **306**(5696): p. 666–669.
230. Kelly, B.T., Physics of graphite. 1981: Applied Science.

18 Problems under Classical and Quantum Pictures

Jheng-Hong Shih, Thi Dieu Hien Nguyen, Wei-Bang Li, Nguyen Thi Han, and Vo Khuong Dien
Department of Physics, National Cheng Kung University, Tainan City, Taiwan

Ming-Fa Lin
Department of Physics and Hierarchical Green-Energy Material (Hi-GEM) Research Center, National Cheng Kung University, Tainan City, Taiwan

CONTENTS

Exercises, which cover the classical and quantum pictures of the basic and applied sciences, are closely related to very interesting topics and certain mainstream materials. They are fully consistent with the development of composite quasiparticles in a series of published books and very suitable for teachings and researches.

18.1 GRAPHITE/GRAPHENE/CARBON NANOTUBES/ GRAPHENE NANORIBBONS

Problem 18.1: There exist strong relations among (a) 3D bulk graphite materials [AA/AB/ABC stacking configurations in Figures 18.1(a)-18.1(c)], (b) 2D layered graphene systems with the distinct layer numbers and stacking/rotation symmetries [1–4], (c) 1D single-walled/coaxial carbon nanotubes characterized by radii and chiralities [5–7], and (d) few-/multi-layer graphene nanoribbons with quantum confinements and open edges [8,9]. The high-symmetry physical environments are fully supported by the planar/normal carbon-honeycomb lattices that are very suitable for the basic science researches under the phenomenological models [10–13]. Generally speaking, the π-electronic bondings [Figure 18.2(a)], which mainly arise from the intralayer and interlayer carbon-$2p_z$ orbital hybridizations [14–17], are sufficient in understanding the low-energy physics for the sufficiently large 1D systems, while it needs to be modified under the opposite case. According

DOI: 10.1201/9781003367215-18

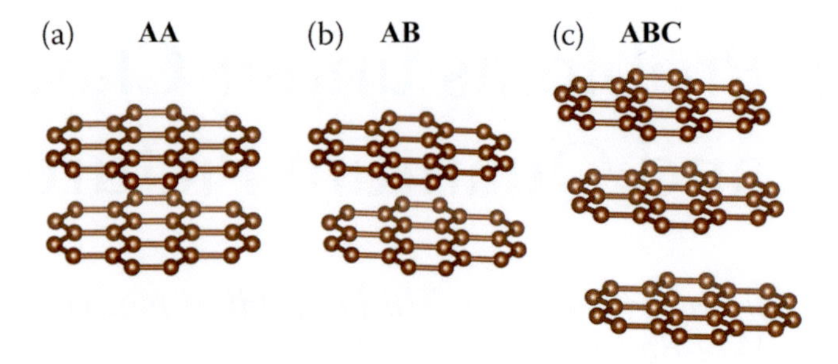

FIGURE 18.1 The (a) AA-. (b) AB, and ABC-stacked bulk graphite systems.

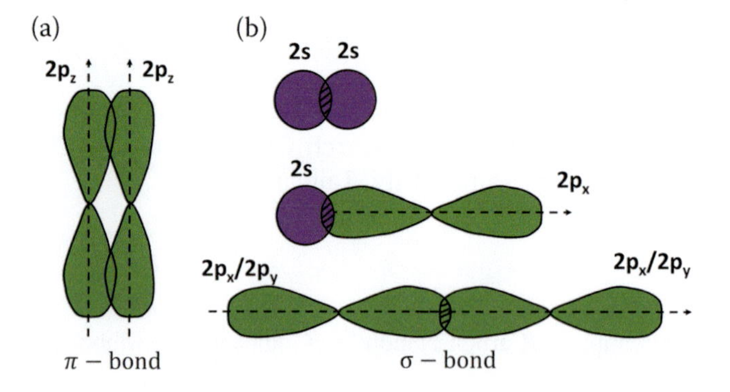

FIGURE 18.2 The (a) π- and (b) σ-electronic bondings, respectively due to $2p_z$-$2p_z$ and 2s-2s/2s-$2p_x$/$2p_x$-$2p_y$.

to the long-term theoretical and experimental studies, the carbon-carbon empirical interaction formulas have been well established in organic materials and very useful for delicate simulations. On the other hand, the σ-electronic bondings [Figure 18.2(b)] due to carbon-[2s, $2p_x$, $2p_y$] orbitals are closely associated with the higher-energy excitation behaviors of $\omega > 3.0$ eV [e.g., the $\omega_p > 5$ eV σ-plasmon modes]. Two kinds of significant bondings are perpendicular to each on the planar lattices, but their hybridizations will gradually become strong in the increase of curvatures, e.g., the smaller carbon nanotubes [18] and curved nanoribbons [19,20]. The $2p_z$-misorientation effects and/or the sp^3 bondings have been successfully utilized to modify the low-lying valence and conduction energy subbands. This leads to the creation of narrow-gap carbon nanotubes under the periodical boundary condition and the curvature-enriched distinct variations of three near-neighbor hopping integrals [21]. The minor changes of band gaps are further verified by the high-resolution measurements of scanning tunneling spectroscopy on the specific single-walled zigzag carbon nanotubes, e.g., the differential conductance versus gate voltage for the narrow-gap (9, 0), (12, 0), (12, 0), (15, 0), and (18, 0) zigzag nanotubes [Figures 18.3(a) and 18.3(b)].

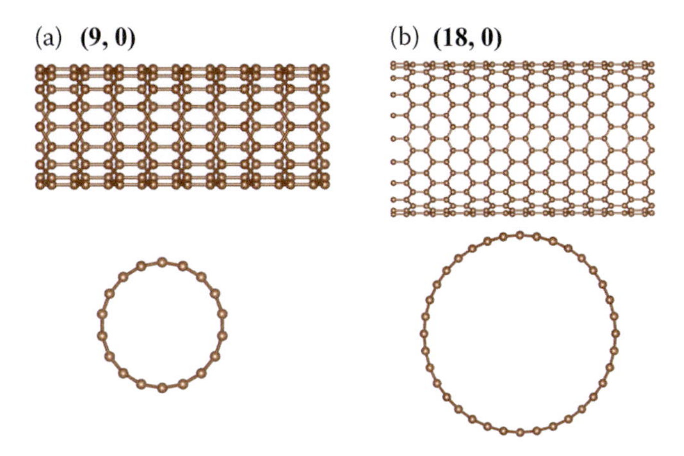

FIGURE 18.3 A single-walled zigzag carbon nanotube: (a) (9, 0), and (18, 0).

Problem 18.2: By using the intralayer and interlayer π-electronic interactions, establish the suitable Hamiltonians for the flowing high-symmetry sp^2-dominated crystal structures: (1) simply hexagonal graphite [Figure 18.1(a)], Bernal graphite [Figure 18.1(b)], and rhombohedral graphite [Figure 18.1(c)], (2) trilayer graphene systems under AAA [Figure 18.4(a)], ABA [Figure 18.4(b)], ABC [Figure 18.4(c)], and AAB stacking configurations [Figure 18.4(d)], (3) the double-wall (5, 5)–(10,10) armchair [Figure 18.5(a)] and (9,0)–(18, 0) zigzag carbon nanotubes [Figure 18.5(b)], and (4) the AA- and AB-stacked bilayer armchair/zigzag graphene nanoribbons with hydrogen passivation on two open edges [22], And then, the exact diagonalization can get the electronic energy spectra and wave functions within the first Brillouin zones. Summarize the diverse band structures and van Hove singularities, mainly owing to the dimensionalities [23,24], layer numbers [2], stacking configurations [1], radius/width, chirality [25,26], and periodical or boundary conditions along the transverse directions. Very interestingly, these physical environments have created the semimetallic behaviors [valence and conduction subband overlaps; [16,17]], as well as, zero- [a intersecting Dirac point; [27]], narrow- [28] and middle-gap semiconductors [29]. Their concise pictures need to propose

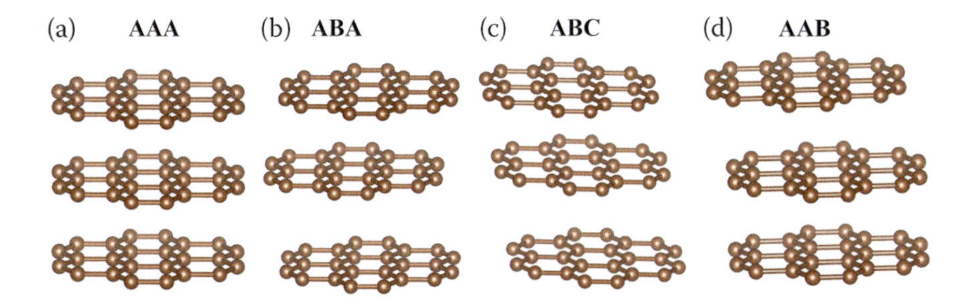

FIGURE 18.4 Trilayer graphene systems with (a) AAA, (b) ABA, (c) ABC and AAB stackings.

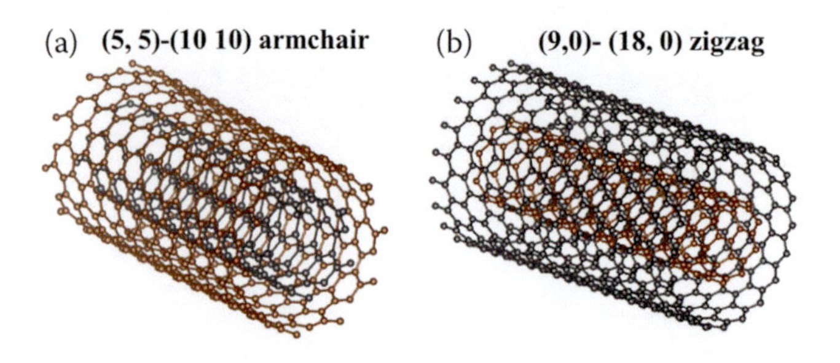

FIGURE 18.5 The double-wall (a) (5, 5)–(10,10) armchair and (b) (9,0)–(18, 0) zigzag carbon nanotubes.

in full understanding of the composite quasiparticles. It should be noticed that the π-electronic bondings could be further modified into the carbon sp^3 ones to clearly reveal the higher-frequency excitation behaviors, such as the single-particle and collective excitations of σ-electronic contributions. The low- and high-frequency essential properties, respectively, arise from the C-$2p_z$ and C-[2s, $2p_x$, $2p_y$] orbitals.

Problem 18.3: The above-mentioned main-stream materials are present in a uniform perpendicular magnetic field [30–32] and gate voltage, leading to the rich magnetic quantization [33] or the periodical Aharonov-Bohm effects [34,35]. The vector potential, which can well characterize the magnetic-field-induced quasiparticle phenomena [36], will drastically change the orbital-dependent hopping integrals of the neighboring atoms through the Peierls phases [37]. The magnetic period is commensurate with that of the crystal potential under the suitable choices of the *B* strengths [e.g., Figure 18.6]. In general, a magnetic unit cell belongs to a giant Moire superlattice for the experimental measurements, e.g., the magnetic Hamiltonian dimensions above 10,000 × 10,000 for $B_z = 1$ T. The exact diagonalization method, being built from the specific rearrangement of the complete set, has been successfully developed in previous works [33,38,39]. As a result, the band-like Hamiltonian matrices are very efficient in solving the Landau-level energy spectra and magnetic wave functions by using their localization features, as clearly illustrated for the

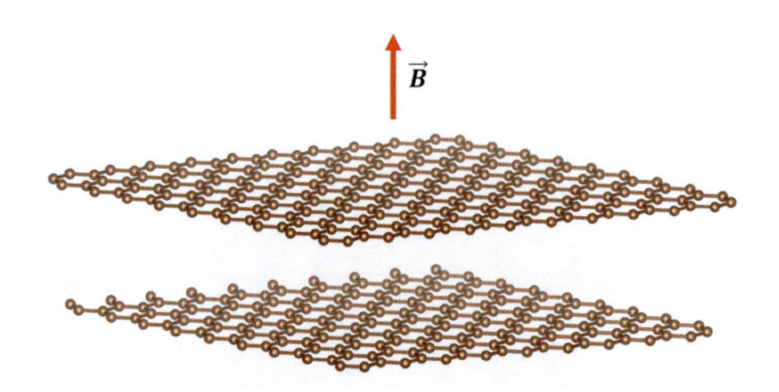

FIGURE 18.6 The AB-stacked bilayer graphene in the presence of a uniform magnetic field composite with a gate voltage, leading to a giant Moire superlattice [a very large rectangle as a unit cell].

graphene-related systems in certain books [details in [33,38,40]]. Clearly illustrate the featured magneto-electronic properties, such as the well-defined quantum number/ energy dispersions/non-crossings/crossings/anti-crossings of Landau levels in trilayer AAA [Figure 18.4(a)], ABA [Figure 18.4(b)], ABC [Figure 18.4(c)], and AAB stackings [Figure 18.4(d)]. A lot of composite quasi-particles are greatly diversified under various physics conditions through the modified methods of phenomenological models. For example, the modified random-phase approximations are successfully developed for the electronic excitations in the layered graphene systems/the coaxial carbon nanotubes [40,41], in which the intralayer/interlayer orbital hybridizations and electron-electron Coulomb interactions, as well as, the external perturbations of magnetic and electric fields, are simultaneously taken into account under a unified framework. The quasiparticle decay rates are also thoroughly investigated through the modified self-energy method [42,43] because of the prominent polarization effects of all neighboring layers [14,40]. Very interestingly, the static charge screenings of metallic and semiconducting systems are totally different from each other within the whole space [44–46], especially for the short and long distances, The former exhibit the unusual superposition behaviors of the Fermi-momentum-based Friedel oscillations in the bilayer graphenes/double-walled carbon nanotubes. How to observe them by scanning tunneling spectroscopy should be the next studying focus.

Problem 18.4: The VASP simulations are very suitable for examining the prominent chemical modification of graphene-related systems [47]. The adatom alkalizations are expected to exhibit the metal-semiconductor/metal-semimetal transitions because of the strong electron transfer from the guest to host atoms under their angle-orbital interlayer hybridizations [48–50]. For example, monolayer graphene [Figure 18.7(a)], armchair graphene nanoribbon [Figure 18.7(b)], and zigzag one [Figure 18.7(c)] clearly illustrate the top-side chemisorptions even under

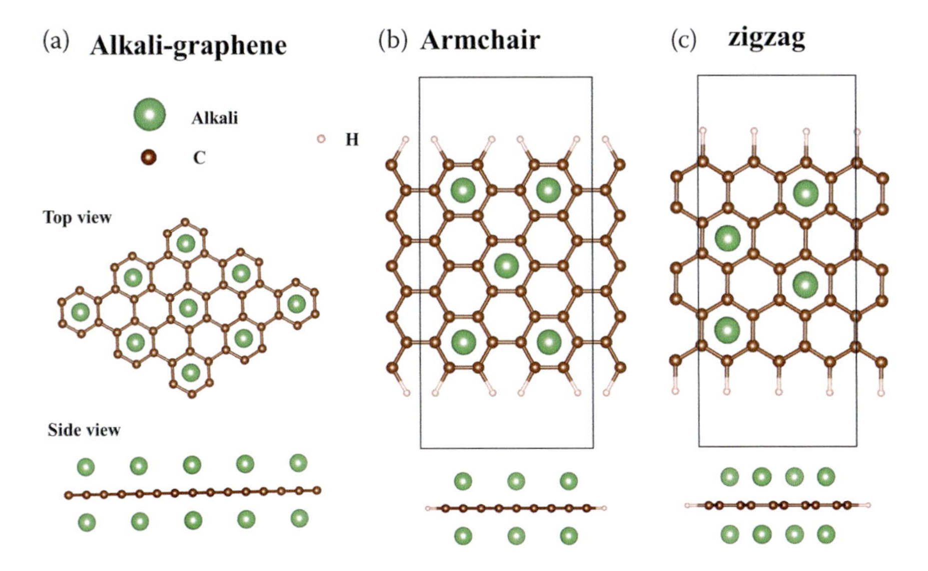

FIGURE 18.7 The double-side alkalization of (a) monolayer graphene, (b) armchair graphene nanoribbon, and (c) zigzag one.

100% double-side cases. Fully explore the main features of optimal geometries, such as the guest-host bond lengths, the planar or buckled honeycomb structures, and open-edge-induced C-C bond length fluctuations through Li-/Na-/K-/Rb-/Cs-adsorptions. Investigate the alkali- and carbon-dominated energy spectra and wave functions and evaluate the existence of the blue shift about the Fermi level. The metallic alkalizations are able to induce the 1D free conduction electrons so that the carrier density is well characterized by the Fermi momenta. Thoroughly examine that it is just proportional to the adatom concentration, being independent of geometric properties and adatom configuration/concentration [22]. This clearly indicates the similar chemical bondings between carbon and all alkali atoms. However, the linear relationships are very difficult to identify in a 2D alkalized monolayer graphene. Propose a concise picture of chemistry and physics in the absence of close partnerships. Similar studies could be generalized to alkalized silicene/germanene/tinene/plumbene materials. The diversified quasiparticles will come to exist frequently in mainstream materials.

Problem 18.5: The up-to-date experimental researches show the high performances of lithium sulfide batteries [51–54], The lithium-, phosphorus-, and sulfur-based material possess special functionalities because of the unusual quasiparticle behaviors. A lot of single-element, binary [55–57], and ternary [54,58,59] compounds have been successfully synthesized by various chemical techniques. Most of the complicated compounds belong to giant Moire superlattices, e.g., those in P_4S_x ($x \leq 10$); [60]]. The various crystal phases are thoroughly identified from the high-resolution X-ray diffraction patterns, in which the position-dependent phosphorus-sulfur bonds exhibit observable fluctuations. The modulation strategy, of course, is the highly non-uniform and anisotropic environments of chemistry, physics, and material engineering. The modulation strategies, which clearly illustrate the diverse quasiparticles, are available for the pristine [12,61,62], chemically modified [e.g., absorptions/substitutions/intercalations/decorations; [63–68]], and physically perturbed materials [applications of gate voltages/magnetic fields/n- or p-type dopings/thermal excitations/mechanical strains; [64,69–73]]. The above-mentioned systems are outstanding candidates for enlarging a unified framework within a series of published books.

Problem 18.6: The diversified quasiparticle phenomena are initiated from the distinct sample syntheses. Their growth processes could be simulated from the molecular dynamics [74,75] under the full assistance of the first-principles methods [76–78]. And then, the latter is available for the delicate examinations of essential physical/chemical/material properties. For example, the optimal 3D crystal structure of binary $P_{32}S_{24}$ compound, as clearly illustrated along (a)/(b)/(c) the different [x, y, z] directions and (d) the first Brillouin zone in Figures 18.8(a)-18.8(d), displays a very complicated rhombohedral cubic. The crystal symmetries are greatly diversified by changing the relative host-atom concentration [e.g., various binary P_4S_x compounds; [57,79,80]], the different atom components [lithium phosphorus/sulfur compounds; [55,56,81]], the various multi-components [single-element and ternary systems], the substituted cases. Obviously, it is very interesting to thoroughly identify chemical bondings through the VASP simulations [78], being very useful and responsible for determining the rich and unique Moire superlattices. The critical mechanisms, as done

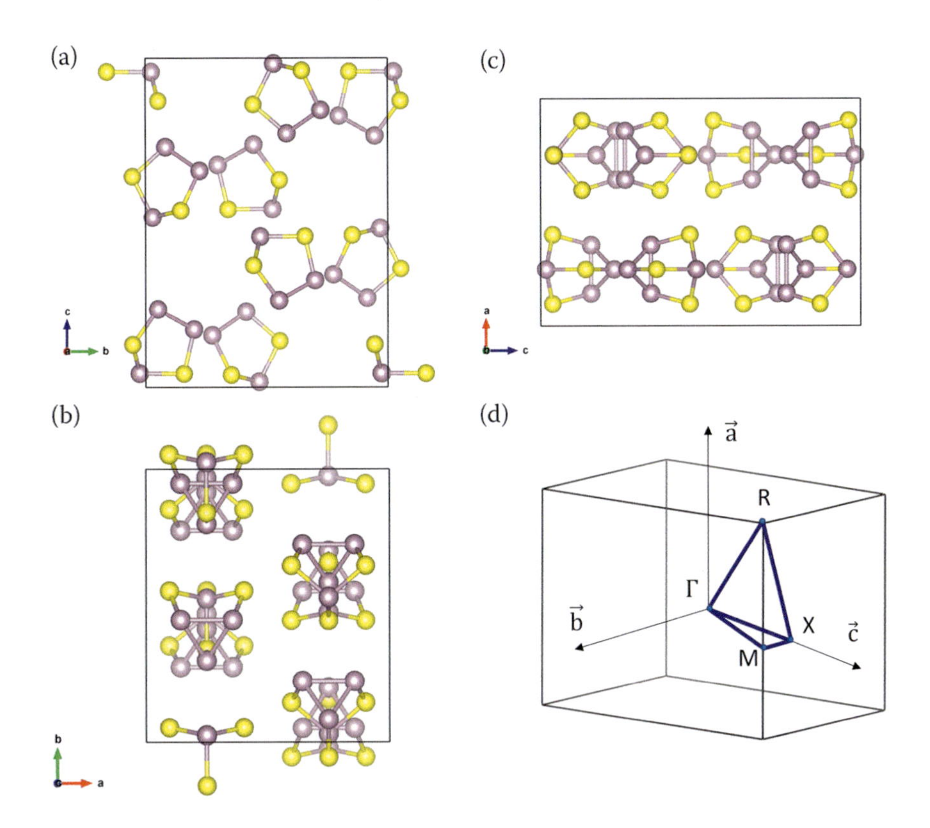

FIGURE 18.8 The optimal 3D crystal structure of binary $P_{32}S_{24}$ compound along (a)/(b)/(c) the different [x, y, z] directions and (d) the first Brillouin zone of the rhombohedral cubic system.

in the previously published books, are achieved from the complicated Moire crystals [82–85], the featured band structures and atom-dominated wave functions, the spatial charge density distributions in the presence of rather prominent multi-orbital mixings, and the obviously merged van Hove singularities due to the atom- and-orbital-projected contributions. That is, the sp^3-sp^3 multi-orbital hybridizations will exhibit diverse behaviors; therefore, the composite quasiparticles are very suitable for research strategies. In addition, whether the low-dimensional materials could survive in experimental laboratories is worthy of theoretical predictions, e.g., the few-layer [86–88], nanotube [89], and nanoribbon structures [90,91].

Problem 18.7: The 3D lithium crystal is an unusual crystal. Use the VASP simulations to identify the bcc symmetry [Figure 18.9(a)] and then analyze the featured geometric properties in a table. Plot the spatial charge density distribution to examine the single-orbital hybridizations in Li-Li bonds and clearly illustrate the 2s-orbital distortions after the crystal formation. Evaluate whether the metallic bonds could be estimated from the high-resolution measurements of the X-ray diffraction patterns [92–94]. Very interestingly, band structure, wave functions

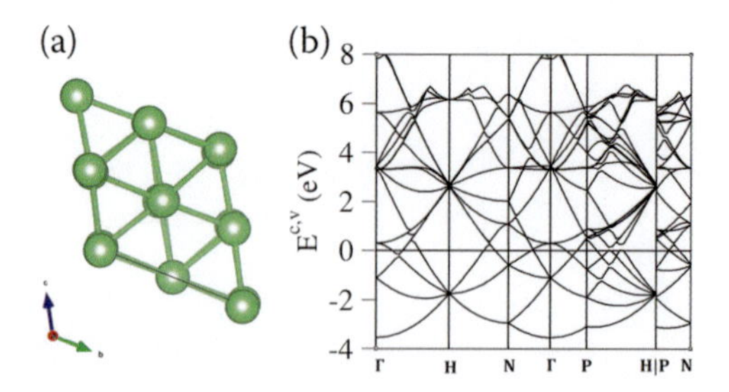

FIGURE 18.9 (a) The optimal lithium crystal and (b) its band structure along the high-symmetry points of the first Brillouin zone.

[Figure 18.9(b)], and van Hove singularities [95,96] show the metallic behaviors in the presence of 2s-2s bondings across the valence and conduction energy subbands. Check that the 3D occupied conduction and valence energy spectra have been successfully verified from the high-precision measurements of angle-resolved photoemission spectroscopy [97–100]. Also, examine its isotropy and the Fermi momenta, being suitable for the estimation of free carrier density or not. Under an electron gas model, roughly estimate the conduction electron density/the isotropic Fermi momenta from a normal crystal structure. Calculate an average kinetic energy per conduction. Such calculations could be generalized to the first-order electron-electron Coulomb interactions [the exchange energy per particle; [101]], of course, the 3D electron gas is an outstanding candidate n thoroughly exploring the diverse [momentum. frequency]-excitation diagrams [e.g., acoustic plasmon modes; [102]], and the quasiparticle lifetimes [62]. On the other side, both 3D phosphorus [Figure 18.10(a)] and sulfur crystals [Figure 18.10(b)] are totally different from lithium one because of the distinct crystal symmetries. Their pure sp^3-sp^3 multi-orbital hybridizations are responsible for all essential properties, such as

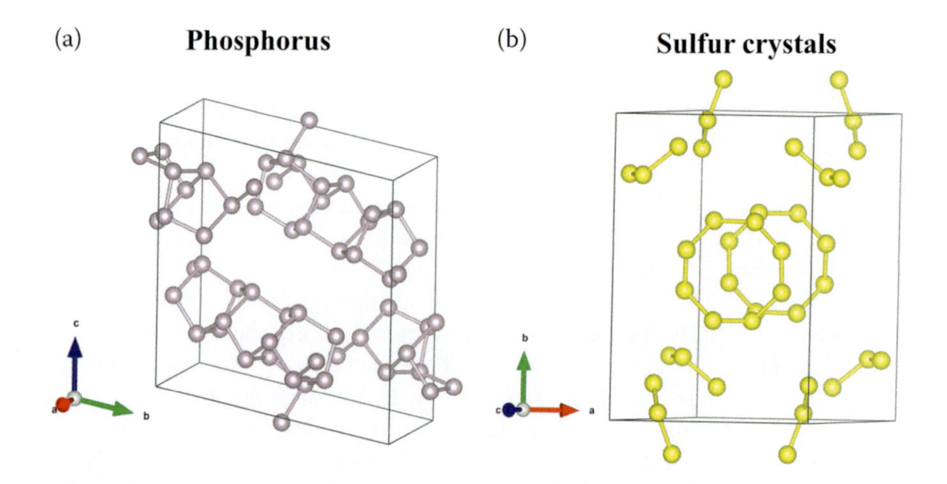

FIGURE 18.10 The 3D (a) phosphorus and (b) sulfur crystals.

semiconducting behaviors, optical plasmon modes [103], threshold absorption frequency, and excitonic effects [104–106].

Problem 18.8: Optical properties of low-dimension materials, being clearly elucidated by the Fermi-golden role or Kubo formula. In addition, the attractive Coulomb interactions of the excited holes and excited electrons, excitonic effects, will also strongly influence the optical absorbance spectra. (a) Achieve the imaginary and the real parts of dielectric functions of monolayer graphene, silicene, germane, MoS_2 under the case of with and without excitonic effects. (b) Clearly characterize the orbital hybridization of the initial states and the final states related to the prominence absorption peaks (c) Explore the other optical properties such as the energy loss function, absorption coefficient, reflection. (d) Similar studies could be done for 1D graphene, silicene, and germanene nanoribbons. Indicate the diversified optical properties and estimate the strength of Coulomb interactions between excited holes and excited electrons.

Problem 18.9: The global impact of greenhouse gases from the internal combustion energies and coalfired power has attracted much research on green energy, e.g., the ingathering and storing of electrical energy from wind and solar power. This situation facilitates the development of energy storage devices. In addition, there exists many potential materials and methods of energy resources, such as alkalization. Take $Li_4Ti_5O_{12}$ as an example, which attracts much attention as an anode material for lithium-ion battery, recent studies display its excellent rate capability and cycle stability [107–109]. The $Li_4Ti_5O_{12}$ is fabricated by in situ synthesis of Li-Ti-O/RGO (reduced graphene oxide) precursor by microwave-assisted alkalization followed by heat treatment. The nanocomposite of $Li_4Ti_5O_{12}$ also performs an excellent high-rate capability even at 20 C-rate over 300 cycles [110]. It is worth to further investigate the detailed physical/chemical properties, such as optimized structures, spatial charge distributions, and the complicated hybridization of orbitals. How to build the complete theoretical framework corresponding to the experiments is quite vital and necessary. Graphene-related systems exhibit a lot of remarkable essential properties arising from the hexagonal symmetry and the nanoscale size, such as high mobility of charge carriers, high transparency, and semiconducting and semi-metal behaviors. The structure-enriched graphenes, graphene oxides, hydrogenated graphene, halogenated graphene, and doped graphene are taken into consideration for emerging materials of electrodes in lithium-ion batteries. The chemical modification on graphene surfaces is a useful tool to engineer new electronic properties. Take the graphane as an example, which is a hydrocarbon compound with extended covalent C-H bonds. This special compound can be oxidized to generate carbon dioxides, water (H_2O), and energy released, so it is a good fuel material. Furthermore, the energy gap of the hydrogenated graphene systems is easily modulated by the distinct concentrations and distributions of H atoms. The hydrogen concentration can be estimated from the intensity ratio of D and G bands in Raman spectroscopy [111,112], and the hydrogen-induced energy gap is verified from the ARPES [113,114]. The delicate theoretical simulations indicate that there exist hydrogen-distribution-dependent non-magnetic and magnetic properties and therein is predicted to possess the splitting of spin-up and spin-down energy bands. Such electronic states in transport measurements are expected to produce the spin-polarized currents which can be applied in spintronic devices. We look forward to much more experiments and applications to support these simulations in the near future

Problem 18.10: A solar cell is an electrical device that transforms light energy into electrical energy via the photovoltaic effect. The advantages of solar cells are not the pollution associated with them, no maintenance cost, and they must last for a long time. Opposite, the disadvantages of the solar cells are the high cost of installation and low efficiency. During cloudy days, energy cannot be produced, and then we will not receive it at night. Though solar cells have several limitations, these problems are likely to be addressed as technology progresses. As technology advances, the cost of solar plates, as well as installation costs, will fall, allowing anybody to effort to build the system. Apparently, perovskite solar cells (PSCs) have recently been widely advocated as a cost-effective and environmentally friendly renewable technology alternative to standard solar cell technologies for solving global concerns in energy generation and climate change. For example, the metal halide perovskites [115,116], with a general formula of ABX_3 (where: A is cation, B is metal cation, and X is halide anion) (Figure 18.11) which are outstanding candidates and are expected to be sketching the diverse quasi-particle phenomena [117–119], have recently emerged as an intriguing class of semiconductors. Such solar-cell materials should be deliberated in the current development of the quasi-particle framework [120,121] as $CsGeX_3$ (X= Cl, Br and I), $BiCrO_3$, and so on.

How to affirm the close relations among the featured properties is worthy of the systematic research. Similarly, reliable verifications of different physical properties require high-precision experimental test. For example, the measurements, such as the X-ray diffraction (XRD) [93,94], the angle-resolved photoemission spectroscopy (ARPES) [122,123] on the occupied energy spectrum can examine the crystal symmetry, the energy gap in band structure, respectively. Besides, the van Hove singularity in the density of states (DOS) can be tested by scanning tunneling

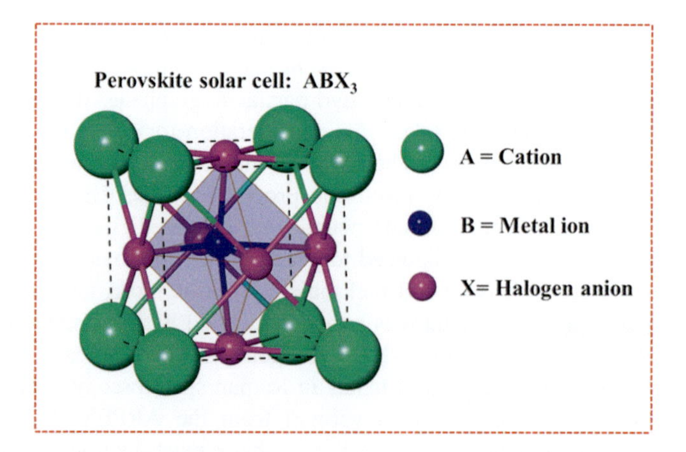

FIGURE 18.11 The optimized structure of the metal halide perovskites: ABX_3 (A= cation, B= metal ion, and X= halogen anion).

spectroscopy (STS). The reflectance, absorption, transmittance, and photoluminescence spectra could be tested by optical spectroscopies [123,124].

Problem 18.11: The harmonic approximation can provide accurate phonon results for a variety of materials, ranging from metal to semiconductors to complex oxides. (a) Express the merits and drawbacks of the harmonic approximation. (b) How to include charge screening effects [valence and conduction carriers] in numerical simulations through the reliable couplings of neighboring oscillators. (c) Getting the effective Hooke force constants within the first, second, third neighboring oscillators of graphene. Whether this strategy can establish the approximate phenomenological models? (d) Express The close relations/the simultaneous progresses of VASP simulations, phenomenological models, and experimental observations under a grand quasiparticle framework.

Problem 18.12: Semiconductor compounds are outstanding candidates for the basic and applied sciences, as well as, even the industry products. In general, according to the atomic configurations in a periodical table [125,126]. Such main-stream materials, which possess the active four orbitals of [s, p_x, p_y, p_z], could be classified into binary III-VI/III-V/IV-IV/II-VI/II-V [127–133] and more complicated ternary and quaternary compounds. Their semiconducting behaviors are able to exhibit diverse phenomena through the significant modifications of chemistry, physics, and material engineering. The main features are suitable for developing functional modulations and thus very useful in providing electronic, optical, transport, energy, and mechanic devices. For example, the first- [134,135], second- [136,137], third- [127,132,138–140], and fourth-chips have been successfully utilized to produce various chips with high-level functionalities, such as CPU. communications, figures, laser optics, solar cells, ion-based batteries energy storage, vehicle controls, electrical resistance, electronic resistors, light modulators, sound sensors, and biological detectors. The large-scale generations of silicon or germanium semiconductors [135], as well as semiconductor compounds [128,129,133], are very successful in industrial companies. Whether there exists another composite way of greatly enhancing merits and reducing drawbacks is the next studying focus. Similar cases appear in the functionalized chips and the integrated engineering. The close partnerships among them can successfully bridge the real gaps between the basic and applied sciences.

Problem 18.13: A lot of main-stream semiconductor compounds could be chosen as model systems in developing composite quasiparticles through a grand framework, as clearly illustrated in the previously published books. For example, both boron nitrogen [Figures 18.12(a)-18.12(b)] and silicon carbon [Figures 18.13(a)-18.13(b)] binary compounds are very suitable in observing the unusual quasiparticle behaviors, such as the insulating behaviors, the abnormal plasmon modes, excellent mechanical coefficients, the abnormal quantum Hall transports [141,142], the excitonic effects/the large threshold absorption frequency, and the rich phonon modes. Using the delicate VASP simulations, determine the rich environments of chemistry/physics/material engineering and analyze the main features of crystal symmetries [78]. And then, how to thoroughly comprehend the close relationships between the position-dependent chemical bonds and the other fundamental properties. The critical multi-orbital hybridizations need to be identified from all

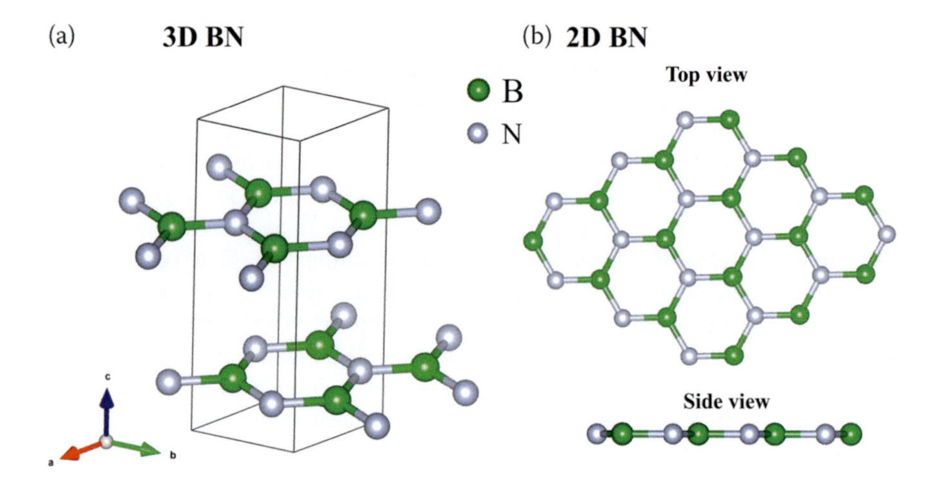

FIGURE 18.12 The optimal geometric structure of (a) 3D and (b) 2D binary boron nitrogen compounds.

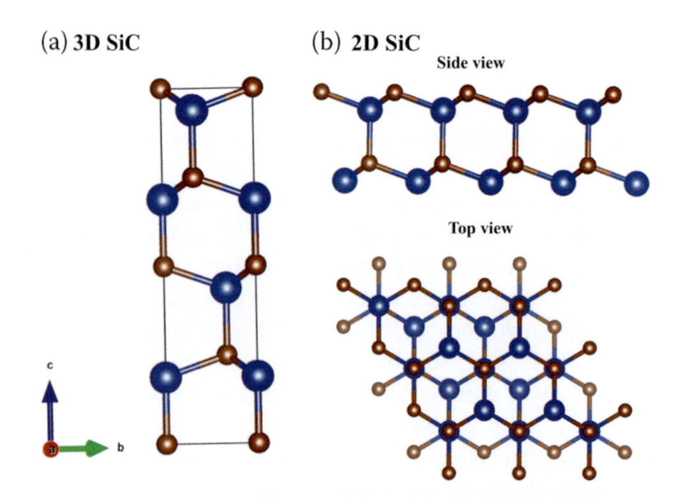

FIGURE 18.13 The crystal structures for (a) 3D and (b) 2D mono-layer silicon carbides.

calculated quantities. Very interesting, 3D boron nitrogen shows a graphite-like layered structure [143], and 2D systems exhibit the few-/multi-layer hexagonal ones without any buckling [144]. The dimensionality crossover should be worthy of systematic investigations [23,24,66]. Illustrate why a 2D composite of boron nitrogen and graphene, with very weak, but significant van der Waals interactions [145,146], is very suitable for the experimental examinations of magnetic quantization phenomenon on the other side, silicon carbide has a 3D quasi-diamond and 2D layered honeycomb lattices. The relative concentrations/arrangements could be modulated by using chemical substitutions [64,147] and thus create dramatic crystal phase transitions. Of course, the diversified quasiparticle phenomena are very sensitive to the variations of impure sp^3-sp^3 multi-orbital hybridizations. Propose that there exist certain important differences between the two materials, being very useful in serving as the available modulation strategy.

Problem 18.14: Many interesting topics could be done through modulation engineering for semiconductor compounds. Under an enlarged framework, the simultaneous developments of phenomenological models [10,148] and first-principles simulations are urgently requested in the near-future researches. The reliable parameters of the tight-binding model/the generalized tight-binding model [149–153] need to be obtained from the well fittings with the lower-lying valence and conduction energy subbands. They cover the atom- and orbital- and configuration-dependent hopping integrals [154,155] and on-site ionization energies. However, the spin-induced single- and many-particle couplings are absent because of the four sp^3 active orbitals. Their magnitudes are able to understand the various bonding strengths in different subgroups, such as the close partnerships among III-V/IV-IV/III-VI/II-V/II-VI from a list of summarized tables. How to utilize them in greatly diversifying the composite quasiparticles must be the studying focus. For example, evaluate whether the wide-gap semiconductors could efficiently generate the exciton decays [104,105] and laser lights, while such materials are not suitable for magnetic quantization/quantum transports [33,156]. Plenty of significant viewpoints, which could be used as modulation strategies, frequently come to exist in basic and applied scientific researches, especially for semiconductor compounds with specific functionalities [157,158].

REFERENCES

1. Fu, H., et al., Stacking-dependent electronic structure of bilayer silicene. *Applied Physics Letters*, 2014. **104**(13): p. 131904.
2. Hao, Y., et al., Probing layer number and stacking order of few-layer graphene by Raman spectroscopy. *Small*, 2010. **6**(2): p. 195–200.
3. Bao, C., et al., Stacking-dependent electronic structure of trilayer graphene resolved by nanospot angle-resolved photoemission spectroscopy. *Nano Letters*, 2017. **17**(3): p. 1564–1568.
4. Do, T.-N., et al., Configuration-enriched magneto-electronic spectra of AAB-stacked trilayer graphene. *Carbon*, 2015. **94**: p. 619–632.
5. Martel, R., et al., Single-and multi-wall carbon nanotube field-effect transistors. *Applied Physics Letters*, 1998. **73**(17): p. 2447–2449.
6. Dresselhaus, G., M.S. Dresselhaus, and R. Saito, Physical properties of carbon nanotubes. 1998: World scientific.
7. Tans, S.J., et al., Individual single-wall carbon nanotubes as quantum wires. *Nature*, 1997. **386**(6624): p. 474–477.
8. Terrones, M., et al., Graphene and graphite nanoribbons: Morphology, properties, synthesis, defects and applications. *Nano Today*, 2010. **5**(4): p. 351–372.
9. Huang, Y., C. Chang, and M.-F. Lin, Magnetic and quantum confinement effects on electronic and optical properties of graphene ribbons. *Nanotechnology*, 2007. **18**(49): p. 495401.
10. Kliros, G.S., A phenomenological model for the quantum capacitance of monolayer and bilayer graphene devices. arXiv preprint arXiv:1105.5827, 2011.

11. Konschuh, S., M. Gmitra, and J. Fabian, Tight-binding theory of the spin-orbit coupling in graphene. *Physical Review B*, 2010. **82**(24): p. 245412.
12. Grüneis, A., et al., Tight-binding description of the quasiparticle dispersion of graphite and few-layer graphene. *Physical Review B*, 2008. **78**(20): p. 205425.
13. Reich, S., et al., Tight-binding description of graphene. *Physical Review B*, 2002. **66**(3): p. 035412.
14. Lin, C.-Y., et al., Electronic and optical properties of graphite-related systems. 2017: CRC Press.
15. Kelly, B.T., Physics of graphite. 1981.
16. Slonczewski, J.C. and P.R. Weiss, Band structure of graphite. *Physical Review*, 1958. **109**(2): p. 272–279.
17. Wallace, P.R., The Band Theory of graphite. *Physical Review*, 1947. **71**(9): p. 622–634.
18. Lu, X. and Z. Chen, Curved Pi-conjugation, aromaticity, and the related chemistry of small fullerenes (<C60) and single-walled carbon nanotubes. *Chemical Reviews*, 2005. **105**(10): p. 3643–3696.
19. Lin, C.-Y., et al., Magneto-optical selection rules of curved graphene nanoribbons and carbon nanotubes. *Carbon*, 2014. **69**: p. 151–161.
20. Li, T., M. Lin, and S. Chang, Modulation of curved graphene nanoribbon optical absorption spectra by an electric field. *Philosophical Magazine*, 2012. **92**(34): p. 4376–4388.
21. Sticlet, D. and F. Piéchon, Distant-neighbor hopping in graphene and Haldane models. *Physical Review B*, 2013. **87**(11): p. 115402.
22. Lin, S.-Y., et al., Structure-and adatom-enriched essential properties of graphene nanoribbons. 2018: CRC press.
23. Ho, C.-H., C.-P. Chang, and M.-F. Lin, Evolution and dimensional crossover from the bulk subbands in ABC-stacked graphene to a three-dimensional Dirac cone structure in rhombohedral graphite. *Physical Review B*, 2016. **93**(7): p. 075437.
24. Ghosh, S., et al., Dimensional crossover of thermal transport in few-layer graphene. *Nature Materials*, 2010. **9**(7): p. 555–558.
25. Neyts, E.C., A.C. Van Duin, and A. Bogaerts, Changing chirality during single-walled carbon nanotube growth: a reactive molecular dynamics/Monte Carlo study. *Journal of the American Chemical Society*, 2011. **133**(43): p. 17225–17231.
26. Ding, J., X. Yan, and J. Cao, Analytical relation of band gaps to both chirality and diameter of single-wall carbon nanotubes. *Physical Review B*, 2002. **66**(7): p. 073401.
27. Castro Neto, A.H., et al., The electronic properties of graphene. *Reviews of Modern Physics*, 2009. **81**(1): p. 109–162.
28. Gunlycke, D., et al., Bandgap modulation of narrow-gap carbon nanotubes in a transverse electric field. *EPL (Europhysics Letters)*, 2006. **73**(5): p. 759.
29. Guo, J., et al., Toward multiscale modeling of carbon nanotube transistors. *International Journal for Multiscale Computational Engineering*, 2004. **2**(2): p. 257–276.
30. Do, T.-N., et al., Peculiar optical properties of bilayer silicene under the influence of external electric and magnetic fields. *Scientific Reports*, 2019. **9**(1): p. 624.
31. Gorbar, E.V., et al., Magnetic field driven metal-insulator phase transition in planar systems. *Physical Review B*, 2002. **66**(4): p. 045108.
32. Wu, J.-Y., Y.-H. Ho, and M.-F. Lin, Magnetic-modulation effects in bilayer Bernal graphene. *The Journal of Physical Chemistry C*, 2010. **114**(27): p. 11940–11945.
33. Lin, C.-Y., et al., Diverse quantization phenomena in layered materials. 2019: CRC Press.

34. Ajiki, H. and T. Ando, Aharonov-Bohm effect in carbon nanotubes. *Physica B: Condensed Matter*, 1994. **201**: p. 349–352.
35. Peshkin, M. and A. Tonomura, The Aharonov-Bohm Effect. Vol. 340. 1989: Springer.
36. Khveshchenko, D., Magnetic-field-induced insulating behavior in highly oriented pyrolitic graphite. *Physical Review Letters*, 2001. **87**(20): p. 206401.
37. Chiu, Y.-H., et al., Electronic structure of a two-dimensional graphene monolayer in a spatially modulated magnetic field: Peierls tight-binding model. *Physical Review B*, 2008. **77**(4): p. 045407.
38. Lin, C.-Y., et al., Optical properties of graphene in magnetic and electric fields. 2017: IOP Publishing.
39. Chen, S.-C., et al., Theory of magnetoelectric properties of 2d systems. 2017: IOP Publishing.
40. Lin, C.-Y., et al., Coulomb excitations and decays in graphene-related systems. 2019: CRC Press.
41. Lin, M.-F. and F.-L. Shyu, Low-frequency electronic excitations in doped carbon nanotubes. *Physica B: Condensed Matter*, 2000. **292**(1–2): p. 117–126.
42. Lin, M. and K.W.-K. Shung, Self-energy of electrons in graphite intercalation compounds. *Physical Review B*, 1996. **53**(3): p. 1109.
43. Ren, X., et al., Random-phase approximation and its applications in computational chemistry and materials science. *Journal of Materials Science*, 2012. **47**(21): p. 7447–7471.
44. Roldán, R. and L. Brey, Dielectric screening and plasmons in AA-stacked bilayer graphene. *Physical Review B*, 2013. **88**(11): p. 115420.
45. Scholz, A., T. Stauber, and J. Schliemann, Dielectric function, screening, and plasmons of graphene in the presence of spin-orbit interactions. *Physical Review B*, 2012. **86**(19): p.195424.
46. Hwang, E. and S.D. Sarma, Screening, Kohn anomaly, Friedel oscillation, and RKKY interaction in bilayer graphene. *Physical Review Letters*, 2008. **101**(15): p. 156802.
47. Tran, N.T.T., et al., Geometric and electronic properties of graphene-related systems: Chemical bonding schemes. 2017: CRC Press.
48. Palser, A.H., Interlayer interactions in graphite and carbon nanotubes. *Physical Chemistry Chemical Physics*, 1999. **1**(18): p. 4459–4464.
49. Koshino, M., Interlayer screening effect in graphene multilayers with A B A and A B C stacking. *Physical Review B*, 2010. **81**(12): p. 125304.
50. Ohta, T., et al., Interlayer interaction and electronic screening in multilayer graphene investigated with angle-resolved photoemission spectroscopy. *Physical Review Letters*, 2007. **98**(20): p. 206802.
51. Kudu, Ö.U., et al., A review of structural properties and synthesis methods of solid electrolyte materials in the Li2S−P2S5 binary system. *Journal of Power Sources*, 2018. **407**: p. 31–43.
52. Dietrich, C., et al., Synthesis, structural characterization, and lithium ion conductivity of the lithium thiophosphate Li2P2S6. *Inorganic Chemistry*, 2017. **56**(11): p. 6681–6687.
53. Homma, K., et al., Crystal structure and phase transitions of the lithium ionic conductor Li3PS4. *Solid State Ionics*, 2011. **182**(1): p. 53–58.
54. Yamane, H., et al., Crystal structure of a superionic conductor, Li7P3S11. *Solid State Ionics*, 2007. **178**(15–18): p. 1163–1167.
55. Zhao, Z., et al., Pressure-induced stable Li5P for high-performance lithium-ion batteries. *The Journal of Physical Chemistry C*, 2017. **121**(39): p. 21199–21205.

56. Hayashi, A., et al., Preparation of Li2S–P2S5 amorphous solid electrolytes by mechanical milling. *Journal of the American Ceramic Society*, 2001. **84**(2): p. 477–479.
57. Tahri, Y., H. Chermette, and G. Hollinger, Electronic structures of phosphorus oxide P4O10 and phosphorus sulfides P4S10 and P4S7. *Journal of Electron Spectroscopy and Related Phenomena*, 1991. **56**(1): p. 51–69.
58. Hakari, T., et al., All-solid-state lithium batteries with Li3PS4 glass as active material. *Journal of Power Sources*, 2015. **293**: p. 721–725.
59. Lepley, N., N. Holzwarth, and Y.A. Du, Structures, Li+ mobilities, and interfacial properties of solid electrolytes Li 3 PS 4 and Li 3 PO 4 from first principles. *Physical Review B*, 2013. **88**(10): p. 104103.
60. Muenow, D. and J. Margrave, Mass spectrometric observations of gaseous phosphorus sulfides and oxysulfides. *Journal of Inorganic and Nuclear Chemistry*, 1972. **34**(1): p. 89–94.
61. Hüser, F., T. Olsen, and K.S. Thygesen, Quasiparticle GW calculations for solids, molecules, and two-dimensional materials. *Physical Review B*, 2013. **87**(23): p. 235132.
62. Giuliani, G.F. and J.J. Quinn, Lifetime of a quasiparticle in a two-dimensional electron gas. *Physical Review B*, 1982. **26**(8): p. 4421.
63. Li, W.-B., et al., Essential electronic properties on stage-1 Li/Li+-graphite-intercalation compounds for different concentrations. arXiv preprint arXiv:2006.12055, 2020.
64. Lv, R., et al., Nitrogen-doped graphene: Beyond single substitution and enhanced molecular sensing. *Scientific Reports*, 2012. **2**(1): p. 586.
65. Kuila, T., et al., Chemical functionalization of graphene and its applications. *Progress in Materials Science*, 2012. **57**(7): p. 1061–1105.
66. Jin, K.-H., S.-M. Choi, and S.-H. Jhi, Crossover in the adsorption properties of alkali metals on graphene. *Physical Review B*, 2010. **82**(3): p. 033414.
67. Panchakarla, L.S., et al., Synthesis, structure, and properties of boron- and nitrogen-doped graphene. *Advanced Materials*, 2009: p. NA-NA.
68. Lin, M., C. Huang, and D. Chuu, Plasmons in graphite and stage-1 graphite intercalation compounds. *Physical Review B*, 1997. **55**(20): p. 13961.
69. Pham, H.D., et al., Rich p-type-doping phenomena in boron-substituted silicene systems. *Royal Society Open Science*, 2020. **7**(12): p. 200723.
70. Tsai, S.-J., et al., Gate-voltage-dependent Landau levels in AA-stacked bilayer graphene. *Chemical Physics Letters*, 2012. **550**: p. 104–110.
71. Fei, Z., et al., Gate-tuning of graphene plasmons revealed by infrared nano-imaging. *Nature*, 2012. **487**(7405): p. 82–85.
72. Wu, J.-Y., et al., The effect of perpendicular electric field on temperature-induced plasmon excitations for intrinsic silicene. *RSC Advances*, 2015. **5**(64): p. 51912–51918.
73. Wong, J.-H., B.-R. Wu, and M.-F. Lin, Strain effect on the electronic properties of single layer and bilayer graphene. *The Journal of Physical Chemistry C*, 2012. **116**(14): p. 8271–8277.
74. Rapaport, D.C. and D.C.R. Rapaport, The art of molecular dynamics simulation. 2004: Cambridge university press.
75. Alder, B.J. and T.E. Wainwright, Studies in molecular dynamics. I. General method. *The Journal of Chemical Physics*, 1959. **31**(2): p. 459–466.
76. Hafner, J., Ab-initio simulations of materials using VASP: Density-functional theory and beyond. *Journal of Computational Chemistry*, 2008. **29**(13): p. 2044–2078.
77. Hafner, J., Materials simulations using VASP—A quantum perspective to materials science. *Computer Physics Communications*, 2007. **177**(1–2): p. 6–13.

78. Hafner, J. and G. Kresse, The vienna ab-initio simulation program VASP: An efficient and versatile tool for studying the structural, dynamic, and electronic properties of materials, *in* Properties of Complex Inorganic Solids. 1997, Springer. p. 69–82.
79. Ozturk, T., E. Ertas, and O. Mert, A Berzelius reagent, phosphorus decasulfide (P4S10), in organic syntheses. *Chemical Reviews*, 2010. **110**(6): p. 3419–3478.
80. Jason, M.E., Transfer of sulfur from arsenic and antimony sulfides to phosphorus sulfides. Rational syntheses of several less-common P4S n species. *Inorganic Chemistry*, 1997. **36**(12): p. 2641–2646.
81. Zhang, L., et al., Revealing the electrochemical charging mechanism of nanosized Li2S by in situ and operando X-ray absorption spectroscopy. *Nano Letters*, 2017. **17**(8): p. 5084–5091.
82. McGilly, L.J., et al., Visualization of moiré superlattices. *Nature Nanotechnology*, 2020. **15**(7): p. 580–584.
83. Chen, G., et al., Tunable correlated Chern insulator and ferromagnetism in a moiré superlattice. *Nature*, 2020. **579**(7797): p. 56–61.
84. Ryu, Y.K., R. Frisenda, and A. Castellanos-Gomez, Superlattices based on van der Waals 2D materials. *Chemical Communications*, 2019. **55**(77): p. 11498–11510.
85. Esaki, L. and R. Tsu, Superlattice and negative differential conductivity in semiconductors. *IBM Journal of Research and Development*, 1970. **14**(1): p. 61–65.
86. Cai, H., et al., Synthesis and emerging properties of 2D layered III–VI metal chalcogenides. *Applied Physics Reviews*, 2019. **6**(4): p. 041312.
87. Xu, K., et al., Synthesis, properties and applications of 2d layered m iii x vi (m= ga, in; x= s, se, te) materials. *Nanoscale*, 2016. **8**(38): p. 16802–16818.
88. Yang, Z. and J. Hao, Recent progress in 2D layered III–VI semiconductors and their heterostructures for optoelectronic device applications. *Advanced Materials Technologies*, 2019. **4**(8): p. 1900108.
89. Hu, P., et al., GaS multi-walled nanotubes from the lamellar precursor. *Applied Physics A*, 2005. **80**(7): p. 1413–1417.
90. Hauchecorne, P., et al., Gallium selenide nanoribbons on silicon substrates for photodetection. *ACS Applied Nano Materials*, 2021. **4**(8): p. 7820–7831.
91. Zhou, W., et al., Tunable half-metallicity and edge magnetism of H-saturated InSe nanoribbons. *Physical Review Materials*, 2018. **2**(11): p. 114001.
92. Bunaciu, A.A., E.G. Udriştioiu, and H.Y. Aboul-Enein, X-Ray diffraction: instrumentation and applications. *Critical Reviews in Analytical Chemistry*, 2015. **45**(4): p. 289–299.
93. Warren, B.E., X-ray diffraction. 1990: Courier Corporation.
94. Whittig, L. and W. Allardice, X-ray diffraction techniques. *Methods of Soil Analysis: Part 1 Physical and Mineralogical Methods*, 1986. **5**: p. 331–362.
95. Li, G., et al., Observation of Van Hove singularities in twisted graphene layers. *Nature Physics*, 2010. **6**(2): p. 109–113.
96. Van Hove, L., The occurrence of singularities in the elastic frequency distribution of a crystal. *Physical Review*, 1953. **89**(6): p. 1189–1193.
97. Sobota, J.A., Y. He, and Z.-X. Shen, Angle-resolved photoemission studies of quantum materials. *Reviews of Modern Physics*, 2021. **93**(2).
98. Cattelan, M. and N. Fox, A perspective on the application of spatially resolved ARPES for 2D materials. *Nanomaterials*, 2018. **8**(5): p. 284.
99. Strocov, V.N., et al., Soft-X-ray ARPES at the Swiss Light Source: From 3D materials to buried interfaces and impurities. *Synchrotron Radiation News*, 2014. **27**(2): p. 31–40.
100. Damascelli, A., Probing the electronic structure of complex systems by ARPES. *Physica Scripta*, 2004. **T109**: p. 61.

101. Perdew, J.P., K. Burke, and Y. Wang, Generalized gradient approximation for the exchange-correlation hole of a many-electron system. *Physical Review B*, 1996. **54**(23): p. 16533.
102. Politano, A. and G. Chiarello, Plasmon modes in graphene: Status and prospect. *Nanoscale*, 2014. **6**(19): p. 10927–10940.
103. Crouse, D. and P. Keshavareddy, Role of optical and surface plasmon modes in enhanced transmission and applications. *Optics Express*, 2005. **13**(20): p. 7760–7771.
104. Mikhnenko, O.V., P.W. Blom, and T.-Q. Nguyen, Exciton diffusion in organic semiconductors. *Energy & Environmental Science*, 2015. **8**(7): p. 1867–1888.
105. Wheeler, D.A. and J.Z. Zhang, Exciton dynamics in semiconductor nanocrystals. *Advanced Materials*, 2013. **25**(21): p. 2878–2896.
106. Mahan, G.D., Excitons in degenerate semiconductors. *Physical Review*, 1967. **153**(3): p. 882–889.
107. Shen, L., et al., In situ growth of Li 4 Ti 5 O 12 on multi-walled carbon nanotubes: novel coaxial nanocables for high rate lithium ion batteries. *Journal of Materials Chemistry*, 2011. **21**(3): p. 761–767.
108. Shen, L., et al., In situ synthesis of high-loading Li 4 Ti 5 O 12–graphene hybrid nanostructures for high rate lithium ion batteries. *Nanoscale*, 2011. **3**(2): p. 572–574.
109. Tang, Y., et al., Synthesis of graphene-supported Li 4 Ti 5 O 12 nanosheets for high rate battery application. *Journal of Materials Chemistry*, 2012. **22**(22): p. 11257–11260.
110. Kim, H.-K., et al., In situ fabrication of lithium titanium oxide by microwave-assisted alkalization for high-rate lithium-ion batteries. *Journal of Materials Chemistry A*, 2013. **1**(47): p. 14849–14852.
111. Elias, D.C., et al., Control of graphene's properties by reversible hydrogenation: evidence for graphane. *Science*, 2009. **323**(5914): p. 610–613.
112. Matis, B.R., et al., Surface doping and band gap tunability in hydrogenated graphene. *ACS Nano*, 2012. **6**(1): p. 17–22.
113. Balog, R., et al., Bandgap opening in graphene induced by patterned hydrogen adsorption. *Nature Materials*, 2010. **9**(4): p. 315–319.
114. Grassi, R., T. Low, and M. Lundstrom, Scaling of the energy gap in pattern-hydrogenated graphene. *Nano Letters*, 2011. **11**(11): p. 4574–4578.
115. Zhang, W., G.E. Eperon, and H.J. Snaith, Metal halide perovskites for energy applications. *Nature Energy*, 2016. **1**(6): p. 1–8.
116. Baranowski, M. and P. Plochocka, Excitons in metal-halide perovskites. *Advanced Energy Materials*, 2020. **10**(26): p. 1903659.
117. Han, N.T., V.K. Dien, and M.-F. Lin, Excitonic effects in the optical spectra of Li 2 SiO 3 compound. *Scientific Reports*, 2021. **11**(1): p. 1–10.
118. Han, N.T., et al., First-principles studies of electronic properties in lithium metasilicate (Li 2 SiO 3). *RSC Advances*, 2020. **10**(41): p. 24721–24729.
119. Dien, V.K., et al., Orbital-hybridization-created optical excitations in Li 2 GeO 3. *Scientific Reports*, 2021. **11**(1): p. 1–10.
120. Van Neck, D., et al., Quasiparticle properties in a density-functional framework. *Physical Review A*, 2006. **74**(4): p. 042501.
121. Bäppler, S.A., et al., Exciton analysis of many-body wave functions: Bridging the gap between the quasiparticle and molecular orbital pictures. *Physical Review A*, 2014. **90**(5): p. 052521.
122. Palczewski, A.D., Angle-resolved photoemission spectroscopy (ARPES) studies of cuprate superconductors. 2010: Iowa State University.

123. Lv, B., T. Qian, and H. Ding, Angle-resolved photoemission spectroscopy and its application to topological materials. *Nature Reviews Physics*, 2019. **1**(10): p. 609–626.
124. French, R.H., H. Müllejans, and D.J. Jones, Optical properties of aluminum oxide: determined from vacuum ultraviolet and electron energy-loss spectroscopies. *Journal of the American Ceramic Society*, 1998. **81**(10): p. 2549–2557.
125. Hargittai, B. and I. Hargittai, Year of the periodic table: Mendeleev and the others. 2019, Springer. p. 1–7.
126. Pyykkö, P., The physics behind chemistry and the periodic table. *Chemical Reviews*, 2012. **112**(1): p. 371–384.
127. Denis, A., G. Goglio, and G. Demazeau, Gallium nitride bulk crystal growth processes: A review. *Materials Science and Engineering: R: Reports*, 2006. **50**(6): p. 167–194.
128. Afzaal, M. and P. O'Brien, Recent developments in II–VI and III–VI semiconductors and their applications in solar cells. *Journal of Materials Chemistry*, 2006. **16**(17): p. 1597–1602.
129. Tamargo, M.C., II-VI semiconductor materials and their applications. Vol. 12. 2002: CRC Press.
130. Walden, R.H., A review of recent progress in InP-based optoelectronic integrated circuit receiver front-ends. *International Journal of High Speed Electronics and Systems*, 1998. **9**(02): p. 631–642.
131. Mohammad, S.N. and H. Morkoç, Progress and prospects of group-III nitride semiconductors. *Progress in Quantum Electronics*, 1996. **20**(5–6): p. 361–525.
132. Harris, G.L., Properties of silicon carbide. 1995: Iet.
133. Turner, W.J., A.S. Fischler, and W.E. Reese, Physical properties of several II-V semiconductors. *Physical Review*, 1961. **121**(3): p. 759–767.
134. Ye, W.N. and Y. Xiong, Review of silicon photonics: History and recent advances. *Journal of Modern Optics*, 2013. **60**(16): p. 1299–1320.
135. Kern, W., Chemical etching of silicon, germanium, gallium arsenide, and gallium phosphide. *RCA Review*, 1978. **39**(2): p. 278–308.
136. Passlack, M., R. Droopad, and G. Brammertz, Suitability study of oxide/gallium arsenide interfaces for MOSFET applications. *IEEE Transactions on Electron Devices*, 2010. **57**(11): p. 2944–2956.
137. Colombo, C., et al., Gallium arsenide p-i-n radial structures for photovoltaic applications. *Applied Physics Letters*, 2009. **94**(17): p. 173108.
138. Baliga, B.J., Gallium nitride devices for power electronic applications. *Semiconductor Science and Technology*, 2013. **28**(7): p. 074011.
139. Izhevskyi, V., et al., Silicon carbide. Structure, properties and processing. *Cerâmica*, 2000. **46**(297): p. 4–13.
140. Casady, J. and R.W. Johnson, Status of silicon carbide (SiC) as a wide-bandgap semiconductor for high-temperature applications: A review. *Solid-State Electronics*, 1996. **39**(10): p. 1409–1422.
141. Ezawa, M., Quantum Hall effects in silicene. *Journal of the Physical Society of Japan*, 2012. **81**(6): p. 064705.
142. Sheng, D.N., L. Sheng, and Z.Y. Weng, Quantum Hall effect in graphene: Disorder effect and phase diagram. *Physical Review B*, 2006. **73**(23).
143. Song, L., et al., Binary and ternary atomic layers built from carbon, boron, and nitrogen. *Advanced Materials*, 2012. **24**(36): p. 4878–4895.
144. Tabert, C.J. and E.J. Nicol, Magneto-optical conductivity of silicene and other buckled honeycomb lattices. *Physical Review B*, 2013. **88**(8): p. 085434.
145. Holstein, B.R., The van der Waals interaction. *American Journal of Physics*, 2001. **69**(4): p. 441–449.

146. Charlier, J.-C., X. Gonze, and J.-P. Michenaud, Graphite interplanar bonding: electronic delocalization and van der Waals interaction. *EPL (Europhysics Letters)*, 1994. **28**(6): p. 403.
147. Pham, H.D., et al., Unusual features of nitrogen substitutions in silicene. *RSC Advances*, 2020. **10**(53): p. 32193–32201.
148. Dodaro, J.F., et al., Phases of a phenomenological model of twisted bilayer graphene. *Physical Review B*, 2018. **98**(7): p. 075154.
149. Hancock, Y., et al., Generalized tight-binding transport model for graphene nanoribbon-based systems. *Physical Review B*, 2010. **81**(24).
150. Schulz, S. and G. Czycholl, Tight-binding model for semiconductor nanostructures. *Physical Review B*, 2005. **72**(16): p. 165317.
151. Sapra, S., N. Shanthi, and D. Sarma, Realistic tight-binding model for the electronic structure of II-VI semiconductors. *Physical Review B*, 2002. **66**(20): p. 205202.
152. Goringe, C.M., D.R. Bowler, and E. Hernández, Tight-binding modelling of materials. *Reports on Progress in Physics*, 1997. **60**(12): p. 1447–1512.
153. Slater, J.C. and G.F. Koster, Simplified LCAO method for the periodic potential problem. *Physical Review*, 1954. **94**(6): p. 1498–1524.
154. Schüler, M., et al., Optimal Hubbard models for materials with nonlocal coulomb interactions: Graphene, silicene, and benzene. *Physical Review Letters*, 2013. **111**(3).
155. Hubbard, J., Electron correlations in narrow energy bands. *Proceedings of the Royal Society of London. Series A. Mathematical and Physical Sciences*, 1963. **276**(1365): p. 238–257.
156. Huang, Y.-K., et al., Feature-rich magnetic quantization in sliding bilayer graphenes. *Scientific Reports*, 2014. **4**(1): p. 1–10.
157. Bent, S.F., Organic functionalization of group IV semiconductor surfaces: principles, examples, applications, and prospects. *Surface Science*, 2002. **500**(1–3): p. 879–903.
158. Chen, O., et al., Surface-functionalization-dependent optical properties of II–VI semiconductor nanocrystals. *Journal of the American Chemical Society*, 2011. **133**(43): p. 17504–17512.

Index

Note: **Bold** page numbers refer to tables and *italic* page numbers refer to figures.